# Elements of
# Power Systems

# Elements of Power Systems

## Pradip Kumar Sadhu • Soumya Das

CRC Press
Taylor & Francis Group
Boca Raton London New York

CRC Press is an imprint of the
Taylor & Francis Group, an **informa** business

CRC Press
Taylor & Francis Group
6000 Broken Sound Parkway NW, Suite 300
Boca Raton, FL 33487-2742

© 2016 by Taylor & Francis Group, LLC
CRC Press is an imprint of Taylor & Francis Group, an Informa business

No claim to original U.S. Government works

Printed on acid-free paper
Version Date: 20150722

International Standard Book Number-13: 978-1-4987-3446-2 (Hardback)

**Visit the Taylor & Francis Web site at**
**http://www.taylorandfrancis.com**

**and the CRC Press Web site at**
**http://www.crcpress.com**

# Contents

# Preface

NUMERICAL MODELING AND SOLUTIONS on digital computers are the only realistic approach to system analysis and planning studies on a present-day power system with its large size and complex and integrated nature. The stage has, therefore, been reached where an undergraduate must be taught the latest techniques of analysis of large-scale power systems. A parallel need also exists in the industry, as practicing power system engineers are constantly faced with the challenges of a rapidly advancing field.

*Elements of Power Systems* is designed as a textbook for engineering degree, diploma, AMIE, or corresponding examinations in electrical power systems in India and overseas. It will also be helpful to students preparing for various competitive examinations. It is equally helpful to practicing engineers for understanding the theoretical aspects of their profession. This book is easy to read and stimulating in its direct approach.

The authors lay no claim to the original research in preparing the book. Liberal use of the materials available in the works of renowned authors has been made. In all modesty, the authors may claim only that they have tried to condense the huge amount of material available from primary and secondary sources into a coherent body of description and analysis.

The authors welcome constructive criticism of the book and will be grateful for any appraisal by readers.

**Pradip Kumar Sadhu**
**Soumya Das**

# Acknowledgments

WE ARE FORTUNATE TO have received many useful comments and suggestions from students, which helped in improving the technical content and clarity of the book. We are grateful to all of them. In particular, Saswata Mukherjee, Saikat Mitra, Anirban Kundu, Sabyasachi Samanta, and Rishabh Das.

We are indebted to many readers in academia and industry worldwide for their invaluable feedback and for taking the trouble to draw our attention to improvements required and to errors in the first edition.

We also thank the reviewers who took time from their busy schedules to send us suggestions.

Most importantly, it was the help and advice of the CRC Press/Taylor & Francis staff that made this whole project a reality. We are thankful to Gagandeep Singh (senior acquisitions editor) for his sincere efforts in handling this project at all stages. We are grateful to the authorities of the Indian School of Mines, Dhanbad, and the University Institute of Technology, Burdwan University, for providing all the facilities required to write this book.

Finally, we are grateful to our families for their love, tolerance, patience, and support throughout this very time-consuming project. Readers of the book are welcome to send their comments and feedback.

**Pradip Kumar Sadhu**
**Soumya Das**

# Authors

**Pradip Kumar Sadhu** earned his bachelor's, postgraduate, and PhD degrees in 1997, 1999, and 2002, respectively, in electrical engineering from Jadavpur University, West Bengal, India. Currently, he is a professor and head of the Electrical Engineering Department of the Indian School of Mines, Dhanbad, India. He has 18 years of experience in teaching and the industry. He has four patents, and has written several journal and conference publications at the national and international levels. He is a principal investigator of some government-funded projects. Dr. Sadhu has guided a large number of doctoral candidates and MTech students. His current areas of interest are power electronics applications, application of high-frequency converter, energy-efficient devices, energy-efficient drives, computer-aided power system analysis, condition monitoring, and light and communication systems for underground coal mines.

**Soumya Das** earned his BTech from the West Bengal University of Technology in 2007 and ME from Jadavpur University in 2010, West Bengal, India. He is presently pursuing a PhD at the Department of Electrical Engineering, Indian School of Mines, Dhanbad, India.

Currently, he is an assistant professor in the Electrical Engineering Department of the University Institute of Technology, Burdwan University, West Bengal, India. Previously, he was an assistant professor in the Electrical Engineering Department at Bengal Institute of Technology and Management, Santiniketan, India. Das has 5 years of teaching experience, and has written several journal publications at the international level, and has guided a large number of BTech and MTech students. His current areas of interest are power system engineering, high-voltage engineering, power electronics applications, computer-aided power system analysis, and solar photovoltaic systems.

# Symbols of Circuit Elements

| S. No. | Circuit Elements | Symbol |
|--------|------------------|--------|
| 1. | Bus bar | |
| 2. | Single-break isolating switch | |
| 3. | Double-break isolating switch | |
| 4. | On load isolating switch | |
| 5. | Isolating switch with earth blade | |
| 6. | Current transformer | |
| 7. | Potential transformer | |
| 8. | Capacitive voltage transformer | |

(*Continued*)

| S. No. | Circuit Elements | Symbol |
|--------|-----------------|--------|
| 9. | Oil circuit breaker | |
| 10. | Air circuit breaker with overcurrent | |
| 11. | Tripping air-blast circuit breaker | |
| 12. | Lightning arrester (valve type) | |
| 13. | Arcing horn | |
| 14. | 3-$\phi$ Power transformer | |
| 15. | Overcurrent relay | |
| 16. | Earth fault relay | |

# CHAPTER 1

# Introductory

## 1.1 INTRODUCTION

Energy is the main reason to progress. The natural resources of a country may be massive but they can only be turned into assets if they are developed, exploited, and interchanged for other goods. This cannot be achieved without energy. Energy exists in different forms in nature but the most important form is electrical energy. Energy is needed for heat, light, motive power, etc. The modern development in science and technology has made it possible to convert electrical energy into any desired form. This has given electrical energy a place of pride in the modern world. The survival of industrial undertakings and our social structures depend primarily upon low cost and continuous supply of electrical energy. In fact, availability of sufficient electrical energy and its proper use in any country can result in its people rising from subsistence level to the highest standard of living.

## 1.2 SIGNIFICANCE OF ELECTRICAL ENERGY

Electrical energy is advanced to all other forms of energy due to the following reasons:

1. *Convenient form.* Electrical energy is a very useful form of energy. It can be easily transformed into other forms of energy. For example, if we want to convert electrical energy into heat, we just need to pass electrical current through a wire of high resistance, for example, a heater. Similarly, electrical energy can be converted into light (e.g., electric bulb), mechanical energy (e.g., electric motors), etc.

2. *Easy control.* The electrically operated machines have simple and easy starting, control, and operation. For instance, an electric motor can be operated by turning on or off a switch. Similarly, with simple arrangements, the speed of electric motors can be easily varied over the desired range.

3. *Greater flexibility.* One important reason for preferring electrical energy is the flexibility that it offers. It can be easily transported from one place to another with the help of conductors.

4. *Cheapness.* Electrical energy is much cheaper than other forms of energy. Thus, it is overall economical to use this form of energy for domestic, commercial, and industrial purposes.

5. *Cleanliness.* Electrical energy is not associated with smoke, fumes, or poisonous gases. Therefore, its use ensures cleanliness and healthy conditions.

6. *High-transmission efficiency.* The consumers of electrical energy are generally situated quite away from the centers of its production. The electrical energy can be transmitted conveniently and efficiently from the centers of generation to the consumers with the help of overhead conductors known as transmission lines.

## 1.3 BASIC CONCEPTS OF A POWER SYSTEM

Generating stations, transmission lines, and the distribution systems are the main components of an electric power system. Generating stations and a distribution system are connected through transmission lines, which also connect one power system (grid, area) to another. A distribution system connects all the loads in a particular area to the transmission lines. For economical and technological reasons, individual power systems are organized in the form of electrically connected areas or regional grids (also called power pools). Each area or regional grid operates independently both technically and economically, but these are eventually interconnected to form a national grid (which may even form an international grid) so that each area is contractually tied to other areas in respect to certain generation and scheduling features. India is now heading for a national grid.

The siting of hydro stations is determined by the natural water power sources. The choice of site for coal-fired thermal stations is more flexible. The following two alternatives are possible.

1. Power stations may be built close to coal mines (called pit head stations), and electric energy is evacuated over transmission lines to the load centers.

2. Power stations may be built close to the load centers, and coal is transported to them from the mines by rail road.

In practice, however, power station siting will depend upon many factors—technical, economical, and environmental. As it is considerably cheaper to transport bulk electric energy over extra high-voltage transmission lines than to transport equivalent quantities of coal over rail road, the recent trends in India (as well as abroad) is to build super (large) thermal power stations near coal mines. Bulk power can be transmitted to fairly long distances over transmission lines of 400/765 kV and above. However, the country's coal resources are located mainly in the eastern belt and some coal-fired stations will continue to be sited in distant western and southern regions. As nuclear stations are not constrained by the problems of fuel transport and air pollution, a greater flexibility exists in their siting, so that these stations are located close to load centers while avoiding high-density pollution areas to reduce the risks, however remote, of radioactivity leakage. In India, as of now, about 75% of electric power used is generated in thermal plants (including nuclear), 23% from mostly hydro stations, and 2% come from renewable and others. Coal is the fuel for most of the steam plants; the rest depends upon oil/natural gas and nuclear fuels.

## 1.4 SINGLE-LINE DIAGRAM OF A POWER SUPPLY NETWORK

The large network of conductors between the power station and the consumers can be broadly divided into two parts viz., transmission system and distribution system. Each part can be further subdivided into two—primary transmission and secondary transmission and primary distribution and secondary distribution. Figure 1.1 shows the layout of a typical AC power supply network by a single line diagram. In Figure 1.1, G.S. represents the generating station where electrical energy is generated by three-phase synchronous generators (alternators). The generation voltages are usually 11 kV. This voltage is too low for transmission over long distance. For economy in the transmission of electric power, the generation voltage is stepped up to 132 kV or more by means of step-up transformer. At that voltage, the electrical energy is transmitted to a bulk power

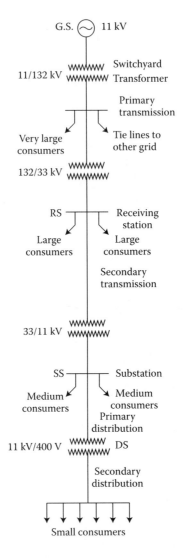

FIGURE 1.1  Schematic diagram depicting power system structure.

substation (receiving station), where energy is supplied from several power stations. The voltage at these substations is stepped down up to 33 kV in India and fed to the subtransmission system for onward transmission to distribution substation. These substations are located in the vicinity of the load centers. The voltage is further stepped down to 11 kV. It may be noted that big consumers (having demand more than 50 kW) are generally supplied power at 11 kV for further handling with their own substations. The voltage is stepped down further by distribution transformers located in residential and commercial areas, where it is supplied to these consumers at the secondary distribution level of 400 V three phase and 230 V single phase.

It is to be noted diagram that it is not necessary that all power schemes should have all the stages shown in the Figure 1.1. For example, in a certain power scheme, there may be no secondary transmission and in another case, the scheme may be so small that there is only distribution and no transmission.

## 1.5 DIFFERENT TYPES OF ENERGY SOURCES

The conversion of energy available in different forms in nature into electrical energy is known as generation of electrical energy. Since electrical energy is produced from energy available in various forms in nature, it is desirable to look into the various sources of energy. Energy classification may be based on its nature, availability, and storing capacity.

*Commercial and noncommercial energy sources*: These are also known as primary energy sources. They are available in nature in raw form, for example, coal, natural gas, and water. The other resources, which are freely available to us like solar energy, agricultural wastes, etc., are known as noncommercial energy sources.

*Conventional and nonconventional energy sources*: Conventional energy sources (also known as commercial sources) are those energy sources which are used traditionally and can be stored. The nonconventional energy sources cannot be easily stored.

*Renewable and nonrenewable energy sources*: Renewable energy sources are those sources which can be used to produce energy again and again, for example, solar energy, geothermal energy, tidal energy, etc. Nonrenewable energy sources cannot be replaced once they are used, for example, coal, oil, gas, etc.

### 1.5.1 Conventional (Nonrenewable) Sources of Electric Energy

Thermal (coal, oil, nuclear) and hydro generations are the main conventional sources of electric energy.

#### 1.5.1.1 Steam Power Station (Thermal Station)

A schematic diagram of a coal-fired thermal plant is shown in Figure 1.2. Coal received in coal storage yard of power station is transferred to the

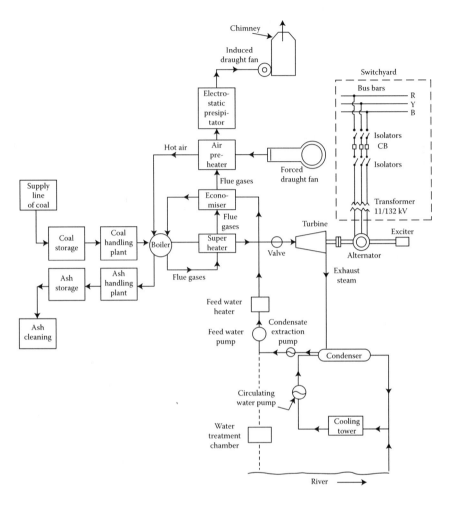

FIGURE 1.2 Schematic diagram of a coal fired thermal plant.

furnace by coal handling unit. Heat produced due to burning of the coal is utilized in converting water in the boiler drum to steam at suitable pressure and temperature. The steam generated is passed through the superheater. Superheated steam then flows through the turbine. In the turbine, the pressure of steam is reduced. Steam leaving the turbine passes through the condenser, where the pressure of steam at the exhaust of turbine is maintained. Steam pressure in the condenser depends upon flow rate and temperature of cooling water and on effectiveness of air removal equipment. Water circulating through the condenser may be taken from the various sources such as river, lake, or sea. If sufficient quantity of water is not available, the hot water is coming out of the condenser may be cooled in cooling tower and circulated again through the condenser. Blade system taken from the turbine at suitable extraction point is sent to low- and high-pressure water heaters. With the help of force draught fan, air is collected from the atmosphere and is first passed through the air preheater, where it is heated by flue gases. The hot air then passes through the furnace. The flue gases after passing through the boiler and superheated tube, flow through economizer, air preheater, and electrostatic precipitator (dust collector), and finally they are exhausted to the atmosphere through the chimney by induced draught fan.

*Merits*

1. The fuel used is quite cheap.

2. Less initial cost as compared to other generating station.

3. It can be installed at any place irrespective of the existence of coal.

4. It requires less space as compared to the hydraulic power station.

5. The cost of generation is lesser than that of the diesel power station.

*Demerits*

1. It pollutes the atmosphere due to the production of large amount of smoke and fumes.

2. Its running cost is costlier than that of the hydraulic plant.

### 1.5.1.2 Gas Turbine Power Plant

With increasing availability of natural gas (methane), prime movers based on gas turbines have been developed on the lines similar to those used in aircraft. Schematic arrangement of gas turbine plant is shown in Figure 1.3. The air at atmospheric pressure is drawn by the compressor via a filter which removes the dust from air. The compressor used in the plant is generally rotatory type. The rotatory blades raise its pressure. Thus air at high pressure is available at the output of the pressure. The exhaust is passed through the regenerator before getting wasted to atmosphere. Regenerator is a device which recovers heat from the exhaust gases of the turbine. A regenerator consists of a nest of tubes contained

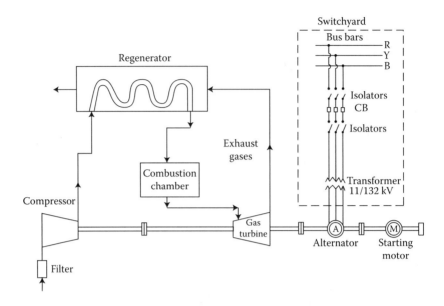

FIGURE 1.3    Schematic arrangement of gas turbine plant.

in shell. The compressed air from the compressor passes through the tubes on its way to the combustion chamber. In this way, compressed air is heated by the exhaust gases. The air at high pressure from the compressor is led to the combustion chamber via regenerator. In this chamber, heat is added to the air by burning oil. The oil is injected through the burner into the chamber at high pressure to ensure atomization of oil and its mixing with air. The result is that the chamber attains very high temperature (about 3000°F). The combustion gases are suitably cooled led to 1300–1500°F and then delivered to the gas turbine. The product of combustion consisting of a mixture of gases at high temperature and pressure is passed to the gas turbine. There gases passing over the turbine blade expand and thus do the mechanical work. The temperature of the exhaust gases from the turbine is about 900°F. The gas turbine is coupled to the alternator. The alternator converts mechanical energy of the turbine into electrical energy. The output from the alternator is given to the bus bars through transformer, circuit breaker, and isolators. Before starting the turbine, the compressor has to be started. For this purpose, electric motor is mounted on the same shaft as that of the turbine. The motor is energized by the batteries. Once the unit starts, a part of mechanical power of the turbine drives the compressor and there is no need of motor now.

*Merits*

1. It is simple in design compared to steam power station since no boilers and their auxiliaries are required.

2. It is much smaller in size as compared to steam power station of the same capacity. This is expected since gas turbine plant does not require boiler, feed water arrangement.

3. The initial and operating costs are much lower than its equivalent steam power station.

4. It requires comparatively less water as no condenser is used.

5. The maintenance charges are quite less.

6. Gas turbines are much simpler in construction and operation than steam turbines.

7. There are no standby losses. However, in a steam power station, losses occur because boiler is kept under operation even when the steam turbine is supplying no load.

*Demerits*

1. There is a problem for starting the unit. It is because before starting the turbine, the compressor has to be operated for which power is required from external source. However, once the unit starts, the external power is not needed as the turbine itself supplies the necessary power to the compressor.

2. Since a greater part of power developed by the turbine is used in driving the compressor, the net output is low.

3. The overall efficiency of such plants is low (about 20%) because the exhaust gases from the turbine contain sufficient heat.

4. The temperature of combustion chamber is quite high (3000°F), so that its life is comparatively reduced.

### 1.5.1.3 Hydroelectric Power Generation

It is known that the chief requirement for hydroelectric power plant is the availability of water in huge quantity at sufficient head and this requirement

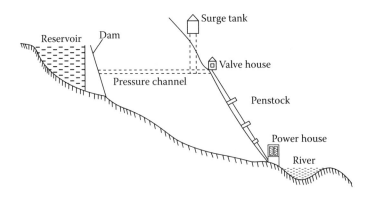

FIGURE 1.4   Schematic arrangement of a hydro plant.

can be met by constructing a conventional dam across a river or lake. A simple schematic arrangement of a hydro plant is given in Figure 1.4. An artificial storage reservoir is formed by constructing a dam across a river and a pressure tunnel is taken off from the reservoir to the valve house at the start of the penstock. The valve house contains main sluice valves for controlling water flow to the power station and automatic isolating valves for cutting off water supply in case of penstock bursts.

A surge tank (open from top) is also provided just before the valve house for better regulation of water pressure in the system. From the reservoir, water is carried to valve house through pressure tunnel and from valve house to the water turbine through pipes of large diameter made of steel or reinforced concrete, called penstock. The water turbine converts hydraulic energy into mechanical energy and the alternator coupled to the water turbine converts mechanical energy into electrical energy. Water after doing useful work is discharged to the tail race.

*Merits*

1. No fuel required. Water is the source of energy. Hence operating costs are low. And there are no problems like handling of the fuel, storage of the fuel, disposal of the ash, etc.

2. The plant is highly reliable and it is the cheapest in operation and maintenance.

3. The load can be varied quickly and the rapidly changing load demand can be met without difficulty.

4. Such plants are robust and have longer life.

4. The efficiency of such plants does not fall with age.

5. Very neat and clean plant as no smoke ash produced.

6. Highly skilled engineers are required only at the time of construction. But later on only a few experienced person will be required.

7. Usually located in remote areas where land is available in cheaper rates.

8. Such plants, in addition to generation of electric power, also serve other purposes such as irrigation, flood control, etc.

*Demerits*

1. Huge area is required.

2. Its construction cost is very high and takes a long time for erection.

3. Long transmission lines are required, as the plants are located in hilly areas which are quite away from the load centers.

4. The output of such plants is never constant owing to vagaries of monsoon and depends on rate of water flow in a river. Long dry season may affect the water.

5. Hydroelectric power plant reservoir submerges huge areas, uproots large population and creates social and other problems.

### 1.5.1.4 Nuclear Power Plant

A nuclear power plant consists of a nuclear reactor (for heat generation), heat exchanger (for converting water into steam by using the heat generated in nuclear reactor), steam turbines, alternators, condenser, etc. As in conventional steam power plant, water for raising steam forms a closed feed system. However, the reactor and the cooling circuit have to be heavily shielded to eliminate radiation hazards. A schematic arrangement of a nuclear power plant is given in Figure 1.5. The tremendous amount of heat energy produced in breaking of atoms of uranium ($U^{235}$) or thorium ($Th^{232}$) of large atomic weight into metals of lower atomic weight by fission process in an atomic reactor is extracted by pumping fluid or molten metal like liquid sodium or gas through the pile. In nuclear fission, the breaking up of nuclei of heavy atoms into two nearly equal parts with release of huge amount of energy. The heated metal or gas is then

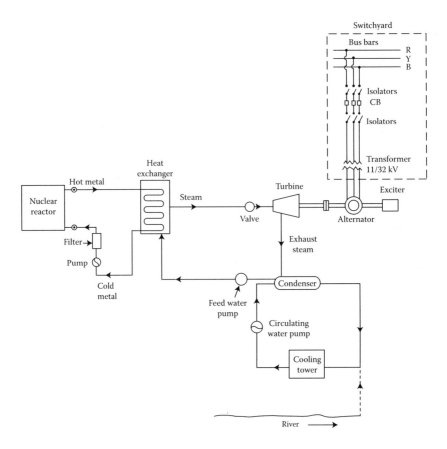

FIGURE 1.5    Schematic arrangement of a nuclear power plant.

allowed to exchange its heat to the heat exchanger by circulation. In heat exchanger, the gas is heated or steam is generated which are utilized to drive gas turbine or steam turbine coupled to an alternator thereby generating electrical energy. After doing useful work in turbine, the steam is exhausted to condenser. The condenser condenses the steam which is fed to heat exchanger through feed water pump. The output of alternator is delivered to bus bars through transformers, circuit breakers, etc.

*Merits*

1. The amount of fuel required is quite small.

2. A nuclear power plant requires less space as compared to any other type of the same size.

3. This plant is very economical for producing bulk electric power.

4. It can be located near the load centers because it does not require large quantities of water and need not be near coal mines. Therefore, the cost of primary distribution is reduced.

5. There are large deposits of nuclear fuels available all over the world. Therefore, such plants can ensure continued supply of electrical energy for thousands of years.

*Demerits*

1. The fuel used is expensive and is difficult to recover.

2. The capital cost on a nuclear plant is very high as compared to other types of plants.

3. The erection and commissioning of the plant require greater technical know-how.

4. The fission by-product is generally radioactive and may cause a dangerous amount of radioactive pollution.

5. The disposal of the by-products, which are radioactive, is a big problem. They have either to be disposed off in a deep trench or in a sea away from sea shore.

6. Maintenance charges are high due to lack of standardization. Moreover high salaries of specially trained personnel employed to handle the plant further raise the cost.

### 1.5.1.5 Diesel Power Plant

In a diesel power plant, generally diesel engine is utilized as the prime mover. Within the engine, diesel is burned and the by-products of this combustion act as the "working fluid" to create mechanical energy. The diesel engine drives the alternator which converts mechanical energy into electrical energy. As the generation cost is considerable due to high price of diesel, such power stations are exclusively used to produce low power. Figure 1.6 shows the formal organization of a typical diesel power station. The plant has the following auxiliaries:

*Fuel supply system*: It consists of a storage tank, strainers, fuel transfer pump, and all day fuel tank. The fuel oil is supplied at the plant site

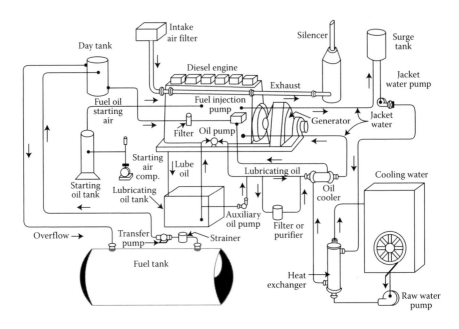

FIGURE 1.6    Schematic arrangement of a typical diesel power station.

by rail or road. This oil is stored in the storage tank. From the storage tank, oil is pumped to smaller all day tank at daily or short intervals. From this tank, fuel oil is passed through strainers to remove suspended impurities. The clean oil is injected into the engine by fuel injection pump.

*Air intake system*: This system supplies necessary air to the engine for fuel combustion. It consists of tubes for the supply of fresh air to the engine manifold. Filters are provided to remove dust specks from the breeze which may act as abrasive in the engine cylinder.

*Exhaust system*: This system leads the engine exhaust gas outside the building and discharges it into the atmosphere. A silencer is usually incorporated in the system to reduce the noise level.

*Cooling system*: The heat liberated by the combustion of fuel in the engine cylinder is partially converted into work. The remainder part of the heat passes through the cylinder walls, piston, rings, etc. and may cause damage to the system. In order to keep the temperature of the engine parts within the safe operating limits, cooling is provided. The cooling system consists of a water source, pump, and cooling towers. The pump circulates water through the piston chamber and head jacket. The water carries away heat from the locomotive and it gets hot. The hot water is cooled by cooling towers and is recirculated for cooling.

*Lubricating system*: This arrangement minimizes the wear of rubbing surfaces of the locomotive. It contains of the lubricating oil tank, pump, filter, and oil cooler. The lubricating oil is drawn from the lubricating oil tank by the pump and is passed through filters to remove impurities. The clean lubricating oil is delivered to the points which require lubrication. The oil coolers incorporated in the system keep the temperature of the oil low.

*Engine starting system*: This is an agreement to rotate the engine initially, while bugging out, until firing starts and the unit runs with its own force. Small sets are taken off manually by handles, but for larger units, compressed air is used for initiating. In the latter case, air at high pressure is admitted to a few of the cylinders, making them to act as reciprocating air motors to turn over the engine shaft. The fuel is admitted to the remaining cylinders which makes the locomotive to take off under its own force.

*Diesel engine generator (alternator)*: The alternator used in diesel electric power plants are of rotating field, salient pole construction, speed ranging from 214 to 1000 rpm (poles 28 to 6), and capacity is ranging from 25 to 5000 kVA at 0.8 pf lagging. Their output voltages are of 440 V in case of small machines and as high as 2200 V in case of large machines. Voltage regulation is about 30%.

They are directly coupled to the diesel engines. They are supplied with automatic voltage regulation and satisfactory parallel operation. The excitation is usually provided at 115 or 230 V from a DC exciter of rating about 2%–4% of the alternator ratings, usually coupled to the engine shaft either directly or through a belt.

*Merits*

1. The design and layout of the plant are quite simple.

2. It occupies less space, as the number and size of the auxiliaries are small.

3. It can be placed at any position.

4. It can be started quickly and can pick up load in a short time.

5. There are no standby losses.

6. It requires less quantity of water for chilling.

7. The overall price is much less than that of steam power station of the same content.

8. The thermal efficiency of the plant is more eminent than that of a steam power station.

9. It requires less operating staff.

*Demerits*

1. The plant has high running charges as the fuel (i.e., diesel) used is costly.

2. The plant does not operate satisfactorily under overload conditions for a longer period.

3. The plant can only generate small power.

4. The cost of lubrication is generally high.

5. The maintenance charges are generally high.

### 1.5.1.6 Magneto Hydrodynamic Generation

In thermal generation of electric energy, the heat released by the fuel is converted to rotational mechanical energy by means of a thermo cycle. The mechanical energy is then applied to rotate the electric generator. Thus, two stages of energy conversion are involved in which the heat to mechanical energy conversion has inherently low efficiency. Besides, the rotating machine has its associated losses and maintenance troubles. In magneto hydrodynamic (MHD) technology, electric energy is directly generated by the hot gases created by the burning of fuel without the demand for mechanical moving components. In an MHD generator, electrically conducting gas at a very high temperature is drawn in a strong magnetic field, thereby generating electricity. High temperature is required to ionize the gas, hence that it has good electrical conductivity. The conducting gas is obtained by burning a fuel and injecting seeding materials such as potassium carbonate in the products of combustion. The principle of MHD power generation is illustrated in Figure 1.7. Approximately, 50% efficiency can be attained if the MHD generator operates in tandem with a conventional steam plant.

### 1.5.2 Nonconventional (Renewable) Sources of Electric Energy

The necessity to conserve fossil fuels has forced scientists and technologists across the world to search for nonconventional sources of electric energy. Some of the sources being explored are solar, wind, and tidal sources. To protect environment and for sustainable development, the importance of renewable energy sources cannot be overemphasized. It is an established and accepted tact that renewable and nonconventional forms of energy will play an increasingly important role in the future, as they are cleaner and easier to use and environmentally benign and bound to become economically more viable with increased use.

#### 1.5.2.1 Solar Energy

The Sun is the primary and main source of energy. The average incident solar energy received on earth's surface is about 600 W/m$^2$ but the actual value varies considerably. It possesses the advantage of being complimentary of cost, non-exhaustible, and completely pollution free. On the other hand, it has several drawbacks—energy density per-unit area is very low, it is available for only a part of the day, and cloudy and hazy atmospheric conditions greatly reduce the energy received. So, harnessing solar energy for electricity generation, challenging technological problems exist, the

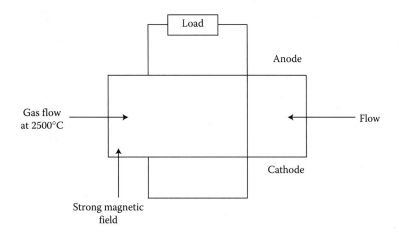

FIGURE 1.7  The principle of MHD power generation.

most important being that of the collection and concentration of solar energy, and its conversion to the electrical form through efficient and relatively economical means.

At present, two technologies are being built up for transition of solar energy to the electrical form.

1.5.2.1.1 Solar–Thermal Energy   In this technology, collectors with concentrators are employed to achieve temperatures high enough (700°C) to operate a heat engine at reasonable efficiency to generate electricity. Nevertheless, there are considerable engineering difficulties in making a single tracking bowl with a diameter exceeding 30 m to get perhaps 200 kW. The scheme calls for large and intricate structures involving a vast capital outlay and as of today is far from being competitive with conventional electricity generation.

1.5.2.1.2 Photovoltaic (PV) Generation   This technology converts solar energy to the electrical form by means of silicon wafer photoelectric cells known as "solar cells." Their theoretical efficiency is about 25% but the practical value is only about 15%. But that does not matter as solar energy is basically free of cost. The main problem is the cost and maintenance of solar cells. With the likelihood of a breakthrough in the large scale production of cheap solar cells with amorphous silicon, this technology may compete with established methods of electricity generation, especially as conventional fuels become scarce.

These systems are of the following two types:

1. *Stand-alone power systems.* In such a system, the PV array is the principal or only source of energy.

2. *Grid-connected power systems.* In this type of arrangement, the load is tied to both a PV power system and an electricity power system.

*1.5.2.2 Wind Power*

This method can be used where wind flows for a considerable length of time. The wind energy is used to run the wind mill which drives a small generator. In order to obtain the electrical energy from a wind mill continuously, the generator is arranged to charge the batteries. These batteries supply the energy when the wind stops. This method has the advantages that maintenance and generation costs are negligible.

In our country, this source of generation of electrical energy will prove economical at a number of places as revealed in a recent study undertaken by National Aerospace Laboratories, Bengaluru.

However, this method is unreliable since the production of electrical energy depends largely upon the pressure of the wind.

### 1.5.2.3 Wave Energy

The energy content of sea waves is very high. In India, with several hundreds of kilometers of coast line, a vast source of energy is available. The power in the wave is proportional to the square of the amplitude and to the period of the motion. Therefore, the long period (~10 s) and large amplitude (~2 m) waves are of considerable interest for power generation, with energy fluxes commonly averaging between 50 and 70 kW/m width of oncoming wave. Though the engineering problems associated with wave power are formidable, the amount of energy that can be harnessed is large and development work is in progress. Sea wave power estimated potential is 20,000 MW.

### 1.5.2.4 Ocean Thermal Energy Conversion

The ocean is the world's largest solar collector. Temperature difference of 20°C between warm, solar absorbing surface water and cooler "bottom" water can occur. This can provide a continually replenished store of thermal power, which is available in the principle conversion to other energy forms. Ocean thermal energy conversion refers to the conversion of some of this thermal energy into work and hence into electricity. Estimated potential energy of ocean thermal power in India is 50,000 MW.

### 1.5.2.5 Biofuels

The material of plants and animals is called biomass, which may be transformed by chemical and biological processes to produce intermediate biofuels such as methane gas, ethanol liquid, or charcoal solid. Biomass is burnt to provide heat for cooking, comfort heat (space heat), crop drying, factory processes, and raising steam for electricity production and transport. In India, potential for bio-energy is 17,000 MW and that for agricultural waste is about 6000 MW.

### 1.5.2.6 Geothermal Energy

Geothermal energy is another energy source that can be harnessed for power generation and thermal applications in near future. Geothermal energy is the natural heat generated from within the earth. The steam and hot water come naturally to the airfoil of the ground at some positions. The primary source of geothermal energy is magma. The extraction of heat form earth's interior needs a natural or artificial heat exchanger. Water is injected by an injection well into hot dry rock inside the ground to extract the hot water and steam from production wells. Water injected into the wall acts as a heat collecting and heat transporting medium.

### 1.5.2.7 Mini and Micro Hydro Plants

In order to match with the present energy crisis partly, a solution is to be developed mini and micro hydro potential in our state. The capacity of micro plant is up to 100 kW and mini plant is from 101 to 100 kW. These applications do not need to conventional dam or retain water to make a hydraulic head; the head is just a few meters, that is, 5–20 m head for mini plants and less than 5 m head for micro plants. Using the current of a river or the naturally occurring tidal flow to create electricity may provide a renewable energy source that will have a minimal impact on the environment.

### EXERCISES

1. Why is electrical energy preferred over other forms of energy?

2. Write a short note on the generation of electrical energy.

3. Discuss the different sources of energy available in nature.

4. Compare the chief sources of energy used for the generation of electrical energy.

5. Discuss the advantages and disadvantages of a steam power station.

6. Draw the schematic diagram of a modern steam power station and explain its operation.

7. Discuss the merits and demerits of a hydroelectric plant.

8. Draw a neat schematic diagram of a hydroelectric plant and explain the functions of various components.

9. Draw the flow diagram of a diesel power station and discuss its operation.

10. Discuss the advantages and disadvantages of a diesel power station.

11. Draw the schematic diagram of a nuclear power station and discuss its operation.

12. Explain the working of a gas turbine power plant with a schematic diagram.

13. Give the comparison of steam power plant, hydroelectric power plant, and nuclear power plant on the basis of operating cost, initial cost, efficiency, maintenance cost, and availability of source of power.

# CHAPTER 2

# Per-Unit Systems

## 2.1 INTRODUCTION

In large interconnected power systems with several voltage levels and various capacity equipments, it has been found very convenient to play with per-unit (pu) systems of quantities for analysis rather than in absolute values of measures. The pu system leads to great simplification of three-phase networks involving transformers. The numerical pu value of any amount is determined as the proportion of its actual value to another arbitrarily chosen value of the quantity of the dimensions assumed as the theme or extension.

$$\text{Per-unit value} = \frac{\text{The actual value of the quantity in any unit}}{\text{The base or reference value of the same unit}} \quad (2.1)$$

For any quantity $K$,

$$K_{pu} = \frac{K_a}{K_b} \quad (2.2)$$

Thus, any quantity can be converted to a pu quantity by dividing the numerical value of a chosen base value of the same proportions. The pu values are dimensionless. Percent quantities differ from pu quantities by a factor of 100. The ratio in percent is 100 times the value in pu.

## 2.2 PER-UNIT REPRESENTATION OF BASIC ELECTRICAL QUANTITIES

In electrical engineering, the three basic quantities are voltage, current, and impedance.

Let $I_{amp}$ the actual current in ampere, $I_b$ the base current in ampere, $V_{volt}$ the actual voltage in volts, $V_b$ the base voltage in volts, $Z_{ohm}$ the actual impedance in ohms, $Z_b$ the base impedance in ohms, $S_{volt-amp}$ the actual volt-ampere, and $S_b$ the base volt-ampere, then

$$\text{Per-unit current} = I_{pu} = \frac{I_{amp}}{I_b} \tag{2.3}$$

$$\text{Per-unit voltage} = V_{pu} = \frac{V_{volt}}{V_b} \tag{2.4}$$

$$\text{Per-unit impedance} = Z_{pu} = \frac{Z_{ohm}}{Z_b} \tag{2.5}$$

$$Z_{ohm} = R_{ohm} + jX_{ohm}$$

$$\therefore Z_{pu} = \frac{Z_{ohm}}{Z_b} = \frac{R_{ohm}}{Z_b} + j\frac{X_{ohm}}{Z_b}$$

or

$$Z_{pu} = R_{pu} + jX_{pu}$$

$$\therefore R_{pu} = \frac{R_{ohm}}{Z_b} \tag{2.6}$$

and

$$X_{pu} = \frac{X_{ohm}}{Z_b} \tag{2.7}$$

Per-unit volt-ampere

$$S_{pu} = \frac{S_{volt-amp}}{S_b} \tag{2.8}$$

$$S = P + jQ = VI^*$$

$$S_{pu} = \frac{S_{volt\text{-}amp}}{S_b} = \frac{P}{S_b} + j\frac{Q}{S_b}$$

$$\therefore S_{pu} = P_{pu} + jQ_{pu}$$

$$\therefore P_{pu} = \frac{P_{watt}}{S_b} \tag{2.9}$$

$$\therefore Q_{pu} = \frac{Q_{vars}}{S_b} \tag{2.10}$$

For a single-phase circuit,

$$Z_b = \frac{V_b}{I_b} \tag{2.11}$$

and

$$S_b = V_b I_b \tag{2.12}$$

The values of the base quantities are selected according to convenience. If any two of the four quantities in Equations 2.11 and 2.12 are specified, the remaining two are fixed automatically.

The base impedance is that which has a voltage drop across which is equal to the base voltage if the current through it is equal to the base current.

$$Z_b = \frac{V_b}{I_b} = \frac{V_b V_b}{V_b I_b} = \frac{V_b^2}{S_b} \text{ ohms} \tag{2.13}$$

$$Y_b = \frac{1}{Z_b} = \frac{S_b}{V_b^2} \text{ siemens}$$

If proper selection of bases is made, the basic circuit relations can be applied to the pu quantities.

$$V_{pu} = Z_{pu} I_{pu} \tag{2.14}$$

$$S_{pu} = V_{pu} I_{pu}^* \qquad (2.15)$$

Equations 2.5 and 2.13 can be combined to give

$$Z_{pu} = Z_{ohm} \frac{S_b}{V_b^2} \qquad (2.16)$$

Also from Equation 2.12, we get

$$I_b = \frac{S_b}{V_b} = \frac{\text{Base kVA}}{\text{Base kV}}$$

where base kW is the numerical value of base kVA and base MW is the numerical value of base MVA.

Let $Y_{Sim}$ be the actual admittance in siemens and $Y_{pu}$ the pu admittance, then

$$Y_{pu} = \frac{1}{Z_{pu}} = \frac{V_b^2}{Z_{ohm} S_b} = Y_{Sim} \frac{V_b^2}{S_b}$$

## 2.3 CHANGE OF BASE

It is sometimes necessary to convert pu quantities from one base to another. Let the base volt-ampere and base voltage in system 1 be represented by $S_{b1}$ and $V_{b1}$, respectively. The corresponding values in system 2 are represented by $S_{b2}$ and $V_{b2}$, respectively.

Base current in base system 1:

$$I_{b1} = \frac{S_{b1}}{V_{b1}} \qquad (2.17)$$

Base current in base system 2:

$$I_{b2} = \frac{S_{b2}}{V_{b2}} \qquad (2.18)$$

The pu value of current $I$ in base system 1:

$$I_{1pu} = \frac{I}{I_{b1}} \qquad (2.19)$$

The pu value of current $I$ in base system 2:

$$I_{2pu} = \frac{I}{I_{b2}} \tag{2.20}$$

Combine Equations 2.18 through 2.20:

$$I_{2pu} = I_{1pu} \frac{I_{b1}}{I_{b2}} = I_{1pu} \frac{S_{b1}}{V_{b1}} \frac{V_{b2}}{S_{b2}} = I_{1pu} \frac{S_{b1}}{S_{b2}} \frac{V_{b2}}{V_{b1}} \tag{2.21}$$

We know

$$Z_{pu} = Z_{ohm} \frac{S_b}{V_b^2}$$

Therefore, the pu value of impedance $Z_{ohm}$ in base system 1 is

$$Z_{1pu} = Z_{ohm} \frac{S_{b1}}{V_{b1}^2} \tag{2.22}$$

Therefore, the pu value of impedance $Z_{ohm}$ in base system 2 is

$$Z_{2pu} = Z_{ohm} \frac{S_{b2}}{V_{b2}^2} \tag{2.23}$$

Elimination of $Z_{ohm}$ from Equations 2.22 and 2.23 gives

$$Z_{2pu} = Z_{1pu} \frac{S_{b2}}{S_{b1}} \left( \frac{V_{b1}}{V_{b2}} \right)^2 \tag{2.24}$$

Equation 2.24 is used for changing the pu impedance from one set of $V$ and $S$ bases to any other set of $V$ and $S$.

So the change of base for the admittance can be written as

$$Y_{2pu} = Y_{1pu} \frac{S_{b1}}{S_{b2}} \left( \frac{V_{b2}}{V_{b1}} \right)^2 \tag{2.25}$$

## 2.4 PER-UNIT QUANTITIES IN A THREE-PHASE SYSTEM

In a star connection,

$$V_1 = \sqrt{3}V_{ph}, \quad V_{lb} = \sqrt{3}V_{phb}$$

$$I_1 = I_{ph}, \quad I_{lb} = I_{phb}$$

$$(V_1)_{pu} = \frac{V_1}{V_{lb}} = \frac{\sqrt{3}V_{ph}}{\sqrt{3}V_{phb}} = (V_{ph})_{pu} \tag{2.26}$$

$$(I_1)_{pu} = \frac{I_1}{I_{lb}} = \frac{I_{ph}}{I_{phb}} = (I_{ph})_{pu} \tag{2.27}$$

where $V_1$ is the line voltage, $V_{ph}$ is the phase voltage, $I_1$ is the line current, and $I_{ph}$ is the phase current in a balance $3\phi$ system.

In a delta connection,

$$V_1 = V_{ph}, \quad V_{lb} = V_{phb}$$

$$I_1 = \sqrt{3}I_{ph}, \quad I_{lb} = \sqrt{3}I_{phb}$$

$$(V_1)_{pu} = \frac{V_1}{V_{lb}} = \frac{V_{ph}}{V_{phb}} = (V_{ph})_{pu} \tag{2.28}$$

$$(I_1)_{pu} = \frac{I_1}{I_{lb}} = \frac{\sqrt{3}I_{ph}}{\sqrt{3}I_{phb}} = (I_{ph})_{pu} \tag{2.29}$$

Thus, it is seen that in both star and delta connections, a pu-phase voltage has the same numerical value as the corresponding pu-line voltage. Also, the pu-phase current has the same numerical value as the corresponding pu-line current.

## 2.5 BASE QUANTITIES IN TERMS OF KV AND MVA

In power systems, it is common practice to specify voltage rating in kilo-volts and the volt-ampere rating in MVA. The results already derived in terms of V and VA can be modified as

$$\text{Base MVA: } S_b = (MVA)_b = V_b I_b \times 10^{-6}$$

Base voltage in kV: $(kV)_b = V_b \times 10^{-3}$

$$Z_b = \left(\frac{V_{b1}}{V_{b2}}\right)^2 \frac{(kV_1)_b^2}{[(MVA)_b]_{3\phi}}$$

Subscript $3\phi$ denotes the three-phase value.

$$Z_{pu} = Z_{ohm}\frac{S_b}{V_b^2} = Z_{ohm}\frac{V_bI_b}{V_b^2} = Z_{ohm}\frac{V_bI_b \times 10^{-6}}{(V_b \times 10^{-3})^2}$$

$$Z_{pu} = Z_{ohm}\frac{(MVA)_b}{(kV)_b^2} \tag{2.30}$$

$$Z_{2pu} = Z_{1pu}\frac{(MVA)_{b2}}{(kV)_{b1}^2}\left[\frac{(kV)_{b1}}{(kV)_{b2}}\right]^2 \tag{2.31}$$

$$Z_{pu} = Z_{ohm}\frac{[(MVA)_b]_{3\phi}}{[(kV_1)_b]^2} \tag{2.32}$$

## 2.6 PER-UNIT IMPEDANCE OF A TRANSFORMER

Consider a single-phase transformer in which the total series impedance of the two windings referred to the primary is $Z_{1e}$. Suppose that the rated values are taken as the base quantities.

In the primary, base current is $I_{b1}$, base voltage is $V_{b1}$, and base impedance, $Z_{b1}$, is $(V_1/I_1)$.

Per-unit impedance of the transformer referred to the primary is

$$Z_{1epu} = \frac{Z_{1e}}{Z_{b1}} \tag{2.33}$$

The total series impedance of the two windings referred to the secondary is

$$Z_{2e} = Z_{1e}\left(\frac{N_2}{N_1}\right)^2 \tag{2.34}$$

where $N_1$ and $N_2$ represent primary and secondary turns, respectively.

In the secondary, base current is $I_{b2}$, base voltage is $V_{b2}$, and base impedance, $Z_{b2}$, is $(V_2/I_2)$.

Per-unit impedance of the transformer referred to the secondary is

$$Z_{2e\,pu} = \frac{Z_{2e}}{V_2/I_2} = \frac{Z_{2e}I_2}{V_2} \tag{2.35}$$

Also

$$I_2 = I_1 \frac{N_1}{N_2} \tag{2.36}$$

and

$$V_2 = V_1 \frac{N_2}{N_1} \tag{2.37}$$

From Equations 2.34 through 2.37,

$$Z_{2e\,pu} = Z_{1e} \left(\frac{N_2}{N_1}\right)^2 \frac{I_1 N_1}{N_2} \frac{N_1}{V_1 N_2} = Z_{1e} \frac{I_1}{V_1} \tag{2.38}$$

From Equations 2.33 and 2.38,

$$Z_{2e\,pu} = Z_{1e\,pu}$$

Thus, the pu impedance of a two-winding transformer referred to either side is the same.

## 2.7 ADVANTAGES OF PU REPRESENTATION

Per-unit system computation has the following advantages:

1. The ordinary parameters (current, impedance, etc.) vary considerably with the variation of physical size, terminal voltage, power rating, etc., while the pu parameters are independent of these quantities over a wide range of the same type of apparatus. In other words, the pu impedance values for apparatus of like ratings lie within a narrow range.

2. Per-unit values provide more meaningful information.

3. The chance of confusion between line and phase values in a three-phase balanced system is reduced.

4. The impedances of machines are specified by the manufacturers in terms of pu values.

5. The pu impedance referred to either side of a single-phase transformer is the same.

6. The computational effort in power systems is very much reduced with the use of pu quantities.

## WORKED EXAMPLES

### EXAMPLE 2.1

A 230-kV transmission line has a series impedance of $(8 + j64)$ ohms and a shunt admittance of $j4 \times 10^{-3}$ s. Using 100 MVA and the line voltage as base values, calculate pu impedance and pu admittance of the line.

**Solution**

$$Z_{pu} = Z_{ohm} = (8 + j64) \times \frac{100}{(230)^2}$$
$$= (0.015 + j0.12) \text{ pu}$$

$$Y_{pu} = j4 \times 10^{-3} \times \frac{(230)^2}{100}$$
$$= j2.116 \text{ pu}$$

### EXAMPLE 2.2

A three-phase, star-connected system is rated at 100 MVA and 110 kV. Express 20,000 kVA of three-phase apparent power as a pu value referred to as

1. The three-phase system kVA as base
2. The per-phase kVA as base

**Solution**

1. For the three-phase base:

$$\text{Base kVA} = 20,000 \text{ kVA} = 1 \text{ pu}$$

$$\text{Base kV} = 110 \text{ kV}$$

$$\text{Per-unit kVA} = \frac{20,000}{100 \times 1000} = 0.2 \text{ pu}$$

2. For the per-phase base:

$$\text{Base kVA} = \frac{1}{3} \times 1,00,000 = 33.33 \text{ MVA} = 1 \text{ pu}$$

$$\text{Base kV} = \frac{110}{\sqrt{3}} = 63.5$$

$$\text{Per-unit kVA} = \frac{1}{3} \times \frac{20,000}{33.33 \times 10^3} = 0.20 \text{ pu}$$

**EXAMPLE 2.3**

A 200-MVA, 11-kV, three-phase generator has a subtransient reactance of 10%. The generator is connected to the motors through transmission lines and transformers. The motors have rated inputs of 20 MVA, 30 MVA, 50 MVA at 20 kV with 10% subtransient reactance. Three-phase transformers are rated at 110 MVA, 13 kV,Δ/110 kV,Y with leakage reactance at 8%. The line has a reactance of 30 Ω. Select the generator rating as the base quantities in other parts of the system and evaluate the corresponding pu values.

**Solution**

Assuming base values as 200 MVA and 11 kV in the generator circuit, the pu reactance of generator will be 10%. The base value of voltage in the line will be

$$11 \times \frac{110}{13} = 93.07 \text{ kV}$$

In this motor circuit,

$$93.07 \times \frac{13}{110} = 11 \text{ kV}$$

The reactance of the transformer given is 8%

Corresponding to 110 MVA, 13 kV

Corresponding to 100 MVA, 11 kV

The pu reactance will be

$$0.08 \times \frac{200}{100} \times \left(\frac{13}{11}\right)^2 = 0.223 \text{ p.u.}$$

$$\text{Per-unit impedance of line} = \frac{30 \times 200}{(93.07)^2} = 0.6926 \text{ p.u.}$$

$$\text{Per-unit reactance for motor 1} = 0.1 \times \frac{200}{20} \times \left(\frac{20}{11}\right)^2 = 3.30 \text{ p.u.}$$

$$\text{Per-unit reactance for motor 2} = 0.1 \times \frac{200}{30} \times \left(\frac{20}{11}\right)^2 = 2.20 \text{ p.u.}$$

$$\text{Per-unit reactance for motor 3} = 0.1 \times \frac{200}{50} \times \left(\frac{20}{11}\right)^2 = 1.32 \text{ p.u.}$$

## EXERCISES

1. Define the terms per-unit voltage, per-unit impedance, and per-unit volt-ampere. Express per-unit impedance in terms of base MVA and base kV for a three-phase system.

2. Derive an expression for per-unit impedance of a given base MVA and base kV in terms of new base MVA and new base kV.

3. Show that the per-unit equivalent impedance of a two-winding transformer is the same whether the calculation is made from the high-voltage side or the low-voltage side.

4. What are the advantages of per-unit representation?

# CHAPTER 3

# Load Characteristics

## 3.1 INTRODUCTION

The primary function of a power station is to serve power to a large number of consumers. Nevertheless, the power needs of consumers are subjected to change depending upon their actions. As a consequence of this variance in demand, the load on a power station is never constant, rather it shifts with time. Due to this reason modern power plant faces a lot of complexities. Regrettably, we cannot store electrical power and, consequently, the power station must create power as and when required to meet the demands of the consumers. On one hand, for maximum efficiency, it is important to run the alternators in the power station at their rated capacity and on the other hand, the requirements of the consumers have wide variances. This makes the design of a power station highly complex.

## 3.2 LOAD

A device that uses electrical energy is said to impose a load on the system. The term load has number of applications such as

- To suggest a device or a collection of devices which consume electrical energy.

- To indicate power required from a given supply circuit.

- To indicate the current passing through a line or a machine.

The load can be resistive, inductive, capacitive, or some combination of them. Load on power systems is split into the following:

- Domestic load—light, fans, refrigerators, heaters, and television
- Commercial load—lighting for shops, fans, and electric appliances used in restaurant
- Industrial load—industrial load consists of load demand by industries
- Municipal load—street lighting, power required for water supply, etc.
- Irrigation load—electric power required for pumps
- Traction loads—tram cars, trolley bus, and railways
- Electronics loads (capacitive loading)—switched-mode power supply and filter circuit

## 3.3 VARIABLE LOAD

The load on the power station changes with time due to uncertain and variable demands of the consumers and is known as *variable load on the station.*

### 3.3.1 Effects of Variable Load

#### 3.3.1.1 Need of Additional Equipment

The variable load on the power station necessitates to have additional equipments. For example, consider a steam power station. Air, coal, and water are the raw materials for this plant. In order to produce variable power, the supply of these materials will be required to be varied correspondingly. For instance, if the power demand on the plant increases, it must be followed by increased flow of coal, air, and water to the boiler in order to meet the increased demand. Therefore, additional equipment has to be installed to accomplish this job. As a matter of fact, in modern plant, there is much equipment devoted entirely to adjust the rates of supply of raw materials in accordance with the power demand made on the plant.

#### 3.3.1.2 Increase in Production Cost

The variable load on the plant increases the cost of production of electrical energy. An alternator operates at maximum efficiency near its rated capacity. If a single alternator is used, it will have poor efficiency during period of light loads on the plant. Therefore, in actual practice, a number of alternators of different capacities are installed so that most of the alternators can be operated at nearly full-load capacity.

However, the use of a number of generating units increases the initial cost per kW of the plant capacity as well as floor area required. This leads to the increase in production cost of energy.

## 3.4 CONNECTED LOAD

Connected load is the sum of continuous ratings of all loads connected to the system.

For instance, if a consumer has connections of five 200 W lamps and power point of 600 W, then connected load of the consumer is $(5 \times 200 + 600 = 1600 \text{ W})$.

## 3.5 DEMAND

The demand of a system is the load that is drawn from the source of supply at a receiving terminal averaged over a suitable and specified interval of time.

The load may be given in kW, kilovar (kVAR), kilovoltampere (kVA), or ampere (A).

## 3.6 DEMAND INTERVAL

Demand interval is the period over which the load is averaged. There are two demands:

1. Instantaneous demand

2. Sustained demand

The former is not very important because all the machines are designed for overloads. The sustained intervals are generally taken as 15 min, 30 min, or even longer. But 30 min is the basic time in India.

## 3.7 MAXIMUM DEMAND OR PEAK LOAD

The maximum demand is the highest demand of load on power station during a given period.

The concept maximum demand should also express the demand interval used to measure it. For example, the specified demand might be maximum of all demands such as daily, weekly, monthly, or annual.

Knowledge of maximum demand helps in determining the installed capacity of a generating station. The generating station must be capable of meeting the maximum demand. Hence, the cost of plant and equipment increases with the increase in maximum demand.

## 3.8 DEMAND FACTOR

The demand factor (DF) is the ratio of the actual maximum demand of the system to the total connected load of the system. Therefore,

$$DF = \frac{\text{Max. demand}}{\text{Total connected load}}$$

The DF can also be found for a part of the system. For example, an industrial or commercial consumer, instead of for the whole system.

## 3.9 AVERAGE LOAD OR AVERAGE DEMAND

The average load occurring on the power station in a given period (day, month, or year) is known as average load or average demand. Therefore,

$$\text{Daily average load} = \frac{\text{No. of units (kWh) generated in a day}}{24 \text{ h}}$$

$$\text{Monthly average load} = \frac{\text{No. of units (kWh) generated in a month}}{\text{No. of hours in that month}}$$

$$\text{Yearly average load} = \frac{\text{No. of units (kWh) generated in a year}}{\text{No. of hours in that year}}$$

## 3.10 LOAD FACTOR

The ratio of average load to the maximum demand during a given period is known as *load factor* (LF).

$$LF = \frac{\text{Average load}}{\text{Max. demand}}$$

If the plant is in operation for $T$ hours,

$$LF = \frac{\text{Average load} \times T}{\text{Max. demand} \times T} = \frac{\text{Units generated in } T \text{ hours}}{\text{Max. demand} \times T \text{ hours}}$$

## 3.11 DIVERSITY FACTOR

The ratio of the individual sum of maximum demands to the maximum demand on power station is known as *diversity factor* ($F_D$).

$$\text{Diversity factor} = \frac{\text{Sum of individual Max. demand}}{\text{Max. demand on the power station}}$$

A power station supplies load to various types of consumers whose maximum demands generally are not the same at the same time. Therefore, the maximum demand on the power station is always less than the sum of maximum individual demands of the consumers.

## 3.12 PLANT CAPACITY FACTOR

It is the ratio of actual energy produced to the maximum possible energy that could have been produced during a given period.

$$\begin{aligned}
\text{Plant capacity factor} &= \frac{\text{Actual energy produced}}{\text{Max. energy that could have been produced}} \\
&= \frac{\text{Average demand} \times T}{\text{Plant capacity} \times T} \\
&= \frac{\text{Average demand}}{\text{Plant capacity}}
\end{aligned}$$

where $T$ denotes the number of hours.

If we consider the period to be 1 year, then

$$\text{Annual plant capacity factor} = \frac{\text{Annual kWh output}}{\text{Plant capacity} \times 8760}$$

The plant capacity factor is a measure of the reserve capacity of the plant. A power station must be designed in such a way that it has some reserve capacity for meeting the increased load demand in future. Therefore, the installed capacity of the plant is always somewhat greater than maximum demand on the plant.

$$\text{Reserve capacity} = \text{Plant capacity} - \text{Max. demand}$$

It is interesting to note that difference between LF and plant capacity factor is an indication of reserve capacity. If the maximum demand on the plant is equal to the plant capacity, then LF and plant capacity factor will have same value. In such case, the plant will have no reserve capacity.

## 3.13 PLANT USE FACTOR

It is the ratio of kWh generated to the product of the plant capacity and the number of hours for which the plant was in operation.

$$\text{Plant use factor} = \frac{\text{Station operation in kWh}}{\text{Plant capacity} \times \text{Hours of use}}$$

Suppose a plant having installed capacity of 20 MW produces annual output of $7.35 \times 10^6$ kWh and remains in operation for 2190 h in a year. Then

$$\text{Plant use factor} = \frac{7.35 \times 10^6}{20 \times 10^3 \times 2190} = 0.167 = 16.7\%$$

## 3.14 UNITS GENERATED PER ANNUM

It is often required to find the kWh generated per annum from maximum demand and LF. The procedure is as follows:

$$\text{LF} = \frac{\text{Average load}}{\text{Max. demand}}$$

$$\text{Average load} = \text{Max. demand} \times \text{LF}$$

$$\text{Units generated/annum} = \text{Average load (in kW)} \times \text{Hours in a year}$$

$$= \text{Max. demand (in kW)}$$

$$\times \text{ LF} \times 8760$$

## 3.15 LOSS FACTOR

It is the ratio of the average power loss ($F_{LS}$) to the peak load power loss during the specified period of time.

$$F_{LS} = \frac{\text{Average power loss}}{\text{Power loss at peak load}}$$

This relationship is applicable for the copper losses of the system but not for the iron losses.

## 3.16 LOAD CURVES

Load curve is a graphical representation between load (in kW or MW) and time (in hours). The curve showing the variation of load on the power station with respect to time is known as load curve. When it is plotted for 24 h a day, it is called daily load curve. If the time considered is 1 year (8760 h), then it is called the annual load curve. Figure 3.1 shows a typical daily load curve of a power station.

It is to be noted that daily load curve of a system is not the same for all days. It differs from day to day and season to season. In practice, two types of curves are drawn—one for summer and the other for winter.

## 3.17 INFORMATION OBTAINED FROM LOAD CURVES

The information listed below are obtained from load curve:

1. Load variation during different hours of the day.

2. The peak load indicated by the load curve gives the maximum demand on the power stations.

3. The area under the load curve gives the total energy generated in the period under consideration.

4. The area under the load curve divided by the total number of hours gives the average load.

5. The ratio of the area under the load curve to the total area of the rectangle in which it is contained gives the LF.

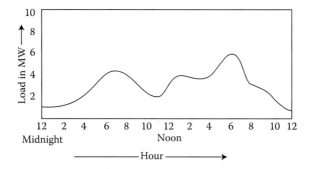

FIGURE 3.1   A typical daily load curve of a power station.

It would be ideal to have a flat-load curve. But in practice, load curves are far from flat. For a flat-load curve, the LF will be higher. Higher LF means more uniform load pattern with less variation in load. This is desirable from the point of view of maximum utilization of associated equipments that are selected on the basis of maximum demand.

## 3.18 LOAD DURATION CURVE

When the load elements of a load curve are arranged in the order of descending magnitudes, the curve thus obtained is called a load duration curve. The load duration curve is derived from the load curve and therefore, represents the same data as that of the load curve but the ordinates are arranged in the order of descending magnitudes. Figure 3.2a and b shows the daily load curve and daily load duration curve, respectively.

## 3.19 INFORMATION AVAILABLE FROM THE LOAD DURATION CURVE

1. It gives minimum load present throughout the given period.

2. It enables the selection of base load and peak load power plants.

3. Any point on the load duration curve gives the total duration in hours for the corresponding load and all loads of greater value.

4. The areas under load curve and corresponding load duration curve are equal. Both areas represent the same associated energy during the period under consideration.

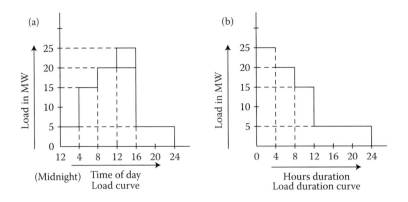

FIGURE 3.2　(a) Daily load curve and (b) daily load duration curve.

5. The average demand during some specified time period such as a day or month or year can be obtained from the load duration curve as follows:

$$\text{Average demand} = \frac{\text{kWh(or MWh) consumed in a given time period}}{\text{Hours in the time period}}$$

$$= \frac{\text{Area under the load duration curve}}{\text{Base of the load duration curve}}$$

## WORKED EXAMPLES

### EXAMPLE 3.1

The peak load on a power station is 60 MW. The load having maximum demand of 30, 20, 15, and 10 MW are connected to the power plant. The capacity of the power plant is 80 MW and the annual LF is 0.80. Estimate (a) the average load on the power plant, (b) the energy supplied per year, (c) the DF, (d) the diversity factor, (e) the utilization factor, (f) the plant capacity factor, and (g) the reverse factor.

**Solution**

(a) Average load = LF × Peak or max. load

$$= 0.8 \times 60 = 48\,\text{MW}$$

(b) Energy supplied per year or $\text{kWh}_{\text{gen}}$ = Average load

$$\times \text{No. of hours in year}$$

$$= 48 \times 10^3 \times 8760$$

$$= 42{,}048 \times 10^4\,\text{kWh}$$

(c) Simultaneous maximum demand of the group of consumers $= (30 + 20 + 15 + 10) = 75$ MW.

Let the connected load be equal to the maximum demand of the group consumers, then

$$\text{DF} = \frac{\text{Max. demand}}{\text{Connected load}} = \frac{60}{75} = 0.8$$

(d) Diversity factor $= \dfrac{\text{Sum of individual consumers max. demand}}{\text{Simultaneous max. demand of the group of consumers}}$

$$= \frac{75}{60} = 1.25$$

(e) Utilization factor $= \dfrac{\text{Max. demand of plant}}{\text{Rated capacity of the plant}} = \dfrac{60}{80} = 0.7$

(f) Plant capacity factor $= \dfrac{\text{Average load}}{\text{Rating or capacity of the plant}}$

$$= \dfrac{48}{80} = 0.6$$

(g) Reverse factor $= \dfrac{\text{LF}}{\text{Capacity factor}} = \dfrac{1}{\text{Utilization factor}}$

$$= \dfrac{1}{0.75} = 1.333$$

## EXAMPLE 3.2

A 150 MW power station delivers 150 MW for 2 h, 75 MW for 8 h, and is shut down for the rest of each day. It is also shut down for maintenance for 50 days each year. Calculate its annual LF.

**Solution**

Energy supplied for each working day $= (150 \times 2) + (75 \times 8)$

$$= 900 \, \text{MWh}$$

Station operates for $= 365 - 50 = 315$ days in a year

Energy supplied/year $= 900 \times 315 = 283{,}500 \, \text{MWh}$

Annual LF $= \dfrac{\text{MWh supplied per annum}}{\text{Max. demand in MW} \times \text{Working hours}} \times 100$

$$= \dfrac{283{,}500}{150 \times (315 \times 24)} \times 100 = 25\%$$

## EXAMPLE 3.3

A power station has a maximum demand of 20,000 kW. The annual LF is 40%, and plant capacity factor is 35%. Determine the reserve capacity of the plant.

**Solution**

Energy generated/annum $=$ Max. demand $\times$ LF $\times$ Hours in a year

$$= (20{,}000) \times (0.4) \times (8760) \, \text{kWh}$$

$$= 70.08 \times 10^{6} \, \text{kWh}$$

$$\text{Plant capacity factor} = \frac{\text{Units generated/annum}}{\text{Plant capacity} \times \text{Hours in a year}}$$

$$\therefore \text{Plant capacity} = \frac{70.08 \times 10^6}{0.35 \times 8760}$$

$$= 22{,}857.14\,\text{kW}$$

$$\text{Reserve capacity} = \text{Plant capacity} - \text{Max. demand}$$

$$= 22{,}857.14 - 20{,}000 = 2857.14\,\text{kW}$$

## EXAMPLE 3.4

A power plant has maximum demand of 80 MW, an LF of 0.7, plant capacity factor of 0.5, and plant use factor of 0.9. Find (a) the daily energy produced, (b) the reverse capacity of the plant, and (c) the maximum energy that could be produced daily if the plant operating schedule is fully loaded when in operation.

### Solution

$$\text{Average load} = \text{LF} \times \text{Peak or max. load}$$

$$= 0.7 \times 80 = 56\,\text{MW}$$

(a)  Daily energy produced $= \text{Average load} \times \text{No. of hours in a day}$

$$= 56 \times 10^3 \times 24 = 1344 \times 10^3\,\text{kWh}$$

(b)  Plant capacity factor $= \dfrac{\text{Average load}}{\text{Capacity of the plant}}$

$$0.5 = \frac{56}{\text{Capacity of the plant}}$$

Therefore,

$$\text{Capacity of the plant} = \frac{56}{0.5} = 112$$

$$\text{Reverse capacity} = \text{Plant capacity} - \text{Max. demand}$$

$$= 112 - 80 = 32\,\text{MW}$$

(c) Plant use factor $= \dfrac{\text{Actual kWh produced}}{\text{Plant capacity (kW)} \times}$

$\text{Actual number of plant operation}$

$0.9 = \dfrac{1344 \times 10^3}{\text{Max. energy that could be produced}}$

Therefore,

$$\text{Max. energy that could be produced} = \frac{1344 \times 10^3}{0.9}$$

$$= 1,493,333.33 \text{ kWh}$$

### EXAMPLE 3.5

Estimate the generation cost per unit of electric energy production from a power plant having the following data:

Output per year $= 5 \times 10^8$ kWh

Annual fixed charges = Rs. 100/kW of installed capacity

Annual running charges = Rs. 0.35/kWh

Annual load factor = 60%

**Solution**

$$\text{Output per charge} = \text{No. of units generated}$$

$$= \text{Average load} \times 8760$$

$$= 5 \times 10^8 \text{ kWh}$$

$$\text{Average load} = \frac{5 \times 10^8}{8760} = 57,077.62 \text{ kW}$$

$$\text{Max. load} = \frac{\text{Average load}}{\text{LF}} = \frac{57,077.62}{0.6}$$

$$= 95,129.37 \text{ kW}$$

Let the installed capacity of the plant be equal to peak load, then

$$\text{Running cost per year} = \text{Annual running charges per kWh}$$

$$\times \text{ Unit number of units}$$

$$= 0.35 \times 5 \times 10^8 = \text{Rs. } 175 \times 10^6$$

Fixed cost = Annual fixed charges per kW × Installed capacity in kW

$$= 100 \times 95{,}129.37 = Rs.\,9{,}512{,}937.6$$

Total annual cost = Fixed cost + Running cost

$$= 9{,}512{,}937.6 + 175 \times 10^6 = Rs.\,1.845 \times 10^8$$

## EXERCISES

1. Why is the load on a power station variable? What are the effects of variable load on the operation of the power station?

2. What do you understand by the load curve? What information are conveyed by a load curve?

3. Define and explain the importance of the following terms in generation:

   a. Connected load

   b. Maximum demand

   c. Demand factor

   d. Average load

4. Explain the terms load factor and diversity factor. How do these factors influence the cost of generation?

5. Explain how load curves help in the selection of size and number of generating units.

6. Discuss the important points to be taken into consideration while selecting the size and number of units.

7. What do you understand about

   a. Base load

   b. Peak load of a power station

8. Write short notes on the following:

   a. Load curves

   b. Load division on hydro-steam system

   c. Load factor

   d. Plant capacity factor

# CHAPTER 4

# Tariffs

## 4.1 INTRODUCTION

The electrical energy that is produced in a power station is delivered to a large number of consumers. The consumers can be convinced to use electrical energy if it is sold at a reasonable price. Here comes the idea of tariffs. A tariff is the schedule of rates structured by the supplier for supplying electrical energy to various types of consumers. The rate at which electric energy is supplied to a consumer is known as tariff. The following elements are engaged into account to determine the tariff:

- Types of load (domestic, commercial, industrial)

- Maximum demand

- Time at which load is required

- Power factor of the load

- Amount of energy used

The way in which consumers pay for electrical energy changes according to their demands. Industrial consumers consume more energy for relatively longer period than domestic consumers. Tariffs should be framed in such a way so that it covers the cost of production, cost of supply, and yet yields some reasonable profit.

The price of energy supplied by a generating station depends on the established capacity of the plant and kWh generated. Maximum demand increases the installed capacity of the generating station.

The instant at which maximum demand occurs is too important in plant economics. If the maximum demand of the consumer and the maximum demand on the system take place simultaneously, additional plant capacity is needed. However, if the maximum demand of the consumer occurs during off-peak hours, then we just need to improve the load factor and no extra plant capacity is needed. Thus, the overall cost per kWh generated is reduced.

Power factor is likewise an important factor from the point of view of plant economics. At a low-power factor, the load current is very high. Therefore, the current to be supplied from the generating station is also large. This high current is also responsible for large $I^2R$ losses in the system and larger voltage drops. Therefore, the regulation becomes poor; in order to supply the consumer's voltage within permissible limits, power factor correction equipment is to be set up. Therefore, the cost of generation increases.

The cost of electrical energy is reduced by using a large amount of energy for a longer period.

## 4.2 OBJECTIVES OF A TARIFF

1. Recovery of cost of producing electrical energy at the power station.

2. Recovery of cost on the capital investment in transmission and distribution systems.

3. Recovery of cost of operation and maintenance of supply of electrical energy. For example, metering equipment, billing, etc.

4. A suitable profit on the capital investment.

## 4.3 DESIRABLE CHARACTERISTICS OF A TARIFF

1. *Proper return.* The tariff should be structured in such a way that it guarantees proper return from each consumer. The total receipts from the consumers must be equal to the cost of producing and supplying electrical energy plus reasonable profit.

2. *Fairness.* The tariff must be fair so that each and every consumer is satisfied with the cost of electrical energy. Thus, a consumer who consumes more electrical energy should be charged at a lower rate than a consumer who consumes little energy. It is because increased energy consumption spreads the fixed charge over a greater number

of units. Hence reducing the overall production cost of electrical energy.

3. *Simplicity.* The tariff should be simple and consumer friendly so that an ordinary consumer can easily understand.

4. *Reasonable profit.* The profit element in the tariff should be reasonable. An electric supply company is a public utility company and generally enjoys the benefits of monopoly.

5. *Attractive.* The tariff should be attractive so that it can attract a large number of consumers to use electricity.

## 4.4 TYPES OF TARIFF

### 4.4.1 Flat-Demand Tariff

This is one of the primitive forms of tariffs used for charging the consumer for consuming electrical energy. In this case, the total demand and the energy consumption are fixed. If $x$ is the number of load connected in kW and $a$ is the rate per lamp or per kW of connected load, then

$$\text{Energy charges} = \text{Rs. } ax$$

### 4.4.2 Simple Tariff

If there is a fixed rate available for per unit of energy consumed, then it is called a simple tariff or uniform rate tariff.

The rate can be delivered as

$$\text{Cost/kWh} = \text{Rs.} \frac{\text{Annual fixed cost} + \text{Annual operating cost}}{\substack{\text{Total number of units supplied to the} \\ \text{consumer per annum}}}$$

*Disadvantages*

- We cannot differentiate various types of consumers (domestic, industrial, bulk) having different load factor, diversity, and power factor.

- The cost per kWh delivered is higher.

- It does not encourage the use of electricity.

### 4.4.3 Flat-Rate Tariff

When different types of consumers are charged at different per-unit rates, it is called a flat-rate tariff. In this type of tariff, the consumers are grouped into various categories, and each type of consumers is charged at a different rate. For instance, the flat rate per kWh for lighting load may be 60 paisa, where as it may slightly less (say 55 paisa) for power load.

*Advantages*

- More fair to different types of consumers.

- Quite simple in calculations.

*Disadvantages*

- It varies with the consumption of electrical energy, and separate meters are required for lighting load, power load, etc. This makes the application of such tariff costly and complex.

- A particular category of consumers are charged at the same rate irrespective of the magnitude of energy consumed. However, big consumers should be charged at a relatively lower rate, as in this case the fixed charges per unit are reduced.

### 4.4.4 Step-Rate Tariff

The step-rate tariff is a group of flat-rate tariffs of decreasing unit charges for higher range of consumption. For example,

- Rs. 4.0/unit if the consumption does not exceed 50 kWh.

- Rs. 3.5/unit if the consumption exceed 50 kW but does not exceed 200 kW.

- Rs. 3.0/unit if the consumption exceeds 200 kW.

*Disadvantage*

- However, by increasing the energy consumption, cost is reduced. Thus, there is a tendency with the consumer, just approaching the limit of the step, to anyhow cross the step and enter the next one in order to reduce the total energy cost.

This drawback is removed in block-rate tariff explained below.

### 4.4.5 Block-Rate Tariff

When a specific block of energy is charged at a specified rate and the succeeding blocks of energy are charged at a progressively reduced rate, it is called a block-rate tariff.

For example, the first 40 units may be charged at 70 paisa/unit, next 35 units at 55 paisa/unit, and remaining additional units at 30 paisa/unit.

This is used for majority of residential and small commercial consumers.

*Advantages*

- Consumers get an incentive to consume more electrical energy.

- This increases the load factor of the system and hence the cost of generation is reduced.

*Disadvantage*

- Its principal defect is that it cannot measure the consumer's demand.

### 4.4.6 Two-Part Tariff

When the rate of electrical energy is charged on the basis of maximum demand of the consumers and the units consumed, then it is called two-part tariff or *Hopkinson demand tariff* (Table 4.1).

In this case, the total cost that is to be charged from the consumer is split into two components:

- *The fixed charges* depend upon the maximum demand of the consumers.

- *The running charges* depend upon the number of units consumed by the consumers.

$$\therefore \text{Total charges} = \text{Rs. } (b \times kW + c \times kWh)$$

where $b$ is the charge per kW of maximum demand and $c$ is the charge per kWh of energy consumed.

- Applicable to industrial consumers who have appreciable maximum demand.

TABLE 4.1 Consumers and Their Tariffs

| Consumers | Examples | Supply Given | Demand Factor | Tariff | Additional Charges |
|---|---|---|---|---|---|
| Domestic | Residential load, light, fan, television, radio, electric irons, domestic pumps, coolers, air conditioners | 1ϕ: supply up to a load of 5 kW 3ϕ: supply for loads exceeding 5 kW | Small consumers (high unity), big consumers (0.5) | 1. Simple 2. Flat rate 3. Block rate | Meter rent and electricity duty |
| Commercial | Shops, business houses, hotels, cinemas, clubs, etc. | 1ϕ: supply up to a load of 5 kW 3ϕ: supply for loads exceeding 5 kW | Fairly high | 1. Simple 2. Flat rate 3. Block rate | Meter rent and electricity duty |
| Agricultural | Tube wells | 3ϕ: power up to 20 kW | Unity | Flat rate | |
| Bulk | Railways, educational institutes, military establishment, hospitals | 3ϕ: power at 415 V or 11 kV depending on their requirement, load exceeding 10 kW | | Flat rate | |
| Industrial (small) | Atta chakkis, small workshop, saw mill, etc. | 3ϕ: power supply at 415 V, load not exceeding 20 kW | Usually high (0.8) | Block tariff | |
| Industrial (medium) | | 3ϕ: power supply at 415 V, load exceeding 20 kW but not exceeding 100 kW | | Two-part tariff | |
| Industrial (large) | | Power supplied at 11 kV or 33 kV, load exceeding 100 kW | 0.5 | KVA maximum demand factor tariff | |

*Advantages*

- Easily understood by the consumers.
- It recovers the fixed charges that depend upon the maximum demand of the consumer but are independent of the units consumed.

*Disadvantages*

- The consumer has to pay the fixed charges irrespective of whether he or she has consumed or not consumed the electric energy.
- There is always error in determining the maximum demand of the consumer.

## 4.4.7 Maximum-Demand Tariff

It is quite similar to two-part tariff; the only difference is that the maximum demand is actually measured by installing maximum demand meter in the premises of the consumer.

*Advantage*

- This eliminates the disadvantage of two-part tariff, where maximum demand is determined merely on the basis of the chargeable value.

This tariff is mostly applied to big consumers.

## 4.4.8 Three-Part Tariff

In three-part tariff, the total charge to be made from the consumer is split into three parts, that is,

- Fixed charges
- Semi-fixed charges
- Running charges

$$\therefore \text{Total charges} = \text{Rs.}\,(a + b \times \text{kW} + c \times \text{kWh})$$

where $a$ is the fixed charge made during each billing period. It includes interest and depreciation on the cost of secondary distribution and labour cost of collecting revenues. *b* is the charge per kW of maximum demand and *c* is the charge per kWh of energy consumed.

The principal objection of this tariff is that the charges are split into three compartments. Generally applied to big consumers.

### 4.4.9 Power Factor Tariff

The tariff in which power factor of the consumer's load is taken into consideration is known as *power factor tariff.*

A low-power factor increases the rating of the station equipment and line losses. Therefore, a consumer having low-power factor must be penalized.

The following are the important types of power factor tariff.

#### 4.4.9.1 kVA Maximum-Demand Tariff

It is a modified form of two-part tariff. The fixed charges are formulated on the basis of maximum demand in kVA, and not in kW. As kVA is inversely proportional to the power factor, a consumer having a low-power factor has to contribute more toward the fixed charges.

*Advantage*

- It encourages the consumers to operate the appliances and machinery at improved power factor.

#### 4.4.9.2 Sliding Scale Tariff

This is known as average power factor tariff. In this case, an average power factor (say 0.8 lagging) is taken as the reference. If the power factor of the consumer falls below this factor, suitable additional charges are made. On the other hand, if the power factor is above the reference, a discount is allowed to the consumers.

#### 4.4.9.3 kW and kVAR Tariff

In this type, both active power (kW) and reactive power (kVAR) supplied are charged separately. A consumer having low-power factor will draw more reactive power and hence shall have to pay more charges.

## WORKED EXAMPLES

### EXAMPLE 4.1

The maximum demand of a consumer is 15 A at 230 V and his/her total energy consumption is 9000 kWh. If the energy is charged at the rate of Rs. 5 per unit for 600 h use of the maximum demand per annum plus Rs. 2 per unit for additional units, calculate (1) annual bill and (2) equivalent flat rate.

**Solution**

Assume the load factor and power factor to be unity.

$$\text{Maximum demand} = \frac{230 \times 15 \times 1}{1000} = 3.45 \text{ kW}$$

1. Units consumed in 600 h = 3.45 × 600 = 2070 kWh

   Charges for 2070 kWh = Rs. 5 × 2070 = Rs. 10,350

   Remaining units = 9000 − 2070 = 6930 kWh

   Charges for 6930 kWh = Rs. 2 × 6930 = Rs. 13,860

   ∴ Total annual bill = Rs. (13,860+10,350) = Rs. 24,210

2. Equivalent flat rate = Rs. 24,210/9000 = Rs. 2.69.

### EXAMPLE 4.2

A consumer has a maximum demand of 150 kW at 50% load factor. If the tariff is Rs. 800 per kW of maximum demand plus Rs. 2 per kWh, find the overall cost per kWh.

**Solution**

$$\text{Units consumed/year} = \text{MD} \times \text{LF} \times \text{Hours in a year}$$
$$= (150) \times (0.5) \times 8760 = 657{,}000 \text{ kWh}$$

$$\text{Annual charges} = \text{Annual MD charges} + \text{Annual energy charges}$$
$$= \text{Rs. } (150 \times 800 + 2 \times 657{,}000)$$
$$= 1{,}434{,}000$$

$$\therefore \text{Overall cost/kWh} = \text{Rs.} \frac{1{,}434{,}000}{657{,}000} = \text{Rs. } 2.18$$

where MD is the maximum demand and LF is the load factor.

### EXAMPLE 4.3

The monthly readings of a consumer's meter are as follows:

$$\text{Maximum demand} = 75\,\text{kW}$$
$$\text{Energy consumed} = 54{,}000\,\text{kWh}$$
$$\text{Reactive energy} = 27{,}500\,\text{kVAR}$$

If the tariff is Rs. 600 per kW of maximum demand plus Rs. 2 per unit plus Rs. 1 per unit for each 1% of power factor below 90%, calculate the monthly bill of the consumer.

**Solution**

$$\text{Average load} = \frac{54{,}000}{24 \times 30} = 75\,\text{kW}$$

$$\text{Average reactive power} = \frac{27{,}500}{24 \times 30} = 38.19\,\text{kVAR}$$

Suppose $\phi$ is the power factor angle,

$$\tan\phi = \frac{\text{kVAR}}{\text{Active power}} = \frac{38.19}{75} = 0.51$$

or

$$\phi = \tan^{-1}(0.51) = 27.02°$$

Therefore, power factor is

$$\cos\phi = \cos(27.02°) = 0.8908$$

$$\text{Power factor surcharge} = \text{Rs.}\,\frac{54{,}000 \times 1}{100} \times (90 - 89.08) = \text{Rs.}\,496.80$$

$$\text{Monthly bill} = \text{Rs.}\,(600 \times 75 + 54{,}000 \times 2 + 496.80)$$
$$= \text{Rs.}\,153{,}496.80$$

## EXAMPLE 4.4

The daily load of an industrial concern is as follows: 150 kW for 10 h, 175 kW for 5 h, 60 kW for 6 h, 50 kW for 3 h, the tariff is Rs. 700 per kW of maximum demand per year plus Rs. 1.20 per kWh. Determine the energy consumption per year and the yearly bill.

**Solution**

$$\text{Energy consumption per day} = (150 \times 10) + (175 \times 5) + (60 \times 6) + (50 \times 3)$$
$$= 2885 \text{ kWh}$$

$$\text{Annual energy consumption} = 2885 \times 365 = 1{,}053{,}025 \text{ kWh}$$

$$\text{Annual cost of energy consumption} = 1{,}053{,}025 \times 1.2 = \text{Rs. } 1{,}263{,}630$$

$$\text{Maximum demand} = 175 \text{ kW}$$

$$\text{Annual maximum demand charges} = \text{Rs. } 175 \times 700 = \text{Rs. } 122{,}500$$

$$\text{Total annual charges} = \text{Rs. } (1{,}263{,}630 + 122{,}500) = \text{Rs. } 1{,}386{,}130.$$

## EXAMPLE 4.5

The following tariffs are offered to a consumer:

1. Rs. 600 per month plus Rs. 1/kWh.
2. Rs. 1.5 for the first 100 units per month and Rs. 1.70 for next 100 units and 1.9 for all the additional units.

Find the energy consumed per year for which the charges due to both tariffs become equal.

**Solution**

Let $x$ be the number of units consumed, where $x > 200$.

$$\text{Annual charges due to first tariff } c_1 = \text{Rs. } (600 + 1x).$$

Annual charges due to second tariff $c_2 =$ Rs. $(1.5 \times 100) + (1.7 \times 100)$
$$+ (1.9)(x - 200)$$
$$= \text{Rs.} \ (1.9x - 60)$$

If $c_1 = c_2$, then

$$600 + 1x = 1.9x - 60$$
$$(1.9 - 1)x = 660$$
$$x = \frac{660}{0.9} = 733.33 \, \text{kWh}$$

## EXERCISES

1. What do you understand by tariff? Discuss the objectives of tariff.

2. Describe the desirable characteristics of a tariff.

3. Describe some of the important types of tariff commonly used.

4. Write short notes on the following:

    a. Two-part tariff

    b. Power factor tariff

    c. Three-part tariff

# CHAPTER 5

# Mechanical Design of Overhead Line

## 5.1 INTRODUCTION

For proper operation of overhead line, protective measures must be considered in the invention of electrical parameters as well as for the mechanical portion. The line should have sufficient current-carrying capacity so that the necessary power transfer can take place without violating allowable voltage drop criterion or overheating. Line losses should be as low as possible and spacing between the line conductors and with the earth should be adequate to cope up with the system voltage. On the other hand, for the mechanical aspects, the line conductors, supports, and the cross arm should have sufficient mechanical strength to cope with the probable weather problems. The mechanical design should estimate the "sag" of the conductor at its mid-span length, as sag is to be allowed to cater the load of the conductor. Adequate distance between the lowest point of the line and the earth must be maintained.

## 5.2 CONDUCTOR MATERIAL

The conductor is one of the important objects, as most of the capital outlay is invested for it. Therefore, choice of material and size of conductor are of utmost importance. The conductor material used for transmission and distribution of electric power should possess the following properties:

1. High electrical conductivity.

2. High tensile strength in order to withstand mechanical strength.

3. Low cost so that it can be used for long distance.

4. Low specific gravity so that weight per-unit volume is small.

All above requirements and specifications are not found in a single material. Therefore, when we select a conductor material, for a particular case, a compromise is made between the cost and the required electrical and mechanical properties.

## 5.2.1 Commonly Used Conductor Materials

The most frequently used conductor materials for overhead lines are copper, aluminum, steel-cored aluminum, galvanized steel, and cadmium copper. The choice of a particular material will depend upon the cost, the required electrical and mechanical properties, and the local conditions.

All conductors used for overhead lines are preferably stranded, in order to increase flexibility. In stranded conductors, there is generally one central wire, and around this, successive layers of wires containing 6, 12, 18, 24, and more wires. Thus if there are $n$ layers, the total number of individual wire is $3n(n + 1) + 1$. In the manufacture of stranded conductors, the consecutive layers of wires are twisted or spiraled in opposite directions so that layers are bound together.

### 5.2.1.1 Copper

Copper is an ideal material for overhead lines due to its high electrical conductivity and greater tensile strength. It is always used in the hard drawn form as stranded conductor. Although hard drawing decreases the electrical conductivity slightly, yet it increases the tensile strength reasonably. Copper has high current density, that is, the current carrying capacity of copper per unit of cross-sectional area is quite large. This leads to two advantages:

1. Smaller cross-sectional area of conductor is required.

2. The area offered by the conductor to wind load is reduced.

Moreover, this metal is quite homogeneous, durable, and has high scrap value. There is no doubt that copper is an ideal material for transmission and distribution of electric power. However, due to its high cost and nonavailability, it is rarely used for these purpose. Nowadays the trend is to use aluminum in place of copper.

### 5.2.1.2 Aluminum

Aluminum is cheap and light as compared to copper but it has much smaller conductivity and tensile strength. The relative comparison of the two materials is given below:

1. The conductivity of aluminum is 60% that of copper. The smaller conductivity of aluminum means that for any particular transmission efficiency, the cross-sectional area of conductor must be larger in aluminum than in copper. For the same resistance, the diameter of aluminum conductor is about 1.26 times the diameter of copper conductor. The increased cross-sectional area of aluminum exposes a greater surface to wind pressure and, therefore, supporting towers must be designed for greater transverse strength. This often requires the use of high towers with consequence of greater sag.

2. The specific gravity of aluminum ($2.71 \text{ g/cm}^3$) is lower than that of copper ($8.9 \text{ g/cm}^3$). Therefore, an aluminum conductor has almost one half the weight of equivalent copper conductor. For this reason, the supporting structures for aluminum need not be made as strong as that of copper conductor.

3. Aluminum conductor being light is liable to greater swings and hence larger crossarms are required.

4. Due to lower tensile strength and higher coefficient of linear expansion of aluminum, the sag is greater in aluminum conductors.

Considering the combined properties of cost, conductivity, tensile strength, weight, etc., aluminum has an edge over copper. Therefore, it is being widely used as a conductor material. It is particularly profitable to use aluminum for high current transmission, where the conductor size is large and its cost forms a major portion of the total cost of complete installation.

### 5.2.1.3 Steel-Cored Aluminum

Due to low tensile strength, aluminum conductors produce greater sag. This prohibits their use for larger spans and makes them unsuitable for long distance transmission. In order to increase the tensile strength, the aluminum conductor is reinforced with a core of galvanized steel wires. The composite conductor thus obtained is known as steel-cored aluminum and is abbreviated as ACSR (aluminum conductor steel reinforced).

Steel-cored aluminum conductor consists of central core of galvanized steel wires surrounded by a number of aluminum strands. Usually, diameter of both aluminum and steel wires is the same. The cross-sections of two metals are generally in the ratio of 1:6 but can be modified to 1:4 in order to get more tensile strength for the conductor. Figure 5.1a and b shows steel-cored aluminum conductor having one steel wire surrounded by six wires of aluminum. The result of this composite conductor is that steel core takes greater percentage of mechanical strength, while aluminum conductors have the following advantages:

1. The reinforcement with steel increases the tensile strength but at the same time keeps the composite conductor light. Therefore, steel-cored aluminum conductors will produce smaller sag and hence larger span can be used.

2. Due to smaller sag with steel-cored aluminum conductors, towers of smaller heights can be used.

### 5.2.1.4 Galvanized Steel
Steel has very high tensile strength. Therefore, galvanized steel conductors can be used for extremely long spans or for short-line sections exposed to abnormally high stresses due to climatic conditions. They are found very suitable in rural areas where cost is the main consideration. Due to poor

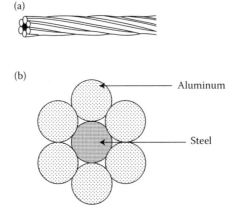

FIGURE 5.1 (a) Conventional ACSR conductor and (b) cross-sectional view of ACSR conductor.

conductivity and high resistance of steel, such conductors are not suitable for transmitting large power over a long distance. However, they can be used to transmit a small power over a small distance where the size of the desired copper conductor would be too small and thus unsuitable for use because of poor mechanical strength.

Nowadays, use of galvanized steel wires is limited to telecommunication lines, stray wires, earth wires, and guard wires.

### 5.2.1.5 Cadmium Copper

The conductor material now being employed in certain cases is copper alloyed with cadmium. An addition of 1% or 2% cadmium to copper increases the tensile strength by 50%, and the conductivity is only reduced by 15% below that of pure copper. Therefore, cadmium copper conductor can be useful for exceptionally long spans. However, due to high cost of cadmium, such conductors will be economical only for lines of small cross-sections, that is, where the cost of conductor material is comparatively small compared with the cost of supports.

### 5.2.1.6 Phosphor Bronze

When harmful gases such as ammonia are present in atmosphere and the spans are extremely long, phosphor bronze is most suitable material for an overhead line conductor. In this conductor, some strands of phosphor bronze are added to the cadmium copper.

## 5.3 LINE SUPPORTS

The supporting structures for overhead line conductors are various types of poles and towers called line supports. In general, the line supports should have the following properties:

1. High mechanical strength to withstand the weight of conductors, wind load, etc.

2. Light in weight without the loss of mechanical strength.

3. Cheaper in cost.

4. Low maintenance cost.

5. Longer life.

6. Easy accessibility of conductors for maintenance.

The line supports used for transmission and distribution of electric power are of various types including wooden poles, steel poles, reinforced cement concrete (RCC) poles, and lattice steel towers. The choice of supporting structure for a particular case depends upon the line span, cross-sectional area, line voltage, cost, and local conditions.

### 5.3.1 Wooden Poles

These are made of seasoned wood (sal or chir) and are suitable for lines of moderate cross-sectional area and of relatively shorter spans, say up to 50 m. Such supports are cheap, easily available, provide insulating properties, and therefore are widely used for distribution purposes in rural areas as an economical proposition. The wooden poles generally tend to rotten below the ground level, causing foundation failure. In order to prevent this, the portion of the pole below the ground level is impregnated with preservative compound like creosote oil. Double pole structures of the "A" or "H" type are often used (Figure 5.2) to obtain a higher transverse strength than could be economically provided by means of a single pole.

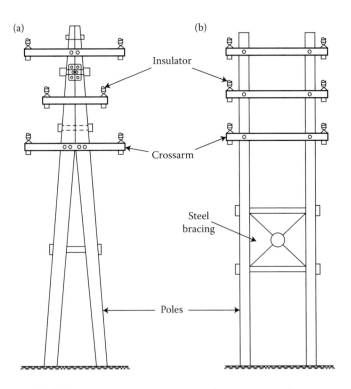

FIGURE 5.2  (a) "A" type wooden poles and (b) "H" type wooden poles.

The main objectives of wooden supports are

- Tendency to rot below the ground level
- Comparatively smaller life (20–25 years)
- Cannot be used for voltages higher than 20 kV
- Less mechanical strength
- Require periodical inspection

### 5.3.2 Steel Poles

The steel poles are often used as a substitute wooden pole. They posses greater mechanical strength, longer life, and permit longer span to be used. Such poles are generally used for distribution purposes in the cities. These types of supports need to be galvanized or painted in order to prolong its life. The steel poles are of three types:

1. Rail poles

2. Tubular poles

3. Rolled steel joints

### 5.3.3 RCC Poles

The reinforced concrete poles usually called the concrete poles are extensively used for low- and high-voltage distribution lines up to 33 kV. They have greater mechanical strength, longer life, and permits longer spans than steel poles. Moreover, they give good outlook, require little maintenance, and have good insulating properties. Figure 5.3 shows RCC poles for single and double circuit. The holes in the poles facilitate the climbing of poles and at the same time reduce the weight of line supports.

The main difficulty with the use of these poles is the high cost of transport owing to their heavy weight; therefore, such poles are often manufactured at the site in order to avoid heavy cost of transmission.

### 5.3.4 Steel Towers

For long distance transmission at higher voltage, steel towers are invariably employed which have greater mechanical strength, longer life, and

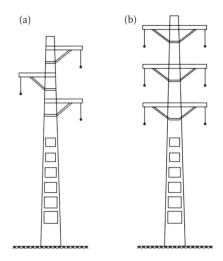

FIGURE 5.3 (a) RCC poles for single circuit and (b) RCC poles for double circuit.

can withstand most severe climate conditions and permit the use of longer spans. The risk of interrupted service due to broken or punctured insulation is considerably reduced owing to longer spans. Tower footings are usually grounded by driving rods into the earth. This minimizes the lightening troubles as each tower acts as a lightening conductor.

Figure 5.4a and c shows a single circuit 110 kV tower and 500 kV tower, respectively. However, at a moderate additional cost, double circuit tower

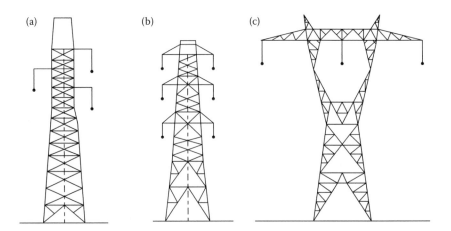

FIGURE 5.4 (a) Single circuit 110 kV steel tower, (b) double circuit steel tower, and (c) single circuit 500 kV steel tower.

can be provided as shown in Figure 5.4b. The double circuit has the advantage that it ensures continuity of supply. In case, there is breakdown of one circuit, the continuity of supply can be maintained by the other circuit.

## 5.4 INDIAN ELECTRICITY RULES (1956) FOR OVERHEAD LINES

Important rules are

- Rule 74—Material and strength

- Rule 75—Joints

- Rule 76—Maximum stress: factor of safety

- Rule 77—Clearance above ground of the lowest conductor

- Rule 85—Maximum interval between supports

- Rule 90—Earthing

- Rule 92—Protection against lightning

## 5.5 SAG IN OVERHEAD LINES

While erecting an overhead line, it is very important that conductors are under safe tension. If the conductors are stretched too much between supports in a bid to save conductor material, the stress in the conductor may reach unsafe value, and in certain cases, the conductor may break due to excessive tension. In order to permit safe tension in the conductors, they are not fully stretched but are allowed to have a dip or sag.

The difference in level between points of supports and the lowest point on the conductor is called sag.

Figure 5.5a shows a conductor suspended between two equilevel supports A and B. The conductor is not fully stretched but is allowed to have a dip. The lowest point on the conductor is O and the sag is S. the following points may be noted:

1. When the conductor is suspended between two supports at the same level, it takes the shape of catenary. However, if the sag is very small compared with the span, the sag–span curve is like a parabola.

FIGURE 5.5    (a) A conductor suspended between two equilevel supports and (b) tension at any point on the conductor.

2. The tension at any point on the conductor acts tangentially. Thus, tension $T_0$ at the lowest point acts horizontally as shown in Figure 5.5b.

3. The horizontal component of tension is constant throughout the length of the wire.

4. The tension at supports is approximately equal to the horizontal tension acting at any point on the wire. Thus, if $T$ is the tension at the support B, then $T = T_0$.

### 5.5.1 Conductor Sag and Tension

This is an important consideration in the mechanical design of overhead lines. The conductor sag should be kept to a minimum in order to reduce the conductor material required and to avoid extra pole height for sufficient clearance above ground level. It is also desirable that tension in the conductor should be low to avoid the mechanical failure of the conductor and to permit the use of less strong supports. However, low conductor supports and minimum sag are not possible. It is because low sag means a tight wire and high tension, whereas low tension means a loose wire and increased sag. Therefore, in actual practice, a compromise is made between the two.

## 5.6 CALCULATION OF SAG

In an overhead line, the sag should be so adjusted that tension in the conductors is within safe limits. The tension is governed by conductor weight, effect of wind, ice loading, and temperature variations. It is a standard practice to keep conductor tension less than 50% of its ultimate tensile strength, that is, minimum factor of safety in respect of conductor tension should be 2. We shall now calculate sag tension of a conductor when (i) supports are at equal level and (ii) supports are at unequal level.

### 5.6.1 When Supports Are at Equal Levels

Consider a conductor between two equilevel supports A and B with O as the lowest point as shown in Figure 5.6. It can be proved that lowest point will be at the mid-span.

Let $l$ be the length of span, $w$, the weight per-unit length of conductor, and $T$, the tension in the conductor.

Consider a point on the conductor. Taking the lowest point O as the origin, let the coordinates of the point P be $x$ and $y$. Assuming that the curvature is so small that curved length is equal to its horizontal projection (i.e., OP = $x$), the forces acting on the portion OP on the conductor are

1. The weight $wx$ of conductor acting at a distance $x/2$ from O.

2. The tension $T$ acting at O.

Equating the moments of above two forces about point O, we get

$$Ty = wx \times \frac{x}{2}$$

or

$$y = \frac{wx^2}{2T}$$

The maximum dip (sag) is represented by the value of $y$ at either of the supports A and B. At support A, $x = l/2$ and $y = S$. Therefore,

$$S = \frac{w(l/2)^2}{2T} = \frac{wl^2}{8T}$$

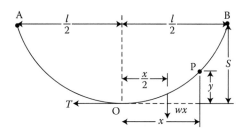

FIGURE 5.6   External forces on the sag between two equilevel supports.

## 5.6.2 When Supports Are at Unequal Levels

In hilly areas, we generally come across conductors suspended between supports at unequal levels. Figure 5.7 shows a conductor suspended between two supports A and B that are at different levels. The lowest point on the conductor is O.

Let $I$ is the span length, $h$ is the difference in levels between two supports, $x_1$ is the distance of support at lower level (i.e., A) from O, $x_2$ is the distance of support at higher level (i.e., B) from O, and $T$ is the tension in the conductor.

If $w$ is the weight per-unit length of the conductor, then

$$\text{Sag } S_1 = \frac{wx_1^2}{2T}$$

$$\text{Sag } S_2 = \frac{wx_2^2}{2T}$$

Also

$$x_1 + x_2 = 1 \tag{5.1}$$

Now

$$S_2 - S_1 = \frac{w}{2T}(x_2^2 - x_1^2) = \frac{w}{2T}(x_2 + x_1)(x_2 - x_1)$$

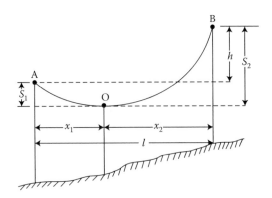

FIGURE 5.7   Two supports at different levels from the ground.

$$\therefore S_2 - S_1 = \frac{wl}{2T}(x_2 - x_1) \quad [\because x_1 + x_2 = 1]$$

But

$$S_2 - S_1 = h$$

$$\therefore h = \frac{wl}{2T}(x_2 - x_1)$$

or

$$x_2 - x_1 = \frac{2Th}{wl} \tag{5.2}$$

Solving Equations 5.1 and 5.2, we get

$$x_1 = \frac{1}{2} - \frac{Th}{wl} \tag{5.3}$$

$$x_2 = \frac{1}{2} + \frac{Th}{wl} \tag{5.4}$$

Having found $x_1$ and $x_2$ values of $S_1$ and $S_2$ can be easily calculated. In Equation 5.3,

$$\text{If } \frac{1}{2} > \frac{Th}{wl}, \quad \text{then } x_1 \text{ is positive}$$

$$\text{If } \frac{1}{2} > \frac{Th}{wl}, \quad \text{then } x_1 \text{ is zero}$$

$$\text{If } \frac{1}{2} < \frac{Th}{wl}, \quad \text{then } x_1 \text{ is negative}$$

If $x_1$ is negative, the lowest point (point O) of the imaginary curve lies outside the actual span as shown in Figure 5.8.

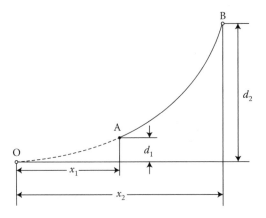

FIGURE 5.8　Case of negative $x_1$.

## 5.6.3 Effect of Wind and Ice Loading

The above formulas for sag are true only in still air and at normal temperature when the conductor is acted by its weight only. However, in actual practice, a conductor may have ice coating and simultaneously subjected to wind pressure. The weight of ice acts vertically downwards, that is, in the same direction as the weight of conductor. The force due to the wind is assumed to act horizontally, that is, at the right angle to the projected surface of the conductor. Hence, the total force on the conductor is the vector sum of horizontal and vertical forces as shown in Figure 5.9.

Total weight of conductor per-unit length is

$$w_t = \sqrt{(w + w_i)^2 + w_w^2}$$

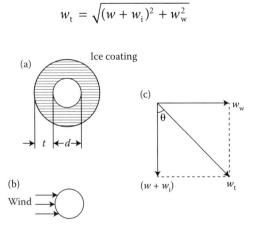

FIGURE 5.9　(a) Ice coating on the conductor, (b) wind pressure on the conductor, and (c) total force on the conductor.

where $w$ is the weight of conductor per-unit length,

$$w = \text{Conductor material density} \times \text{Volume per-unit length}$$

$w_i$ is the weight of ice per-unit length,

$$w_i = \text{Density of ice} \times \text{Volume of ice per-unit length}$$
$$= \text{Density of ice} \times \frac{\pi}{4}[(d + 2t)^2 - d^2] \times 1$$
$$= \text{Density of ice} \times \pi t(d + t)$$

and $w_w$ is the wind force per-unit length,

$$w_w = \text{Wind pressure per-unit area} \times \text{Projected area per-unit length}$$
$$= \text{Wind pressure} \times [(d + 2t) \times 1]$$

## 5.7 SAG TEMPLATE

For perfect design and maintaining economic balance, location of structures of profile with a template is very essential. Sag template is a suitable device, which is often used in designing a transmission line to determine the location and height of the structures. Sag template can be a reliable option which provides the following:

1. Economic layout

2. Minimum errors in design and layout

3. Proper grading of structures

4. Controls excessive insulator swing

Generally two types of towers are used:

1. The standard or straight run or intermediate tower

2. The angle or anchor or tension tower

The straight run towers are used for straight runs and normal conditions. The angle towers are designed to withstand heavy loading as

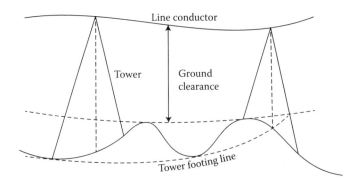

FIGURE 5.10   Sag template for locating towers.

compared to standard towers because angle towers are used at angles, terminals, and other points, where a large unbalanced pull may be thrown on the supports.

For standard towers, normal or average spans, the sag and the nature of the curve (catenary or parabola) the line conductor occupies under expected loading conditions are evaluated and plotted on the template. Template will also show the required minimum ground clearance by plotting a curve parallel to the conductor shape curve. For the standard tower and same height, the tower footing line can also be plotted on the template. Tower footing line is used for locating the position of towers, and minimum ground clearance is maintained throughout. Figure 5.10 shows the sag template used for locating towers.

## 5.8  STRINGING CHART

For use in the field work of stringing the conductors, temperature–sag and temperature–tension charts are plotted for the given conductor and loading conditions. Such curves are called stringing charts (Figure 5.11). These charts are very helpful while stringing overhead lines.

## WORKED EXAMPLES

### EXAMPLE 5.1

A 132-kV transmission line has the following data: weight of conductor = 700 kg/km, length of span = 250 m, ultimate strength = 2800 kg, and safety factor = 2.

Calculate the height above ground at which the conductor should be supported. Ground clearance required is 15 m.

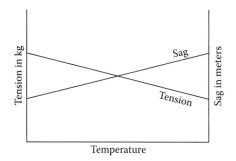

FIGURE 5.11   Stringing charts.

### Solution

- Weight of conductor/meter run,

$$w = \frac{700}{1000} = 0.7\,\text{kg}$$

- Working tension,

$$T = \frac{\text{Ultimate strength}}{\text{Safety factor}} = \frac{2800}{2} = 1400\,\text{kg}$$

- Span length, $l = 250$ m. Therefore,

$$\text{Sag} = \frac{wl^2}{8T} = \frac{0.7 \times 250^2}{8 \times 1400} = 3.90\,\text{m}$$

∴ Conductor should be supported at a height of $15 + 3.9 = 18.9$ m.

### EXAMPLE 5.2

A transmission line has a span of 250 m between level supports. The conductor has an effective diameter of 1.44 cm and weighs 0.900 kg/m. Its ultimate strength is 7520 kg. If the conductor has ice coating of radial thickness 1.43 cm and is subjected to a wind pressure of 3.8 g/cm² of projected area, calculate sag for a safety factor of 2. Weight of 1 cm³ of ice is 0.91 g.

**Solution**

Span length, $l = 250$ m
Weight of conductor/meter length, $w = 0.900$ kg
Conductor diameter, $d = 1.44$ cm
Ice-coating thickness, $t = 1.43$ cm
Working tension is

$$T = \frac{7520}{2} = 3760\,\text{kg}$$

Volume of ice per meter length of conductor is

$$= \pi t(d + t) \times 100\,\text{cm}^3$$
$$= \pi \times 1.43 \times (1.44 + 1.43) \times 100 = 1289.34\,\text{cm}^3$$

Weight of ice per meter length of conductor is

$$w_i = 0.91 \times 1289.34 = 1173.3\,\text{g} = 1.173\,\text{kg}$$

Wind force/meter length of conductor is

$$w_w = [\text{Pressure}] \times [(d + 2t) \times 100]$$
$$= [3.8] \times (1.44 + 2 \times 1.43) \times 100\,\text{g} = 1634\,\text{g} = 1.634\,\text{kg}$$

Total weight of conductor per meter length of conductor is

$$w_t = \sqrt{(w + w_i)^2 + (w_w)^2}$$
$$= \sqrt{(0.900 + 1.173)^2 + (1.634)^2}$$
$$= 2.639\,\text{kg}$$

$$\therefore \text{Sag} = \frac{w_t l^2}{8T} = \frac{2.639 \times (250)^2}{8 \times 3760} = 5.48\,\text{m}$$

**EXAMPLE 5.3**

A transmission line conductor at a river crossing is supported from two towers at a height of 50 and 80 m above water level. The horizontal distance between the towers is 300 m. If the tension in the conductor is 2000 kg, find the clearance between the conductor and water level at a point midway between the towers. Weight of the conductor per meter is 0.844 kg. Assume that the conductor takes the shape of parabolic curve.

**Solution**

Difference in level between the two supports: $h = 80 - 50 = 30$ m

Distance of lowest point of conductor from the support of low level: $x = (L/2) - (Th/wL)$

$$\therefore x = \frac{300}{2} - \frac{2000 \times 30}{0.844 \times 300} = -86.967 \text{ m}$$

The negative value of the $x$ shows that the support A is on the same side of O as support B.

Distance of midpoint P from O is

$$\frac{L}{2} - x = \frac{300}{2} - (-86.967)$$
$$= 236.967 \text{ m}$$

Distance of point B from O is

$$L - x = 300 + 86.967 = 386.967 \text{ m}$$

Height of midpoint P above O is

$$S_{mid} = \frac{w \times ((L/2) - x)^2}{2T}$$
$$= \frac{0.844 \times (236.967)^2}{2 \times 2000}$$
$$= 11.848 \text{ m}$$

Height of point B from O is

$$S_2 = \frac{w \times (L - x)^2}{2T}$$
$$= \frac{0.844 \times (386.967)^2}{2 \times 2000}$$
$$= 31.596 \text{ m}$$

Hence, midpoint P is (31.596 − 11.848), that is, 19.748 m below point B or (80 − 19.748), that is, 60.252 m above the water level.

## EXERCISES

1. Name the important components of an overhead transmission line.

2. Discuss the various conductor materials used for overhead lines. What are their relative advantages and disadvantages?

3. Discuss the various types of line supports.

4. What is a sag in overhead lines? Discuss the disadvantages of providing too small or too large sag on a line.

5. Deduce an approximate expression for sag in overhead lines when

   a. Supports are at equal levels

   b. Supports are at unequal levels

# CHAPTER 6

# Overhead Line Insulators

## 6.1 INTRODUCTION

The overhead line conductors are open and do not have any insulated coating over them. Those conductors should be supported on the poles or towers in such a way that current from conductors do not flow to earth through supports, that is, line conductors must be properly insulated. This is accomplished by connecting line conductors to a support with the help of insulators. The insulator provides necessary insulation between line conductors and supports and thus prevents any leakage current from conductors to ground. Insulators also provide support to the conductor.

In general, the insulator should have the following desirable properties:

- High mechanical strength in order to withstand conductor load, wind load, etc.

- High electrical resistance of insulator material in order to avoid leakage current to earth.

- High relative permittivity of insulator material in order that dielectric strength is high.

- The insulator material should be nonporous, free from impurities, otherwise permittivity will be lowered.

- High ratio of puncture strength to flashover.

## 6.2 INSULATOR MATERIALS

The following three materials are widely used in the manufacture of insulator units:

- Porcelain

- Glass

- Synthetic resin

The most commonly used materials for overhead line is porcelain. Porcelain is a ceramic material. It is produced by firing at high temperature a mixture of kaolin, feldspar, and quartz. The metal parts within the insulator are made of malleable cast iron with galvanizing. It is mechanically stronger than glass, gives less trouble from leakage, and is less affected by temperature change.

Glass is also used as an insulator material instead of porcelain. However, glass insulators are mainly used for EHV AC and DC systems. The glass is toughened by heat treatment. Though it is more brittle, its transparency, cracks, and defects within the insulator material can be detected easily by visual inspection. The glass insulators, on the other hand, are disadvantageous from the point of view that moisture condensation is more likely on the insulator surface causing higher leakage of current.

Synthetic insulators are mostly used in various indoor applications. They contain compounds of silicon, rubber, resin, etc. Synthetic insulators have high strength and lower weight. However, leakage current is higher and longevity is low. On the other hand, they are comparatively cheaper and have applications in bushings mainly.

## 6.3 TYPES OF INSULATORS

Various types of insulators used for overhead transmission, and distribution lines are described below.

### 6.3.1 Pin-Type Insulators

The part section of a pin-type insulator is shown in Figure 6.1. As the name suggests, the pin-type insulator is secured to the crossarm on the pole. There is a groove on the upper end of the insulator for housing the conductor. The conductor passes through this groove and is bound by the annealed wire of the same material as that of the conductor.

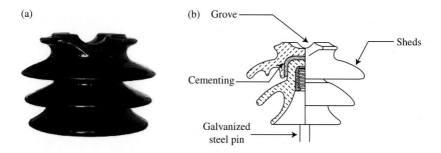

FIGURE 6.1    (a) Pin-type insulator and (b) cross-sectional view of pin-type insulator.

Pin-type insulators are used for transmission and distribution of electric power at voltages up to 33 kV. Beyond operating voltage of 33 kV, pin-
type insulators become too bulky and hence uneconomical.

### 6.3.1.1 Causes of Insulator Failure

Insulators are required to withstand both mechanical and electrical
stresses. The latter type is primarily due to line voltage and may cause
breakdown of the insulator. The electrical breakdown of the insulator can
occur due to flashover or puncture. In flashover, an arc occurs between the
line conductor and the insulator pin (i.e., earth) and the discharge jumps
across the air gaps, following shortest distance. Figure 6.2 shows the arcing distance (i.e., $a + b + c$) for the insulator. In this case, the insulator will
continue to act in its proper capacity, unless extreme heat produced by the
arc destroys the insulator.

In case of puncture, the discharge occurs from conductor to pin through
the body of the insulator. In case of puncture, insulator is permanently
destroyed due to excessive heat. To avoid the puncture, sufficient thickness

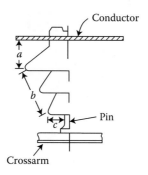

FIGURE 6.2    Arcing distance.

of porcelain is provided in the insulator. The ratio of puncture strength to flashover voltage is known as safety factor, that is,

$$\text{Safety factor of the insulator} = \frac{\text{Puncture strength}}{\text{Flash} - \text{Over voltage}}$$

It is desirable that the value of safety factor is high so that flashover takes place before the insulator gets punctured. For pin insulators, the value of safety factor is about 10.

### 6.3.2 Suspension Type Insulators

As the working voltage increases, the cost of pin-type insulator increases rapidly. Therefore, this type of insulator is not economical beyond 33 kV. For high voltages (>33 kV), it is a usual practice to use suspension type insulators shown in Figure 6.3. They consist of a number of porcelain disks connected in series by metal links in the form of a string. The conductor is suspended at the bottom end of this string, while the other end of the string is secured to the crossarm of the tower. Each unit or disk is designed for low voltage (say 11 kV). The number of disk would obviously depend upon the working voltage. For instance, if the working voltage in 66 kV, then six disks in series will be provided on the string.

*Advantages*

1. Cheaper than insulators for voltages beyond 33 kV.

2. Each unit or disk is designed for low voltage (say 11 kV). The number of disk would obviously depend upon the working voltage.

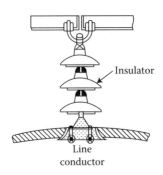

FIGURE 6.3   Suspension type insulators.

3. If any disk is damaged, the whole string does not become useless because the damaged disk can be replaced by the sound one.

4. The arrangement provides greater flexibility to the line. The connection at the crossarm is such that insulator is free to swing in any direction and can take up the position, where the mechanical stresses are minimum.

5. The suspension type insulators are generally used with steel towers. As the conductors run below the earthed crossarm of the tower, therefore, this arrangement provides partial protection from lightning.

6. In case of increased demand on the transmission line, it is found more satisfactory to supply greater demand by raising the line voltage, than to provide another set of conductor. The additional insulation required for the raised voltage can be easily obtained in the suspension arrangement by adding the desired number of disks.

### 6.3.2.1 Types of Suspension Insulators
The types of suspension insulators in use are

1. Cap-and-pin type

2. Hewlett or interlink type

The first type is more common. A galvanized cast iron or forged-steel cap and galvanized forged-steel pin are connected to porcelain in the cap-and-pin type construction. The units are joined together either by ball and socket or clevis–pin connections. Cap-and-pin type construction is given in Figure 6.4.

The interlink type unit (Figure 6.5) employs porcelain having two curved channels with planes at right angles to each other. U-shaped level covered steel links pass through these channels and serve to connect the

FIGURE 6.4 Cap-and-pin type construction.

FIGURE 6.5   Interlink type insulator.

units. Interlink type insulator is mechanically stronger than the cap-and-pin type unit. The metal links continue to support the line if the porcelain between the links breaks. Thus, the supply is not interrupted. The Hewlett type of insulator suffers from the disadvantage that the porcelain between links is highly stressed electrically and, therefore, its puncture strength is lesser as compared to other types.

### 6.3.3  Strain Insulators

When there is a dead end of the line or there is corner or sharp curve, the line is subjected to greater tension. In order to relieve the line of excessive tension, strain insulators are used. For low-voltage lines (<11 kV), shackle insulators are used as strain insulators. However, for high-voltage transmission lines, strain insulator consists of an assembly of suspension lines; strain insulator consists of an assembly of suspension insulators as shown in Figure 6.6. The disks of strain insulators are used in the vertical plane. When the tension in the line is exceeding high, as at long river spans, two or more strings are used in parallel.

### 6.3.4  Shackle Insulators

In early days, the shackle insulators were used as strain insulators. But now a days, they are frequently used for low-voltage distribution lines

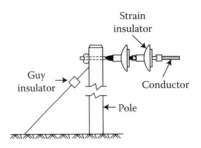

FIGURE 6.6   Strain insulator.

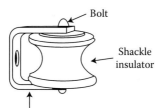

FIGURE 6.7    Shackle insulator.

FIGURE 6.8    Stay insulator.

(<11 kV). Such insulators can be used either in horizontal position or in vertical position. They can be directly fixed to the pole with a bolt or to the crossarm. Figure 6.7 shows a shackle insulator fixed to the pole. The conductor in the groove is fixed with a soft binding wire.

### 6.3.5  Stay Insulators

These kind of insulators are of egg shape, also called strain or guy insulators, and are used in guy cables, where it is very important to insulate the lower portion of the guy cable from the pole for the safety of human beings and animals on the ground. This type of insulator comprises of a porcelain piece pierced with two holes at right angles to each other through which two ends of the guy wires are looped. This compresses the porcelain between the two loops in and the guy wire remains in the same position even if the insulator breaks due to any reason. Figure 6.8 shows a stay insulator.

## 6.4  POTENTIAL DISTRIBUTION OVER SUSPENSION INSULATOR STRING

A string of suspension insulator consists of a number of porcelain disks connected in series through metallic links. Figure 6.9a shows three disks string of suspension insulators. The porcelain portion of each disk is in between two metal links. Therefore, each disk forms a capacitor $C$ known as mutual capacitance or self-capacitance. If there were mutual capacitance alone, then charging current would have been the same through all the disks and consequently voltage across each unit would have been same, that is, $V/3$ as shown in Figure 6.9b. However, in actual practice, capacitance also exists between metal fitting of each disk and tower or earth. This is known as shunt capacitance $C_1$. Due to shunt capacitance, charging current is not same through all the discs of the string shown in Figure 6.9c. Therefore, voltage across each disk will be different. Obviously, the disk nearest to the line conductor will have the maximum voltage. Therefore, $V_3$ will be much more than $V_1$ or $V_2$.

The following points may be noted regarding the potential distribution over a string of suspension insulators:

1. The voltage impressed on a string of suspension insulators does not distribute itself uniformly across the individual disks due to the presence of shunt capacitance.

2. The disk nearest to the conductor has maximum voltage across it. As we move toward the crossarm, the voltage across each disk goes on decreasing.

3. The unit nearest to the conductor is under maximum electric stress and is likely to be punctured.

4. If the voltage impressed across the string were DC, then voltage across each unit would be the same. It is because insulator capacitances are ineffective for DC.

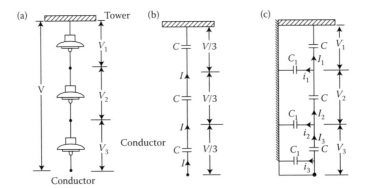

FIGURE 6.9   (a) Three discs string of suspension insulators; (b) voltage distribution shown in the presence of mutual capacitances; (c) voltage distribution shown in the presence of both mutual and shunt capacitances.

## 6.5 STRING EFFICIENCY

As stated above, the voltage applied across the string of suspension insulators is not uniformly distribute across various units or disks. The disk nearest to the conductor has much higher potential than the other disks. The unequal potential distribution is undesirable and usually expressed in terms of string efficiency.

The ratio of voltage across the whole string to the product of the number of disks and the voltage across the disk nearest to the conductor is known as *string efficiency.*

$$\text{String efficiency} = \frac{\text{Voltage across the string}}{n \times \text{Voltage across the disk nearest to the conductor}}$$

where $n$ is the number of disks in the string.

String efficiency is an important consideration since it decides the potential distribution along the string. Greater the string efficiency, the more uniform is the voltage distribution. Thus, 100% string efficiency is an ideal case for which the voltage across the disk will be exactly the same. Although it is impossible to achieve 100% string efficiency, yet efforts should be made to improve it as close to this value as possible.

## 6.6 MATHEMATICAL EXPRESSION

Figure 6.10 shows the equivalent circuit for a three-disk string. Let us suppose that self-capacitance of each disk is $C$. Let us further assume

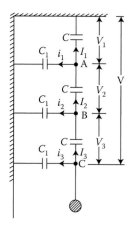

FIGURE 6.10   Equivalent circuit for a three-disk string.

that shunt capacitance $C_1$ is some fraction $K$ of self-capacitance, that is, $C_1 = KC$. Starting from the crossarm or tower, the voltage across each unit is $V_1$, $V_2$, and $V_3$ as shown.

Applying Kirchhoff's current law to node A, we get

$$I_2 = I_1 + i_1$$

or

$$V_2 \omega C = V_1 \omega C + V_1 \omega C_1$$

or

$$V_2 \omega C = V_1 \omega C + V_1 \omega KC$$

or

$$V_2 = V_1(1 + K) \tag{6.1}$$

Applying Kirchhoff's current law to node B, we get

$$I_3 = I_2 + i_2$$

or

$$V_3 \omega C = V_2 \omega C + (V_1 + V_2) \omega C_1$$

or

$$V_3 \omega C = V_2 \omega C + (V_1 + V_2) \omega KC$$

or

$$\begin{aligned}
V_3 &= V_2 + (V_1 + V_2)K \\
&= KV_1 + (1 + K)V_2 \\
&= KV_1 + V_1(1 + K)^2 \quad [\because V_2 = V_1(1 + K)] \\
&= V_1[K + (1 + K)^2]
\end{aligned}$$

$$\therefore V_3 = V_1[1 + 3K + K^2] \tag{6.2}$$

Voltage between conductor and earth (i.e., tower) is

$$V = V_1 + V_2 + V_3$$
$$= V_1 + V_1(1 + K) + V_1[1 + 3K + K^2]$$
$$= V_1 + (3 + 4K + K^2)$$

$$\therefore V = V_1(1 + K)(3 + K) \tag{6.3}$$

From Equations 6.1 through 6.3, we get

$$\frac{V_1}{1} = \frac{V_2}{1 + K} = \frac{V_3}{[1 + 3K + K^2]} = \frac{V}{(1 + K)(3 + K)}$$

Therefore, voltage across the top unit is

$$V_1 = \frac{V}{(1 + K)(3 + K)}$$

Voltage across the second unit from top is

$$V_2 = V_1(1 + K)$$

Voltage across the third unit from top is

$$V_3 = V_1[1 + 3K + K^2]$$

$$\% \text{ Age string efficiency} = \frac{\text{Voltage across the string}}{n \times \text{Voltage across the disk nearest to the conductor}} \times 100$$

$$= \frac{V}{3 \times V_3} \times 100$$

The following points may be noted from the above mathematical analysis:

1. Disk nearest to the conductor has maximum voltage across it. The voltage across other disk decreasing progressively as the crossarm in approached.

2. The greater value of $K$ ($=C_1/C$), the more nonuniform is the potential across the disks and lesser is the string efficiency.

3. The inequality in voltage distribution increases with the increase of number of disks in the string. Therefore, shorter string has more efficiency than the larger one.

## 6.7 METHODS OF IMPROVING STRING EFFICIENCY

For satisfactory performance, it is essential that the voltage distribution across the units of the string should be uniform. Different methods have been attempted to get uniform distribution of voltage along the insulators in order to exploit its insulation strength fully. The following methods may be utilized to achieve the uniformity.

### 6.7.1 Use of a Longer Crossarm

The value of string efficiency depends upon the value of $K$, that is, ratio of shunt capacitance to mutual capacitance. The leaser the value of $K$, the greater is the string efficiency and more uniform is the voltage distribution. The value of $K$ can be decreased by reducing the shunt capacitance. In order to reduce shunt capacitance, the distance of conductor from tower must be increased, that is, a longer crossarm should be used. However, limitation of cost and strength of tower do not allow the use of very long crossarms. In practice, $K = 0.1$ is the limit that can be achieved by this method (Figure 6.11).

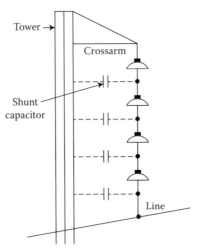

FIGURE 6.11   Insulator string with longer crossarm.

## 6.7.2 Grading of Units

It is seen that nonuniform distribution of voltage across an insulator string is due to leakage current from the insulator pin to the supporting structure. This current cannot be eliminated. However, it is possible that disks of different capacities are used such that the product of their capacitive reactance and the current flowing through the respective unit is same. This can be achieved by grading the mutual capacitance of the insulator units, that is, by having lower units of more capacitance—maximum at the line unit and minimum at the top unit, nearest to the crossarm—it can be shown that by this method complete equality of voltage across the units of an insulator string can be obtained but this method needs a large number of different-sized insulator units. This involves maintaining spares of all varieties of insulator disks which is contrary to the tendency of standardization. So this method is not used in practice below 200 kV.

## 6.7.3 Use of Guard Ring

The potential across each unit in a string can be equalized by using a guard ring, which is a metal ring electrically connected to the conductor and surrounding the bottom insulator as shown in Figure 6.12. The guard ring introduces capacitance between the metal fittings and the line conductor. The guard ring is connected in such a way that shunt capacitance current $i_1$, $i_2$, etc., are equal to metal fitting line capacitance current $i_1'$, $i_2'$, etc., resulting in the flow of same charging current $I$ through each unit of string. Consequently, there will be uniform potential distribution across the units.

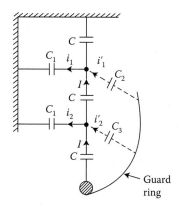

FIGURE 6.12   Insulator string with guard ring.

## 6.8 EFFECTS OF RAIN ON STRING EFFICIENCY

In the rainy season, insulators are naturally wet. Due to this, capacitive reactance decreases and mutual capacitance value increases. Thus, the ratio of shunt to mutual capacitance, that is, $K$ decreases. This results in uniform voltage distribution. Hence, in rainy season, string efficiency is higher.

## WORKED EXAMPLES

### EXAMPLE 6.1

The three bus-bar conductors in an outdoor substation are supported by units of post-type insulators. Each unit consists of a stack of three pin-type insulators fixed one on the top of the other. The voltage across the lowest insulator is 11.3 kV and that across the next unit is 10.2 kV. Find the bus-bar voltage of the station.

**Solution**

The equivalent circuit of insulators is the same as shown in Figure 6.13. It is given that $V_3 = 11.3$ kV and $V_2 = 10.2$ kV. Let $K$ be the ratio of shunt capacitance to self-capacitance of each unit.

Applying Kirchhoff's current law to junctions A and B, we can easily derive the following equations.

$$V_2 = V_1(1 + K)$$

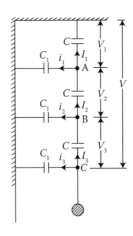

FIGURE 6.13 Equivalent circuit of string insulators.

or

$$V_1 = \frac{V_2}{(1 + K)} \tag{6.4}$$

and

$$V_3 = V_2 + (V_1 + V_2)\, K \tag{6.5}$$

Putting the value of $V_1 = V_2/(1 + K)$ in Equation 6.5, we get

$$V_3 = V_2 + \left( \frac{V_2}{(1 + K)} + V_2 \right) K$$

or

$$V_3(1 + K) = V_2(1 + K) + [V_2 + V_2(1 + K)]\, K$$

or

$$V_3(1 + K) = V_2[(1 + K) + K + (K + K^2)]$$
$$= V_2[1 + 3K + K^2]$$

Therefore,

$$11.3\,(1 + K) = 10.2\,[1 + 3K + K^2]$$

or

$$11.3 + 11.3K = 10.2 + 30.6K + 10.2K^2$$

or

$$10.2K^2 + (30.6 - 11.3)K + (10.2 - 11.3) = 0$$

or

$$10.2K^2 + 19.3K - 1.1 = 0$$

or

$$K = \frac{-19.3 \pm \sqrt{(19.3)^2 + (4 \times 10.2 \times 1.1)}}{2 \times 10.2} \quad \text{or} \quad \frac{-19.3 \pm 20.42}{2 \times 10.2}$$

$$= 0.05$$

Therefore,

$$V_1 = \frac{V_2}{(1 + K)} = \frac{10.2}{1 + 0.05} = 9.71$$

Voltage between line and earth $= V_1 + V_2 + V_3 = 9.71 + 10.2 + 11.3 = 31.21$ kV. Therefore, voltage between bus bars (i.e., line voltage) $= 31.21 \times \sqrt{3} = 54.06$ kV.

### EXAMPLE 6.2

An insulator string consists of three units, each having a safe working voltage of 13 kV. The ratio of self-capacitance to shunt capacitance of each unit is 10:1. Find the maximum safe working voltage of the string. Also find the string efficiency.

### Solution

The equivalent circuit of string insulators is the same as shown in Figure 6.13. The maximum voltage will appear across the lowest unit in the string.

$$V_3 = 13 \text{ kV}, \quad K = \frac{1}{10} = 0.1$$

Applying Kirchhoff's current law to junction A, we get

$$V_2 = V_1(1 + K)$$

or

$$V_1 = \frac{V_2}{(1 + K)} = \frac{V_2}{1 + 0.1} = 0.909 \, V_2 \tag{6.6}$$

Applying Kirchhoff's current law to junction B, we get

$$V_3 = V_2 + (V_1 + V_2)K = V_2 + (0.909 \, V_2 + V_2) \times 0.1$$

or

$$V_3 = 1.190 \, V_2 \qquad\qquad (6.7)$$

Therefore, voltage across middle unit is

$$V_2 = \frac{V_3}{1.190} = \frac{13}{1.190} = 10.916 \text{ kV}$$

Voltage across top unit is $V_1 = 0.909$ and $V_2 = 0.909 \times 10.916 = 9.92$ kV.
Voltage across the string is $V_1 + V_2 + V_3 = 9.92 + 10.916 + 13 = 33.84$ kV.

$$\text{String efficiency} = \frac{33.84}{3 \times 13} \times 100\% = 86.76\%$$

## EXAMPLE 6.3

In a transmission line, each conductor is at 30 kV and is supported by a string of three suspension insulators. The air capacitance between each cap–pin junction and tower is one-eighth of the capacitance of each insulator unit. A guard ring, effective only over the line-end insulator unit is fitted so that the voltages on the two units nearest the line-end are equal. (a) Calculate the voltage on the line-end unit. (b) Calculate the value of capacitance $C_x$ required (Figure 6.14).

**Solution**

Voltage between unit conductor and earth $V = 30$ kV.

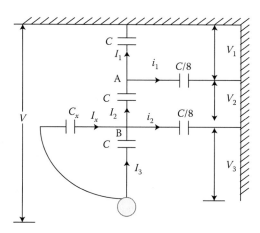

FIGURE 6.14   Equivalent circuit of string insulators with guard ring.

Mutual capacitance = $C$

Shunt capacitance = $C/8 = 0.125C$

Applying Kirchhoff's first law to node A:

$$I_2 = I_1 + i_1$$

$$\omega C V_2 = \omega C V_1 + \omega \times 0.125 C V_1$$

or

$$V_2 = 1.125 V_1$$

also

$$V_3 = V_2$$

and

$$V = V_1 + V_2 + V_3$$

or

$$30 = V_1 + V_2 + V_3$$

or

$$30 = V_1 + 1.125 V_1 + 1.125 V_1$$

or

$$V_1 = \frac{30}{1 + 1.125 + 1.125} = 9.23 \text{ kV}$$

or

$$V_3 = 1.125 \times V_1 = 10.38 \text{ kV}$$

a. Voltage on the line-end unit $\geq V_3 = 10.38$ kV
b. Applying Kirchhoff's first law to node B, we have:

$$I_3 + i_x = I_2 + i_2$$

or

$$\omega C V_3 + \omega C_x V_3 = \omega C V_2 + \omega \frac{C}{8}(V_1 + V_2)$$

or

$$C \times 1.125V_1 + C_x \times 1.125V_1 = C \times 1.125V_1 + \frac{C}{8} 2.125V_1$$

or

$$C + C_x = \frac{1.39}{1.125} C \text{ or } C_x = 0.236C \text{ farads}$$

## EXAMPLE 6.4

In a string of three identical suspension insulator units supporting a transmission line conductor, if the self-capacitance of each unit is denoted as $C$ farads, the capacitance of each connector pin to ground can be taken as $0.2C$ farads. Determine the voltage distribution across the string if the maximum possible voltage par unit is given as 30 kV. Also determine the string efficiency.

**Solution**

Number of units, $n = 3$. Ratio of shunt capacitance to mutual capacitance is

$$K = \frac{0.2C}{C} = 0.2$$

Voltage across the bottom most unit, $V_3 =$ safe working voltage $= 30$ kV. Hence, voltage across top most unit is

$$V_1 = \frac{V_3}{1 + 3K + K^2} = \frac{30}{1 + 0.6 + 0.04} = 18.29 \text{ kV}$$

Voltage across middle unit is

$$V_2 = V_1(1 + K) = 18.29 \times 1.2 = 21.95 \text{ kV}$$

Maximum safe working voltage of the string is

$$V = V_1 + V_2 + V_3 = 18.29 + 21.95 + 30 = 70.24 \text{ kV}$$

$$\text{String efficiency} = \frac{V}{nV_n} = \frac{70.25}{3 \times 30} \times 100 = 78.04\%$$

## EXERCISES

1. Why are insulators used with overhead lines? Discuss the desirable properties of insulators.

2. Discuss the advantages and disadvantages of (i) pin-type insulators and (ii) suspension-type insulators.

3. Explain how the electrical breakdown can occur in an insulator.

4. What is a strain insulator and where is it used? Give a sketch to show its location.

5. Give reasons for unequal potential distribution over a string of suspension insulators.

6. Define and explain string efficiency. Can its value be equal to 100%?

7. Show that in a string of suspension insulators, the disk nearest to the conductor has the highest voltage across it.

8. Explain various methods of improving string efficiency.

9. Explain why the voltage across the insulators of a simple insulator string are not equal?

# CHAPTER 7

# Corona

## 7.1 INTRODUCTION

The use of high-voltage supply has become necessary in order to fulfill the rapidly increasing demand of power. With extra high-voltage transmission lines of more than 230 kV coming into prominence, the corona characteristics of conductors have gained great importance. Below this size of the conductor is mainly determined by corona loss and radio noise. A series of experiments carried out in many countries reveal that it is now possible to predict the extent of corona performance of a line under different operating conditions with a fair degree of accuracy.

## 7.2 THE PHENOMENON OF CORONA

When an alternating potential difference is applied across two conductors whose spacing is large as compared to their diameters, there is no apparent charge in the condition of atmospheric air surrounding the wires if the applied voltage is low. However, when applied voltage exceeds a certain value, called critical disruptive voltage, the conductors are surrounded by a faint violet or yellowish-blue glow called corona.

The phenomenon of corona is accompanied by a hissing sound, production of ozone, power loss, and radio interference. The higher the voltage is raised, the larger and higher the luminous envelope becomes, and greater are the sound, the power loss, and the radio noise. If the applied voltage increased to breakdown value, a flashover will occur between the conductors due to the breakdown of air insulation. Corona occurs when the electrostatic stress in the air around the conductor exceeds 30 kV max/cm or 21.21 kV rms/cm.

The phenomenon of violet glow, hissing noise, and production of ozone gas in an overhead transmission line is known as corona.

If the conductors are polished and smooth, the corona glow will be uniform throughout the length of the conductor; otherwise, the rough points will appear brighter. With DC voltage, there is difference in the appearance of the two wires. The positive wire has uniform glow around it, while the negative conductor has spotty glow.

## 7.3 THEORY OF CORONA FORMATION

The electrons and ions are always present to a small extent in the atmospheric air due to cosmic rays, ultra violet radiations, and radioactivity. Therefore, under normal conditions, the air around the conductors contains some ionized particles and neutral molecules. When potential difference is applied between the conductors, potential gradient is set up in the air which will have maximum value at the conductor surfaces. Under influence of potential gradient, the existing free electrons acquire greater velocities. The greater the applied voltage, the greater the potential gradient, and more is the velocity of free electrons.

When the potential gradient at the conductor surface reaches about 30 kV/cm (maximum value), the velocity acquired by the electrons is sufficient to strike a neutral molecule with enough force to dislodge one or more electrons from it. This produces another ion and one or more free electrons, which in turn are accelerated until they collide with other neutral molecules, thus, producing other ions. The other process of ionization is cumulative. The result of this ionization is that either corona is formed or spark takes place between the conductors.

## 7.4 FACTORS AFFECTING CORONA

The phenomenon of corona is affected by the factors given below:

1. *Conductor.* It has been observed that the corona very much depends upon the shape and condition of conductor. A rough and irregular surfaced conductor gives more corona, that is, near rough and dirty surface, corona glow is intensified because unevenness of the surface decreases the value of breakdown voltage. For example, stranded conductor gives more corona than a single solid conductor (without strand).

The corona depends upon the following:

a. Size (diameter)

b. Shape (solid/stranded)

c. Surface condition (clean/dirty)

Corona decreases with increase in diameter of conductor.

2. *Spacing between conductors.* With increase in distance between the conductors, the corona effect is reduced considerably. It is because larger distances between the conductors reduce the electrostatic stresses at the conductor surface, thus avoiding corona formation.

3. *Line voltage.* The line voltage greatly affects corona. A low-voltage corona is not observed. At higher voltage or at very high electrostatic stress, air gets ionized, which gives rise to corona effect and it is the cause of occurrence of corona.

4. *Atmosphere.* Corona is caused by the ionization of air surrounds the conductor. A stormy and foggy weather has more ions and therefore gives rise to more corona as compared to fair, dry, and clean weather.

## 7.5  ADVANTAGES OF CORONA

1. Due to corona formation, the air surrounding the conductor become conducting and hence virtual diameter of the conductor is increased. The increased diameter reduces the electromagnetic stresses between the conductors.

2. Corona reduces the effects of transients produced by surges.

## 7.6  DISADVANTAGES OF CORONA

1. Corona is accompanied by a loss of energy. This affects the transmission efficiency of the line.

2. Ozone is produced by corona and may cause corrosion of the conductor due to chemical action.

3. The current drawn by the line due to corona is non-sinusoidal, and hence non-sinusoidal voltage drop occurs in the line. This may cause inductive interference with neighboring communication line.

## 7.7 METHODS OF REDUCING CORONA EFFECT

1. *By increasing conductor size.* By increasing conductor size, the voltage at which corona occurs is raised and hence corona effects are considerably reduced. This is one of the reasons that ACSR conductors that have larger cross-sectional area are used in transmission line.

2. *By increasing conductor spacing.* Between conductors, the voltage at which corona occurs is raised, and hence corona effects can be eliminated. However, spacing cannot be increased too much otherwise the cost of supporting structure (e.g., bigger crossarms and supports) may increase to a considerable extent.

## 7.8 CRITICAL DISRUPTIVE VOLTAGE

It is the minimum phase to neutral voltage at which corona occurs.

Consider two conductors (Figure 7.1) of radius $r$ (cm) and spaced $d$ (cm) apart. If $V$ is the phase-neutral potential, then the potential gradient at the conductor surface is given by

$$g = \frac{V}{r \log_e(d/r)} \text{ V/cm}$$

In order that corona is formed, the value of $g$ must be made equal to the breakdown strength of air. The breakdown strength of air at 76 cm pressures and temperature of 25°C is 30 kV/cm (max) or 21.21 kV/cm (rms) and is denoted by $g_0$. If $V_c$ is the phase-neutral potential required under these conditions, then

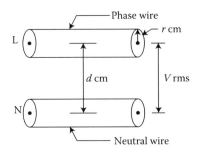

FIGURE 7.1 One-phase two-wire line.

$$g_0 = \frac{V_c}{r \log_e (d/r)}$$

$g_0$ = Breakdown strength of air at 76 cm of mercury and 25°C
= 30 kV/cm (max) or 21.21 kV/cm (rms)

Therefore, critical disruptive voltage is

$$V_c = g_0 r \log_e \frac{d}{r}$$

The above expression for disruptive voltage is under standard condition, that is, at 76 cm of Hg and 25°C. However, if these conditions vary, the air density also changes, thus altering the value of $g_0$. The value $g_0$ is directly proportional to air density. Thus, the breakdown strength of air at a barometric pressure of $b$ cm of mercury and temperature of $t$°C becomes $\delta g_0$, where

$$\delta = \text{Air density factor} = \frac{3.92b}{273 + t}$$

Under standard conditions, the value of $\delta = 1$, and critical disruptive voltage is

$$V_c = g_0 \delta r \log_e \frac{d}{r}$$

Correction must also be made for the surface conditions of the conductor. This is accounted for by multiplying the above expression by irregularity factor $m_0$. Then, the critical disruptive voltage is

$$V_c = m_0 g_0 \delta r \log_e \frac{d}{r} \text{ kV/phase}$$

where $m_0$ is 1 for polished conductors, 0.98 to 0.92 for dirty conductors, 0.87 to 0.8 for stranded conductors, and 0.90 (approximately) for large cable more than seven strands.

## 7.9 VISUAL CRITICAL VOLTAGE

It is the minimum phase to neutral voltage at which corona glow appears all along the line conductors.

It has been seen that in case of parallel conductors, the corona glow does not begin at the disruptive voltage $V_c$ but at a higher voltage $V_v$, called visual critical voltage.

When the voltage applied is made equal to the critical disruptive voltage, corona phenomenon occurs, but it remains invisible because the charged ions in the air must gain some finite energy for the occurrence of further ionization by collisions. For a radial field, it must reach a gradient $g_v$ at the surface of the conductor to cause a gradient $g_0$ a finite distance away from the surface of the conductor. The distance between $g_v$ and $g_0$ is called the energy distance. According to Peek, this distance is equal to $(r + 0.301\sqrt{r})$ for two parallel conductors and $(r + 0.308\sqrt{r})$ for coaxial conductors. From this, it is clear that $g_v$ is not constant but $g_0$ is, and is a function of the size of the conductor. For two wires in parallel,

$$g_v = g_0\delta\left(1 + \frac{0.3}{\sqrt{r\delta}}\right)\text{kV/cm}$$

Also, if $V_v$ is the critical visual disruptive voltage, then

$$V_v = g_v r \ln\frac{d}{r}$$

or

$$g_v = \frac{V_v}{r\log_e(d/r)} = g_0\delta\left(1 + \frac{0.3}{\sqrt{r\delta}}\right)$$

or

$$V_v = rg_0\delta\left(1 + \frac{0.3}{\sqrt{r\delta}}\right)\log_e\frac{d}{r}\text{kV}$$

In case the irregularity factor is taken into account,

$$V_v = g_0 m_v \delta r \left[ 1 + \frac{0.3}{\sqrt{r\delta}} \right] \log_e \frac{d}{r}$$

$$= 21.21 \, m_v \delta r \left[ 1 + \frac{0.3}{\sqrt{r\delta}} \right] \log_e \frac{d}{r} \, \text{kV rms}$$

The irregularity factor value $m_v$ is 1 for polished wires, 0.98 to 0.93 for rough conductor exposed to atmospheric severities, and 0.72 for local corona on stranded conductors.

## 7.10 POWER LOSS DUE TO CORONA

Formation of corona is always accompanied by energy loss which is dissipated in the form of light, heat, sound, and chemical action. When disruptive voltage is exceeded, the power loss due to corona is given by

$$P = 242.2 \left( \frac{f+25}{\delta} \right) \sqrt{\frac{r}{d}} (V - V_c)^2 \times 10^{-5} \, \text{kW/km/phase}$$

where $f$ is the supply frequency in Hertz, $V$ is the phase to neutral voltage (rms), and $V_c$ is the disruptive voltage (rms) per phase.

## 7.11 RADIO INTERFERENCE

Radio interference is one of the adverse effects caused by corona on wireless broadcasting. The corona discharges radiation which may introduce noise signal in the communication lines, radio, and television receivers. It is primarily because of the brush discharges on the surface irregularities of the conductor during positive half cycles. This causes corona to occur at voltages below the critical voltages. The negative discharges are less troublesome for radio reception. Radio interference is considered as a field measured in microvolts per meter at any distance from the transmission line and is significant only at voltages greater than 200 kV. There is gradual increase in radio interference (RI) level until the voltage reaches a value which causes corona to take place. Above this voltage, there is rapid increase in RI level. The rate of increase is more for smooth and large diameter conductors. The amplitude of RI level varies inversely as the frequency at which the interference is measured. Thus, the services in the higher frequency band, for example, television, frequency-modulated broadcasting, microwave relay, radar, etc., are less affected. Radio interference is one of the very important factors while designing a transmission line.

## 7.12 INDUCTIVE INTERFERENCE BETWEEN POWER AND COMMUNICATION LINES

It is general practice to run communication lines along the same route as the power lines, since the user of electrical energy is also the user of electrical communication system. The transmission lines transmit bulk power at relatively higher voltages. These lines give rise to electromagnetic and electrostatic fields of sufficient magnitude which induce currents, and voltages in the neighboring communication lines. The effects of extraneous currents and voltages on communication systems include interference with communication service, for example, superposition of extraneous currents on the true speech currents in the communication wires, hazards to person, and damage to apparatus due to extraneous voltages. In extreme cases, the effect of these fields may take it impossible to transmit any message faithfully and may raise the potential of the apparatus above the ground to such an extent as to render the handling of the telephone receiver extremely dangerous.

### 7.12.1 Electromagnetic Effects

Figure 7.2A–C shows the power conductors of a three-phase single circuit line on a transmission tower, and D and E, the conductors of a neighboring communication line running on the same transmission towers as the power conductors or a neighboring separate line. Let the distances between power conductors and communication conductors be $d_{AD}$, $d_{AE}$, $d_{BD}$, $d_{BE}$, $d_{CD}$, and $d_{CE}$, and the currents through power conductors be $I_A$, $I_B$, and $I_C$, such that $I_A + I_B + I_C = 0$. The flux linkage to conductor D due to current $I_A$ in conductor A will be $\psi_{AD} = 2 \times 10^{-7} I_A \log_e(\infty/d_{AD})$. Similarly, the flux linkage to conductor E due to current $I_A$ in a conductor A.

$$\psi_{AE} = 2 \times 10^{-7} I_A \log_e \frac{\infty}{d_{AE}}$$

Therefore, mutual flux linkage between conductor D and conductor E due to current $I_A$ will be

$$\psi_{AD} - \psi_{AE} = 2 \times 10^{-7} I_A \log_e \frac{d_{AE}}{d_{AD}}$$

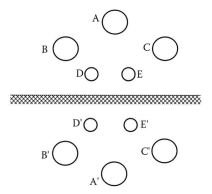

FIGURE 7.2 Three-phase single-circuit power line, communication line, and their images.

or mutual inductance

$$M_A = \frac{\Psi_{AD} - \Psi_{AE}}{I_A} = 2 \times 10^{-7} \log_e \frac{d_{AE}}{d_{AD}} \text{ Hz/m}$$

Similarly, the mutual inductances $M_B$ and $M_C$ between conductor B and loop $d_e$ and between conductor C and loop $d_e$, respectively, are given as

$$M_B = 2 \times 10^{-7} \log_e \frac{d_{BE}}{d_{BD}} \text{ Hz/m}$$

$$M_C = 2 \times 10^{-7} \log_e \frac{d_{CE}}{d_{CD}} \text{ Hz/m}$$

These mutual inductances are due to fluxes that have a phase displacement of 120°; therefore, the net effect of the magnetic field will be

$$M = M_A + M_B + M_C$$

where $M$ is the net mutual inductance which is the phasor sum of the three inductances.

If $I$ is the current in the power conductors and $f$ is the supply frequency, the voltage induced in the communication conductors D and E will be $V = 2\pi f M I$ V/m.

It is to be noted that larger the distance between the power conductors and the communication conductors, smaller is the value of mutual inductance and since the current through the power conductors is displaced by 120°, there is appreciable amount of cancellation of the power frequency voltages. But the presence of harmonics and multiple of third harmonics will not cancel, as they are in phase in all the power conductors and, therefore, are dangerous for the communication circuits. Also, since these harmonics come within audio frequency range, they are dangerous for the communication circuits.

### 7.12.2 Electrostatic Effects

Consider again Figure 7.2. Let $Q$ be the charge per-unit length of the power line. The voltage of conductor D due to charge on conductor can be obtained by considering the charge on conductor A and its image on the ground. Let conductor A be at a height $h_A$ from the ground. Therefore, the voltage of conductor D will approximately be

$$V_{AD} = \frac{Q}{2\pi\epsilon_0} \int_{h_A}^{d_{AD}} \left[ \frac{1}{x} + \frac{1}{(2h_A - x)} \right] dx$$

$$= \frac{Q}{2\pi\epsilon_0} \left[ \ln \frac{2h_A - x}{x} \right]_{d_{AD}}^{h_A} = \frac{Q}{2\pi\epsilon_0} \left[ \log_e \frac{2h_A - d_{AD}}{d_{AD}} \right]$$

Now from the geometry, the voltage of conductor A is

$$V_A = \frac{Q}{2\pi\epsilon_0} \ln \frac{2h_A}{r}$$

where $r$ is the radius of conductor A.

Substituting for $Q$ in the equation for $V_{AD}$ above, we get

$$V_{AD} = \frac{2\pi\epsilon_0 V_A}{\log_e(2h_A/r)} \frac{1}{2\pi\epsilon_0} \ln \frac{2h_A - d_{AD}}{d_{AD}}$$

$$= V_A \frac{\log_e[(2h_A - d_{AD})/d_{AD}]}{\log_e(2h_A/r)}$$

Similarly, we can obtain the potential of conductor D due to conductors B and C, and hence the potential of conductor D due to conductors A through C will be

$$V_D = V_{AD} + V_{BD} + V_{CD}$$

Similarly, the potential of conductor E due to conductors A through C can be obtained.

## WORKED EXAMPLES

### EXAMPLE 7.1

A three-phase transmission line consists of 1 cm radius conductors spaced symmetrically 3 m apart. Dielectric strength of air is 30 kV/cm. Determine the unit voltage for commencing of corona (Irregularity factor = 0.9, temperature 20°C, barometric pressure 72.2 cm of mercury).

**Solution**

$$d = 3 \text{ m} = 300 \text{ m}$$
$$r = 1 \text{ cm}$$

Dielectric strength of air: $v_0 = \dfrac{30}{\sqrt{2}} \text{ kV/cm} = 21.21 \text{ kV/cm}$

Air density factor $\delta = \dfrac{3.92b}{273 + t} = \dfrac{3.92 \times 72.2}{273 + 20} = 0.966$

Line voltage for commencing corona $= \sqrt{3}V_c = \sqrt{3}g_0 \cdot m_0 r \log_e \dfrac{d}{r}$

$$= \sqrt{3} \times 21.21 \times 0.966 \times 0.96 \times 1 \times \log_e \dfrac{300}{1} = 194.31 \text{ kV}$$

### EXAMPLE 7.2

A three-phase, 220-kV, 50-Hz transmission line consists of 1.15 radius conductor spaced 3 m apart in equilateral triangular formation, if temperature is 20°C and atmospheric pressure is 72.2 cm, calculate the corona loss per kilometer of the line, take $m_0 = 0.96$.

**Solution**

The corona loss is given by

$$P = \frac{242.2}{\delta}(f + 25)\sqrt{\frac{r}{d}}(V - V_c)^2 \times 10^{-5} \text{ kW/km/phase}$$

Now

$$\delta = \frac{3.92b}{273 + t} = \frac{3.92 \times 72.2}{273 + 20} = 0.966$$

Assuming, $g_0 = 21.21$ kV/cm (rms). So, critical disruptive voltage per phase is

$$V_c = m_0 g_0 \delta \, r \log_e \frac{d}{r} \text{kV}$$

$$= 0.966 \times 21.21 \times 0.96 \times 1.15 \times \log_e \frac{300}{1.15}$$

$$= 125.85 \text{ kV}$$

Supply voltage for phase $= (220/\sqrt{3}) = 127$ kV

Substituting the above values, we have corona loss as

$$P = \frac{242.2}{0.966}(50 + 25)\sqrt{\frac{1.15}{300}}(127 - 125.8)^2 \times 10^{-5}$$

$$= 0.015 \text{ kW/km/phase}$$

Therefore, total corona loss per kilometer for three phases $= 3 \times 0.015 = 0.045$ kW.

**EXERCISES**

1. What is corona? What are the factors which affect corona?

2. Discuss the advantages and disadvantages of corona.

3. Explain the following terms with reference to corona:

   a. Critical disruptive voltage

   b. Visual critical voltage

   c. Power loss due to corona

4. Describe the various methods for reducing corona effect in an overhead transmission line.

# CHAPTER 8

# Transmission Line Parameters

## 8.1 INTRODUCTION

An electric transmission line consists of four parameters, namely resistance, inductance, capacitance, and shunt conductance. Shunt conductance, which is mostly due to leakage over line insulators, is almost always neglected in overhead transmission lines. The electrical design and operation of a line are dependent on these parameters. These four parameters are uniformly distributed on the whole distance of the cable. The communication channel parameters are functions of the communication channel geometry, building material, and operational frequency. The line resistance and inductance form the series impedance. The capacitance and conductance form the shunt admittance.

## 8.2 LINE INDUCTANCE

When an alternating current flows through a conductor, a changing flux is set up. With the variation of current in the circuit, the number of lines of flux also changes and an electromotive force (emf) is induced in it. The magnitude of the self-induced emf is directly proportional to the rate of change of flux linkage, and its direction is such as to oppose the cause, that is, the change of current which produces it.

Mathematically, the induced emf is given by

$$|e| = \frac{d}{dt}(\phi N) = N\frac{d\phi}{dt}\,\text{Volts}$$

(8.1)

where ($\phi N$) is the number of flux linkages of the circuit in Weber-turns. Flux linkages mean the sum of flux lines linking with each turn of the circuit, so that the number of flux linkage is equal to the product of the flux and the number of turns of the circuit linked.

The change in the circuit current causes a change in flux linkages proportionately provided the permeability of the medium in which the magnetic field produced is assumed to be constant. The self-induced emf will, therefore, be proportional to the rate of change of current, that is,

$$|e| = L\frac{di}{dt}\text{Volts} \tag{8.2}$$

where $L$ is the constant of proportionality and is known as the self-inductance of the circuit.

Equating the two values of induced emf from Equations 8.1 and 8.2,

$$N\frac{d\phi}{dt} = L\frac{di}{dt}, \quad L = N\frac{d\phi}{di} \tag{8.3}$$

If the permeability of the magnetic circuit is assumed to be constant,

$$\frac{d\phi}{di} = \frac{\phi}{i}$$

and

$$L = \frac{\phi N}{i}\text{H} \tag{8.4}$$

which shows that the self-inductance of an electric circuit is numerically equal to the flux linkage of the circuit per unit of current:

$$L = \frac{\phi}{i}\text{H} \tag{8.5}$$

Because only one current path links the flux.

## 8.3 FLUX LINKAGE DUE TO A SINGLE CURRENT-CARRYING CONDUCTOR

Consider a long straight cylindrical conductor of radius $r$ meters and carrying a current $I$ amperes (rms) as shown in Figure 8.1a. This current will set up magnetic field. The magnetic lines of force will exist inside the conductor as well as outside the conductor. Both these fluxes will contribute to the inductance of the conductor.

### 8.3.1 Flux Linkage Due to Internal Flux

The cross-section of the conductor is shown magnified clarity (Figure 8.1b).

The current inside a line of force of radius $x$,

$$I_x = \frac{I}{\pi r^2} \pi x^2 = \frac{I x^2}{r^2}$$

The magnetic field intensity at a point $x$ meters from the center is given by

$$H_x = \frac{\text{Current}}{2\pi \times \text{Distance}}$$

$$\therefore H_x = \frac{I x^2}{r^2} \times \frac{1}{2\pi x} = \frac{I x}{2\pi r^2} \text{ AT/m}$$

Flux density, $B_x = \mu_0 \mu_r H_x$ Wb/m² and $\mu(=\mu_0\mu_r)$ is the permeability of the medium, and for nonmagnetic material $\mu_r = 1$.

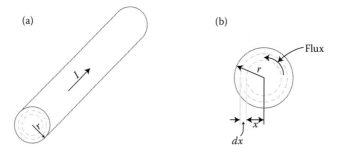

FIGURE 8.1 (a) Single current carrying conductor and (b) flux linkage due to internal flux (cross-sectional view).

$$\therefore B_x = \mu_o H_x \text{ Wb/m}^2 = \frac{\mu_o I x}{2\pi r^2} \text{ Wb/m}^2$$

Now flux $d\phi$ through a cylindrical shell of radial thickness $dx$ and axial length 1 m is given by

$$d\phi = B_x \times 1 \times dx = \frac{\mu_o x I}{2\pi r^2} dx \text{ Wb}$$

This flux links with current $I_x = ((\pi x^2/\pi r^2)I)$ only.
Therefore, flux linkage per meter length of the conductor is

$$d\Psi = \frac{\pi x^2}{\pi r^2} d\phi = \frac{\mu_o I x^3}{2\pi r^4} dx \text{ Wb}$$

Total flux linkage from center up to the conductor surface is

$$\Psi_{int} = \int_0^r \frac{\mu_o x^3 I}{2\pi r^4} dx = \frac{\mu_o I}{8\pi} \text{ Wb/m}$$

$$= \frac{4\pi \times 10^{-7} \times I}{8\pi} \text{ Wb/m} \tag{8.6}$$

$$= \frac{1}{2} I \times 10^{-7} \text{ Wb/m}$$

### 8.3.2 Flux Linkage of a Conductor Due to External Flux

Consider two points 1 and 2 at distance $d_1$ and $d_2$ from the center of the conductor (Figure 8.2). Since the flux paths are concentric circles around

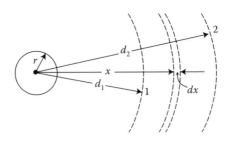

FIGURE 8.2 Flux linkages due to external flux.

the conductor, whole of the flux between points 1 and 2 lies within the concentric cylindrical surface passing through these points 1 and 2.

The field strength at any distance $x$ from the center of the conductor $(x > r)$,

$$H_x = \frac{I}{2\pi x} \text{ AT/m}$$

Flux density,

$$B_x = \mu_o H_x \text{ Wb/m}^2 = \frac{\mu_o I}{2\pi x} \text{ Wb/m}^2$$

So, the flux through a cylindrical shell of radial thickness $dx$ and axial length 1 m.

$$d\phi = B_x \times 1 \times dx = \frac{\mu_o I}{2\pi x} dx \text{ Wb/m}$$

Now flux linkages per meter is equal to $d\phi$, since flux external to conductor links all the current in the conductor only once

$$d\Psi = \frac{\mu_o I}{2\pi x} dx \text{ Wb/m}$$

Total flux linkage between points 1 and 2 is

$$\Psi_{ext} = \int_{d_1}^{d_2} \frac{\mu_o I}{2\pi x} dx = \frac{4\pi \times 10^{-7}}{2\pi} I \left[ \log_e x \right]_{d_1}^{d_2} \text{ Wb/m}$$

$$= 2 \times 10^{-7} I \log_e \frac{d_2}{d_1} \text{ Wb/m}$$

(8.7)

## 8.4 INDUCTANCE OF A SINGLE-PHASE TWO-WIRE LINE

Consider a single-phase line consisting of two parallel conductors A and B of radii $r_1$ and $r_2$ spaced $d$ meters apart as shown in Figure 8.3. Conductors A and B carry the same current (i.e., $I_A = I_B$) in magnitude but opposite in

FIGURE 8.3   Single-phase two-wire line.

directions, as one forms the return path for the other. The inductance of each conductor is due to internal flux linkages and external flux linkages, and the following points are to be noted regarding external flux linkages:

1. A line of flux produced due to current in conductor A at a distance equal to or greater than $(d + r_2)$ from the center of conductor A links with a zero net current, as the current following in the two conductors A and B are equal in magnitude but opposite in directions.

2. Flux lines at a distance $(d - r_2)$ link with a current $I$ and those between $(d + r_2)$ and $(d - r_2)$ link with a current varying from $I$ to zero.

As a simplifying assumption, it can be assumed that all the flux produced by current in conductor A links all the current up to the center of conductor B and that the flux beyond the center of conductor B does not link any current.

The above assumption simplifies the calculations and results obtained are quite accurate specifically when $d$ is much greater than $r_1$ and $r_2$ as is usually the case in overhead lines.

Based on the above assumption flux linkages of conductor A due to external flux can be determine from Equation 8.7 by substituting $d_2 = d$ and $d_1 = r_1$. Thus flux linkage of conductor A due to external flux only is

$$\Psi_{A\,ext} = 2 \times 10^{-7} I \, \log_e \frac{d}{r_1} \, \text{Wb/m}$$

Flux linkage of conductor A due to internal flux only is

$$\Psi_{A\,int} = \frac{1}{2} \times I \times 10^{-7} \, \text{Wb/m}$$

Total flux linkage of conductor is

$$\Psi_A = \Psi_{A\,ext} + \Psi_{A\,int} \left( 2 \times 10^{-7} I \log_e \frac{d}{r_1} + \frac{1}{2} \times I \times 10^{-7} \right) Wb/m$$

$$= \left( 0.5 + 2 \times \log_e \frac{d}{r_1} \right) I \times 10^{-7} \; Wb/m$$

Total inductance of conductor A is

$$L_A = \frac{\Psi_A}{I} = \frac{\left( 0.5 + 2\log_e (d/r_1) \right)}{I} \times I \times 10^{-7} \; H/m$$

$$= 2 \times 10^{-7} \left( 0.25 + \log_e \frac{d}{r_1} \right) H/m$$

$$= 2 \times 10^{-7} \left( \log_e e^{1/4} + \log_e \frac{d}{r_1} \right) H/m$$

$$= 2 \times 10^{-7} \log_e \frac{d}{r_1 e^{-1/4}} \; H/m$$

The product $(r_1 e^{-1/4})$ is known as geometric mean radius (GMR) of the conductor and is equal to 0.7788 times the radius of the conductor. Let it be represented by $r_1'$, where

$$r_1' = 0.7788 r_1$$

$$\therefore L_A = 2 \times 10^{-7} \log_e \frac{d}{r_1'} \; H/m$$

Similarly, inductance of conductor B is

$$L_B = 2 \times 10^{-7} \log_e \frac{d}{r_2'} \; H/m$$

Loop inductance of the line is

$$L = L_A + L_B$$

$$= 2 \times 10^{-7} \left( \log_e \frac{d}{r_1'} + \log_e \frac{d}{r_2'} \right)$$

If $r_1' = r_2' = r'$, the loop inductance of the line is given as

$$L = 4 \times 10^{-7} \times \log_e \frac{d}{r'} \, H/m$$

$$= 0.4 \times \log_e \frac{d}{r_1 e^{-1/4}} \, mH/km$$

The idea of replacing the original conductor of radii $r$ by a fictitious conductor of radii $r'$ is quite attractive because streamlined equation for inductance can be developed without bogging down in accounting for the internal flux.

## 8.5 FLUX LINKAGES OF ONE CONDUCTOR IN A GROUP OF CONDUCTORS

Consider a group of parallel conductors 1, 2, 3,..., $n$ carrying current $I_1$, $I_2$, $I_3$,..., $I_n$ respectively, as illustrated in Figure 8.4. Let it be assumed that the sum of the current in various conductors is zero, that is, $I_1 + I_2 + I_3 + \cdots + I_n = 0$.

Theoretically, the flux due to a conductor extends from the center of the conductor right up to infinity but let us assume that the flux linkages extend up to a remote point P and the respective distances are as marked in Figure 8.4.

The current in each conductor sets up a certain flux due to its own current. The sum of all these fluxes is the total flux of the system, and the total flux linkages of any one conductor is the sum of its linkages with all the individual fluxes set up by the conductor of the system.

Now let us determine the flux linkages of conductor 1 due to current $I_1$ carried by the conductor itself and flux linkages to conductor 1 due to current carried by other conductors (2, 3,..., $n$).

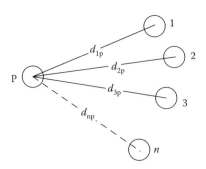

FIGURE 8.4 Arbitrary group of $n$ parallel conductors carrying current.

The flux linkage of conductor 1 due to its own current $I_1$ (internal and external), up to point P is

$$\Psi_{1p1} = 2 \times 10^{-7} I_1 \log_e \frac{d_{1p}}{r_1'} \text{ Wb/m}$$

The flux linkage of conductor 1 due to current in conductor 2 is

$$\Psi_{1p2} = 2 \times 10^{-7} I_2 \log_e \frac{d_{2p}}{d_{12}}$$

Flux due to conductor 2 that lies between conductor 2 and conductor 1 does not link conductor 1 and therefore the distances involved are $d_{2p}$ and $d_{12}$.

Thus the expression for flux linkages of conductor 1 due to current in all conductors can be written as

$$\Psi_{1p} = 2 \times 10^{-7} \left[ I_1 \log_e \frac{d_{1p}}{r_1'} + I_2 \log_e \frac{d_{2p}}{d_{12}} + I_3 \log_e \frac{d_{3p}}{d_{13}} + \cdots + I_n \log_e \frac{d_{np}}{d_{1n}} \right]$$
Wb/m

The above equation may be written as

$$\Psi_{1p} = 2 \times 10^{-7} \left[ I_1 \log_e \frac{1}{r_1'} + I_2 \log_e \frac{1}{d_{12}} + I_3 \ln \frac{1}{d_{13}} + \cdots + I_n \log_e \frac{1}{d_{1n}} \right]$$
$$+ 2 \times 10^{-7} [I_1 \log_e d_{1p} + I_2 \log_e d_{2p} + I_3 \log_e d_{3p} + \cdots + I_n \log_e d_{np}]$$
(8.8)

To account for the total flux linkage to conductor 1, the point P must be approach infinity and in this condition

$$d_{1p} \approx d_{2p} \approx d_{3p} \approx \cdots \approx d_{np} \approx d(\text{say})$$

Then,

$$\lim_{d \to \infty} [I_1 + I_2 + I_3 + \cdots + I_n] \log_e d = 0, \quad \because I_1 + I_2 + I_3 + \cdots + I_n = 0$$

This simplifies the Equation 8.8 and the equation for the flux linkages to conductor 1 becomes

$$\Psi_1 = 2 \times 10^{-7} \left[ I_1 \log_e \frac{1}{r_1'} + I_2 \log_e \frac{1}{d_{12}} + I_3 \log_e \frac{1}{d_{13}} + \cdots + I_n \log_e \frac{1}{d_{1n}} \right] \text{Wb/m}$$

(8.9)

### 8.5.1 Inductance of Composite Conductor Lines—Self and Mutual Geometric Mean Distances

Consider a single-phase line consisting of two parallel conductor A and B, conductor A consisting of $x$ and conductor B of $y$ strands (Figure 8.5).

Let the conductors A and B carry current $I$ and $-I$, respectively. Assuming uniform current density in both the conductors, the current carried by each strand of conductor A will be $I/x$ while that carried by each strand of conductor B will be $-I/y$.

Using Equation 8.9, the flux linkages of strand 1 in conductor A can be written as

$$\Psi_1 = 2 \times 10^{-7} \frac{I}{x} \left[ \log_e \frac{1}{r_1'} + \log_e \frac{1}{d_{12}} + \log_e \frac{1}{d_{13}} + \cdots + \log_e \frac{1}{d_{1x}} \right]$$

$$-2 \times 10^{-7} \frac{I}{y} \left[ \log_e \frac{1}{d_{11'}} + \log_e \frac{1}{d_{12'}} + \log_e \frac{1}{d_{13'}} + \cdots + \log_e \frac{1}{d_{1y}} \right]$$

$$\Psi_1 = 2 \times 10^{-7} I \log_e \frac{\sqrt[y]{(d_{11'} d_{12'} d_{13'} \cdots d_{1y})}}{\sqrt[x]{r_1' d_{12} d_{13} d_{14} \cdots d_{1x}}} \text{Wb/m}$$

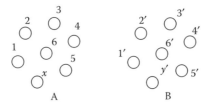

FIGURE 8.5  Single-phase line consisting of two composite conductors.

Inductance of strand 1 of conductor A

$$L_1 = \frac{\Psi_1}{I/x} = 2 \times 10^{-7} x \log_e \frac{\sqrt[y]{(d_{11'} d_{12'} d_{13'} \dots d_{1y})}}{\sqrt[x]{r' d_{12} d_{13} d_{14} \dots d_{1x}}} \qquad (8.10)$$

In above expression, numerator of argument of $\log_e$ is written $y$th root of distances $d_{11'}, d_{12'}, d_{13'}, \dots, d_{1y}$ multiplied together where distances $d_{11'}, d_{12'}, d_{13'}, \dots, d_{1y}$ are the distances of strands 1′, 2′, 3′, 4′,..., $y$ (all segments of conductor B) from segment 1 under consideration.

The denominator of argument ln is $x$th root of distances $d_{12}, d_{13}, \dots, d_{1x}$, and $r'$ multiplied together where distances $d_{12}, d_{13}, \dots, d_{1x}$ are the distances of strands 2, 3, 4,..., $n$ (all strands of conductor A) from strands 1. $r'$ can also be represented by distance $d_{11}$, and the expression for inductance for conductor A becomes

$$L_1 = \frac{\Psi_1}{I/x} = 2 \times 10^{-7} x \log_e \frac{\sqrt[y]{(d_{11'} d_{12'} d_{13'} \dots d_{1y})}}{\sqrt[x]{d_{11} d_{12} d_{13} d_{14} \dots d_{1x}}} \text{ H/m} \qquad (8.11)$$

Similarly, the expression for inductance for strand 2 can be written as

$$L_2 = \frac{\Psi_2}{I/x} = 2 \times 10^{-7} x \log_e \frac{\sqrt[y]{(d_{21'} d_{22'} d_{23'} \dots d_{2y})}}{\sqrt[x]{d_{21} d_{22} d_{23} d_{24} \dots d_{2x}}} \text{ H/m} \qquad (8.12)$$

Thus, we see that the different strands of a conductor have different inductances.

Average inductance of strands of conductor A is

$$L_{av} = \frac{L_1 + L_2 + L_3 + \dots + L_x}{x}$$

Since $x$ such strands of conductor A are electrically parallel, inductance of conductor A, therefore, is

$$L_A = \frac{L_{av}}{x} = \frac{L_1 + L_2 + L_3 + \dots + L_x}{x^2}$$

$$= 2 \times 10^{-7} \log_e \left[ \frac{\sqrt[xy]{(d_{11'} d_{12'} d_{13'} \dots d_{1y})(d_{21'} d_{22'} d_{23'} \dots d_{2y}) \dots (d_{x1'}, d_{x2'}, d_{x3'} \dots d_{xy})}}{\sqrt[x^2]{(d_{11} d_{12} d_{13} \dots d_{1x})(d_{21} d_{22} d_{23} \dots d_{2x}) \dots (d_{x_1} d_{x_2} d_{x_3} \dots d_{xx})}} \right]$$

$$(8.13)$$

In the above expression, the numerator of argument of $\log_e$ is called geometric mean distance (often called the mutual GMD) between conductor A and conductor B, and the denominator of argument $\log_e$ is called GMR (often called the self GMD). GMD and GMR are denoted by $D_m$ and $D_s$, respectively.

$$L_A = 2 \times 10^{-7} \log_e \frac{D_m}{D_{sA}} \text{ H/m}$$

Similarly

$$L_B = 2 \times 10^{-7} \log_e \frac{D_m}{D_{sB}} \text{ H/m}$$

If conductors A and B are identical,

$$D_{sA} = D_{sB} = D_s \text{(say)}$$

Loop inductance,

$$L = L_A + L_B = 2 \times 10^{-7} \log_e \frac{D_m}{D_s} + 2 \times 10^{-7} \log_e \frac{D_m}{D_s}$$

$$= 4 \times 10^{-7} \log_e \frac{D_m}{D_s} \text{ H/m} \quad \text{or} \quad 0.4 \log_e \frac{D_m}{D_s} \text{ mH/km}$$

(8.14)

## 8.6 INDUCTANCE OF A THREE-PHASE OVERHEAD LINE WITH UNSYMMETRICAL SPACING

Consider a 3-ϕ line with conductors A, B, and C; each of radius $r$ meters (Figure 8.6). Let the spacing between them be $d_1$, $d_2$, and $d_3$ and current flowing through them be $I_A$, $I_B$, and $I_C$, respectively.

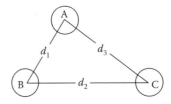

FIGURE 8.6   Cross-sectional view of a three-phase overhead line with unsymmetrical spacing.

The flux linkages of conductor A due to its own current $I_A$ and other conductor current $I_B$ and $I_C$.

$$\Psi_A = 2 \times 10^{-7} \left[ I_A \log_e \frac{1}{r'} + I_B \log_e \frac{1}{d_1} + I_C \log_e \frac{1}{d_3} \right] \text{Wb/m}$$

Similarly

$$\Psi_B = 2 \times 10^{-7} \left[ I_B \log_e \frac{1}{r'} + I_A \log_e \frac{1}{d_1} + I_C \log_e \frac{1}{d_2} \right] \text{Wb/m}$$

and

$$\Psi_C = 2 \times 10^{-7} \left[ I_C \log_e \frac{1}{r'} + I_A \log_e \frac{1}{d_3} + I_B \log_e \frac{1}{d_2} \right] \text{Wb/m}$$

If the system is balanced, $I_A = I_B = I_C = I$ (say) in magnitude.

Taking $I_A$ as a reference phasor, the current are represented, in symbolic form as

$$I_A = I, \quad I_B = I(-0.5 - j0.866), \quad I_C = I(-0.5 + j0.866)$$

Substituting these values of $I_B$ and $I_C$ in the expression for $\psi_A$, we get

$$\Psi_A = 2 \times 10^{-7} \left[ I \log_e \frac{1}{r'} + I(-0.5 - j0.866) \log_e \frac{1}{d_1} + I(-0.5 + j0.866) \log_e \frac{1}{d_3} \right]$$

$$= 2 \times 10^{-7} I \left[ \log_e \frac{1}{r'} + \log_e \sqrt{d_1 d_3} + j\sqrt{3} \log_e \sqrt{\frac{d_1}{d_3}} \right]$$

and

$$L_A = \frac{\Psi_A}{I_A} = 2 \times 10^{-7} \left[ \log_e \frac{1}{r'} + \log_e \sqrt{d_1 d_3} + j\sqrt{3} \log_e \sqrt{\frac{d_1}{d_3}} \right] \text{H/m}$$

$$(8.15)$$

Similarly

$$L_B = 2 \times 10^{-7} \left[ \log_e \frac{1}{r'} + \log_e \sqrt{d_1 d_2} + j\sqrt{3} \log_e \sqrt{\frac{d_2}{d_1}} \right] \text{H/m} \quad (8.16)$$

$$L_C = 2 \times 10^{-7} \left[ \log_e \frac{1}{r'} + \log_e \sqrt{d_2 d_3} + j\sqrt{3} \log_e \sqrt{\frac{d_3}{d_2}} \right] \text{H/m} \qquad (8.17)$$

When three-phase line conductors are not equidistant from each other, the conductor spacing is said to be unsymmetrical. Under such conditions, the flux linkages and inductance of each phase are not same. A different inductance in each phase results in unequal voltage drop in three phases even if the current in the conductors are balanced. Therefore, the voltage at the receiving end will not be same for all phases. In order that voltage drops are equal in all conductors, we generally interchange the positions of the conductor at regular intervals along the line, so that each conductor occupies the original position of every other conductor over an equal distance. Such an exchange of position is known as transposition, as shown in Figure 8.7. In practice, the conductors are so transposed that each of the three possible arrangement of conductors exist for one-third of the total length of the line.

1. The effect of transposition is that each conductor has the same average inductance, which is given as

$$L = \frac{1}{3}[L_A + L_B + L_C]$$

$$= 2 \times 10^{-7} \left[ \log_e \frac{1}{r'} + \frac{1}{3} \left( \log_e \sqrt{d_1 d_3} + \log_e \sqrt{d_1 d_2} + \log_e \sqrt{d_2 d_3} \right) \right.$$

$$\left. + \frac{j\sqrt{3}}{3} \left( \log_e \sqrt{\frac{d_1}{d_3}} + \log_e \sqrt{\frac{d_2}{d_1}} + \log_e \sqrt{\frac{d_3}{d_2}} \right) \right] \text{H/m} \qquad (8.18)$$

$$= 2 \times 10^{-7} \left[ \log_e \frac{1}{r'} + \log_e \sqrt[3]{d_1 d_2 d_3} + 0 \right]$$

$$= 2 \times 10^{-7} \log_e \sqrt[3]{\frac{d_1 d_2 d_3}{r'}} \text{H/m}$$

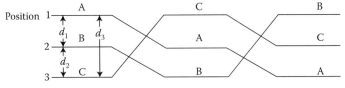

FIGURE 8.7   A complete transposition cycle.

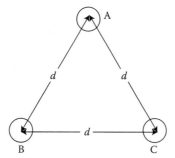

FIGURE 8.8   Three-phase line with equilateral spacing.

2. If the conductors are equispaced (let the spacing be equal to $d$), as shown in Figure 8.8, the inductance of each conductor will be same and can be obtained by substituting $d_1 = d_2 = d_3$ in Equations 8.15 through 8.17. So inductance of each conductor is

$$L = 2 \times 10^{-7} \log_e \sqrt[3]{\frac{d_1 d_2 d_3}{r'}} \text{ H/m}$$

$$= 2 \times 10^{-7} \log_e \frac{d}{r'} \text{ H/m}$$

For stranded conductor, $r'$ will be replaced by $D_s$ (self-GMD)

3. When the conductors of three-phase transmission line are in the same plane, as shown in Figure 8.9.
In this position, $d_1 = d_2 = d$ and $d_3 = 2d$.
In general, Equations 8.15 through 8.17 for $L_A$, $L_B$, and $L_C$, we get

$$L_A = 2 \times 10^{-7} \left[ \log_e \frac{1}{r'} + \log_e \sqrt{2d \times d} + j\sqrt{3} \log_e \sqrt{\frac{d}{2d}} \right] \text{H/m}$$

$$= 2 \times 10^{-7} \left[ \log_e \frac{d}{r'} + \frac{1}{2} \log_e 2 + j0.866 \log_e 2 \right] \text{H/m}$$

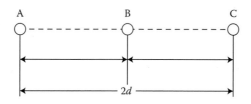

FIGURE 8.9   Horizontally spaced line conductors.

$$L_B = 2 \times 10^{-7} \left[ \log_e \frac{1}{r'} + \log_e \sqrt{d \times d} + j\sqrt{3} \log_e \sqrt{\frac{d}{d}} \right] \text{H/m}$$

$$= 2 \times 10^{-7} \log_e \frac{d}{r'} \text{H/m}$$

$$L_C = 2 \times 10^{-7} \left[ \log_e \frac{1}{r'} + \log_e \sqrt{2d \times d} + j\sqrt{3} \log_e \sqrt{\frac{2d}{d}} \right] \text{H/m}$$

$$= 2 \times 10^{-7} \left[ \log_e \frac{d}{r'} + \frac{1}{2} \log_e 2 + j0.866 \log_e 2 \right] \text{H/m}$$

4. When the conductors are at the corners of right-angled triangle, as shown in Figure 8.10. In this position,

$$d_1 = d_2 = d \quad \text{and} \quad d_3 = \sqrt{2}d$$

Substituting this values in Equations 8.15 through 8.17 for $L_A$, $L_B$, and $L_C$, respectively, we get

$$L_A = 2 \times 10^{-7} \left[ \log_e \frac{1}{r'} + \log_e \sqrt{\sqrt{2}d \times d} + j\sqrt{3} \log_e \sqrt{\frac{d}{\sqrt{2}d}} \right] \text{H/m}$$

$$= 2 \times 10^{-7} \left[ \log_e \frac{1}{r'} + \frac{1}{2} \log_e \sqrt{2} - j0.866 \log_e \sqrt{2} \right] \text{H/m}$$

$$L_B = 2 \times 10^{-7} \left[ \log_e \frac{1}{r'} + \log_e \sqrt{d \times d} + j\sqrt{3} \log_e \sqrt{\frac{d}{d}} \right] \text{H/m}$$

$$= 2 \times 10^{-7} \log_e \frac{d}{r'} \text{H/m}$$

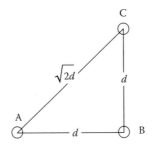

FIGURE 8.10   Conductors are at the corners of a right-angled triangle.

$$L_C = 2 \times 10^{-7} \left[ \log_e \frac{1}{r'} + \log_e \sqrt{\sqrt{2d} \times d} + j\sqrt{3} \log_e \sqrt{\frac{\sqrt{2d}}{d}} \right] \text{H/m}$$

$$= 2 \times 10^{-7} \left[ \log_e \frac{d}{r'} + \frac{1}{2} \log_e \sqrt{2} + j0.866 \log_e \sqrt{2} \right] \text{H/m}$$

## 8.7 INDUCTANCE OF A THREE-PHASE LINE WITH MORE THAN ONE CIRCUIT

It is usual practice to run three-phase transmission lines with more than one circuit in parallel on the same tours, because it gives greater reliability and a higher transmission capacity. If such circuits are so widely separated that the mutual inductance between them becomes negligible, the inductance of the equivalent single circuit would be half of each of the individual circuits considered alone. But in actual practice, the separation is not very wide and the mutual inductance is not negligible. GMD method is used for the determination of inductance per phase by considering the various conductors collected in parallel as strands of one composite conductor.

It is desirable to have a configuration that provides minimum inductance so as to have maximum transmission capacity. This is possible only with low GMD and high GMR. Therefore, the practice is to have the individual conductor of a phase widely separated to provide high GMR and the distance between the phases small to give low GMD. Thus, in the case of a double circuit in vertical formation, the arrangement of conductor would be as illustrated in Figure 8.11a and not as illustrated in Figure 8.11b, because the arrangement of conductors given in Figure 8.11a results in low inductance in compression to that given by the arrangement illustrated in Figure 8.11b.

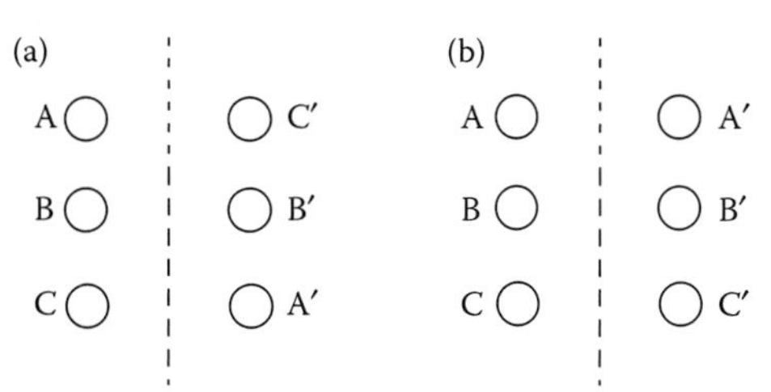

FIGURE 8.11 Arrangements of conductors in a double circuit three-phase line.

### 8.7.1 Inductance of a Three-Phase Double Circuit Line with Symmetrical Spacing

Consider a three-phase double circuit line connected in parallel conductors A, B, and C forming one circuit and conductors A′, B′, and C′ forming the other one, as illustrated in Figure 8.12.

Flux linkage of phase A conductor is

$$\Psi_A = 2 \times 10^{-7} \left[ I_A \left( \log_e \frac{1}{r'} + \log_e \frac{1}{2d} \right) + I_B \left( \log_e \frac{1}{d} + \log_e \frac{1}{\sqrt{3d}} \right) \right.$$
$$\left. + I_C \left( \log_e \frac{1}{\sqrt{3d}} + \log_e \frac{1}{d} \right) \right]$$

$$= 2 \times 10^{-7} \left[ I_A \log_e \frac{1}{2dr'} + (I_B + I_C) \log_e \frac{1}{\sqrt{3d^2}} \right]$$

$$= 2 \times 10^{-7} \left[ I_A \log_e \frac{1}{2dr'} - I_A \log_e \frac{1}{\sqrt{3d^2}} \right] \quad \because I_A + I_B + I_C = 0$$

$$= 2 \times 10^{-7} I_A \log_e \frac{\sqrt{3d}}{2r'} \text{ Wb/m}$$

Inductance of conductor A is, $\quad L_A = \dfrac{\Psi_A}{I_A} = 2 \times 10^{-7} \log_e \dfrac{\sqrt{3d}}{2r'}$ H/m $\quad$ (8.19)

Similarly, inductance of remaining conductors can be worked out, which will be the same as $L_A$. This is due to the fact that the conductors of different phases are symmetrically placed.

Since conductors are electrically in parallel, inductance of each phase is

$$= \frac{1}{2} L_A = 1 \times 10^{-7} \log_e \frac{\sqrt{3d}}{2r'} \text{ H/m} \tag{8.20}$$

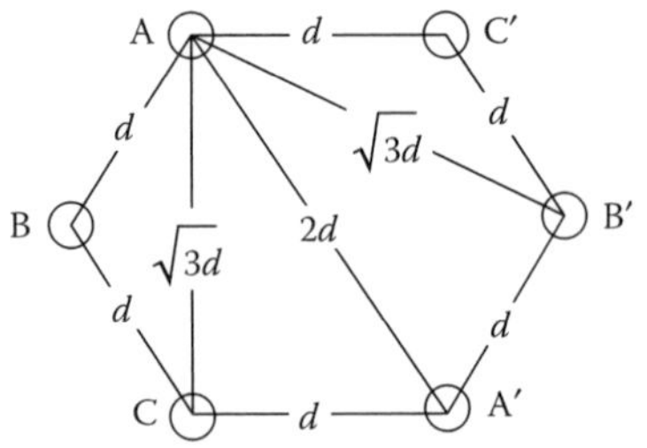

FIGURE 8.12  Three-phase double circuit line with symmetrical spacing.

### 8.7.2 Inductance of a Three-Phase Double Circuit with Unsymmetrical but Transposed

Now consider a three-phase double circuit connected in parallel—conductors A, B, and C forming one circuit and conductors A′, B′, and C′ forming the other one, as illustrated in Figure 8.13 (conductors unsymmetrically spaced and transposed).

Since the conductors are thoroughly transposed, the conductor situations in the transposition cycle would be as illustrated in Figure 8.13a–c.

Flux linkages with conductor A in position (a)

$$
\Psi_{A1} = 2 \times 10^{-7} \left[ I_A \left( \log_e \frac{1}{r'} + \log_e \frac{1}{\sqrt{4d_1^2 + d_2^2}} \right) + I_B \left( \log_e \frac{1}{d_1} + \log_e \frac{1}{\sqrt{d_1^2 + d_2^2}} \right) \right.
$$
$$
\left. + I_C \left( \log_e \frac{1}{2d_1} + \log_e \frac{1}{d_2} \right) \right]
$$

Similarly flux linkages with conductor A in positions (b) and (c) are

$$
\Psi_{A2} = 2 \times 10^{-7} \left[ I_A \left( \log_e \frac{1}{r'} + \log_e \frac{1}{d_2} \right) + I_B \left( \log_e \frac{1}{d_1} + \log_e \frac{1}{\sqrt{d_1^2 + d_2^2}} \right) \right.
$$
$$
\left. + I_C \left( \log_e \frac{1}{d_1} + \log_e \frac{1}{\sqrt{d_1^2 + d_2^2}} \right) \right]
$$

$$
\Psi_{A3} = 2 \times 10^{-7} \left[ I_A \left( \log_e \frac{1}{r'} + \log_e \frac{1}{\sqrt{4d_1^2 + d_2^2}} \right) + I_B \left( \log_e \frac{1}{2d_1} + \log_e \frac{1}{d_2} \right) \right.
$$
$$
\left. + I_C \left( \log_e \frac{1}{d_1} + \log_e \frac{1}{\sqrt{d_1^2 + d_2^2}} \right) \right]
$$

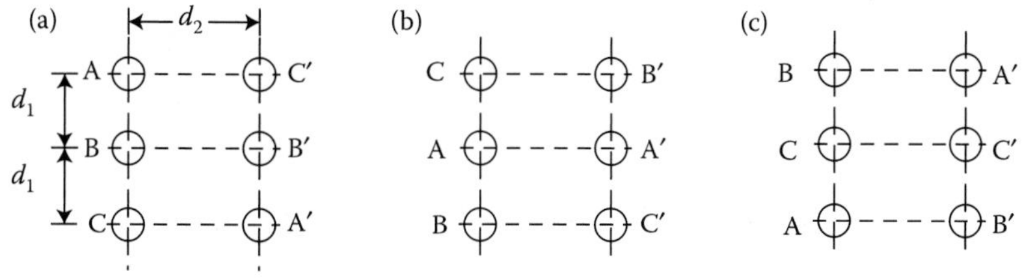

FIGURE 8.13 Three-phase double-circuit line with unsymmetrical spacing (fully transposed).

Average flux linkages with conductor A

$$\Psi_A = \frac{\Psi_{A1} + \Psi_{A2} + \Psi_{A3}}{3}$$

$$= \frac{2 \times 10^{-7}}{3} \left[ I_A \left( 3\log_e \frac{1}{r'} + \log_e \frac{1}{d_2} + \log_e \frac{1}{4d_1^2 + d_2^2} \right) \right.$$

$$+ I_B \left( 2\log_e \frac{1}{d_1} + \log_e \frac{1}{2d_1} + \log_e \frac{1}{d_1^2 + d_2^2} + \log_e \frac{1}{d_2} \right)$$

$$\left. + I_C \left( \log_e \frac{1}{2d_1} + \log_e \frac{1}{d_1^2 + d_2^2} + \log_e \frac{1}{d_2} + 2\log_e \frac{1}{d_1} \right) \right.$$

$$= \frac{2 \times 10^{-7}}{3} \left[ I_A \left( 3\log_e \frac{1}{r'} + \log_e \frac{1}{d_2} + \log_e \frac{1}{4d_1^2 + d_2^2} \right) \right.$$

$$\left. + (I_B + I_C) \left( \log_e \frac{1}{2d_1} + \log_e \frac{1}{d_1^2 + d_2^2} + \log_e \frac{1}{d_2} + 2\log_e \frac{1}{d_1} \right) \right]$$

$$= \frac{2 \times 10^{-7}}{3} \left[ I_A \left( 3\log_e \frac{1}{r'} + \log_e \frac{1}{d_2} + \log_e \frac{1}{4d_1^2 + d_2^2} \right) \right.$$

$$\left. - I_A \left( \log_e \frac{1}{2d_1} + \log_e \frac{1}{d_1^2 + d_2^2} + \log_e \frac{1}{d_2} + 2\log_e \frac{1}{d_1} \right) \right]$$

$$\because I_A + I_B + I_C = 0$$

$$\Psi_A = \frac{2 \times 10^{-7}}{3} I_A \log_e \frac{2d_1^3 (d_1^2 + d_2^2) d_2}{(r')^3 d_2 (4d_1^2 + d_2^2)}$$

$$\text{or} \quad \Psi_A = 2 \times 10^{-7} I_A \log_e \frac{2^{1/3} d_1 (d_1^2 + d_2^2)^{1/3}}{r' (4d_1^2 + d_2^2)^{1/3}} \, \text{H/m}$$

And inductance, $L_A = \dfrac{\Psi_A}{I_A}$

$$= 2 \times 10^{-7} \log_e 2^{1/3} \frac{d_1}{r'} \left( \frac{d_1^2 + d_2^2}{4d_1^2 + d_2^2} \right)^{1/3} \, \text{H/m} \tag{8.21}$$

Inductance of each phase $L = \dfrac{1}{2} \times$ Inductance per conductor $= \dfrac{1}{2} L_A$

$$= 2 \times 10^{-7} \log_e 2^{1/6} \left( \dfrac{d_1}{r'} \right)^{1/2} \left( \dfrac{d_1^2 + d_2^2}{4d_1^2 + d_2^2} \right)^{1/6} \text{H/m}$$

(8.22)

If the distance $d_2$ is too large as compared to $d_1$, $(d_1^2 + d_2^2)/(4d_1^2 + d_2^2)$ would tend to be unity and inductance per phase,

$$L = 2 \times 10^{-7} \log_e 2^{1/6} \left( \dfrac{d_1}{r'} \right)^{1/2} \text{H/m} \quad \text{or} \quad L = 2 \times 10^{-7} \log_e 2^{1/6} \left( \dfrac{d_1}{r'} \right)^{1/2} \text{mH/km}$$

(8.23)

## 8.8 CAPACITANCE

We know that any two conductors separate by an insulating material constitute a capacitor. As any two conductors of an overhead transmission line are separated by air which acts as insulation. Therefore, capacitance exists between any two overhead line conductors. The capacitance between the conductors is the charge per-unit potential difference.

$$\text{Capacitance } C = \dfrac{Q}{V} \text{ Farad}$$

where $Q$ is the charge on the line in coulomb, and $V$ is the potential difference between the conductors in volts.

## 8.9 POTENTIAL AT A CHARGED SINGLE CONDUCTOR

The electric potential at a point due to a charge is measured by the work done in bringing a unit positive charge from infinite distance to that point. Electric potential is an extremely important factor for determining the capacitance of a circuit since it is defined as the charge per-unit potential. We shall now discuss in details the electric potential due to some important conductor arrangements.

Consider a long straight, isolated conductor carrying a charge $+Q$ coulomb/m. The charge is uniformly distributed over the surface of the conductor. The electric lines of flux will be straight, radial, and uniformly spaced. The points which are equally placed from the conductor will be at the same potential and have the same flux density. All the cylinders concentric with the conductor will be equipotential surfaces. The electric flux density $D_x$ at a point $x$ meters from the axis of the conductor is the quotient of flux leaving the conductor per meter length and the curved surface of a cylinder 1 m long and having a radius $x$ meters.

$$D_x = \frac{Q}{2\pi x \times 1} \, \text{C/m}^2$$

The electric field intensity or potential gradient at the point considered is

$$E_x = \frac{Q}{2\pi\varepsilon_0\varepsilon_r x} \, \text{V/m}$$

Taking air as medium, that is, $\varepsilon_r = 1$, $\varepsilon_0 =$ permittivity of free space.

The potential difference between two points A and B (Figure 8.14) kept at distances $d_1$ and $d_2$ meters from the conductor is same as the integral of the electric field intensity over a radial path between the equipotential surfaces passing through A and B. It does not matter whether A and B lie on the same radial line or not.

$$V_{AB} = \int_{d_1}^{d_2} E_x \, dx = \int_{d_1}^{d_2} \frac{Q}{2\pi\varepsilon_0 x} \, dx$$

$$V_{AB} = \frac{Q}{2\pi\varepsilon_0} \log_e \frac{d_2}{d_1} \tag{8.24}$$

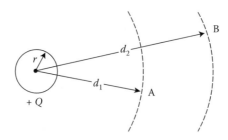

FIGURE 8.14   Electric field of a long straight conductor.

Equation 8.24 is very useful specially in determining charges and capacitances of a system of conductors.

## 8.10 SYSTEM OF CONDUCTORS

Consider a system of $n$ conductors (Figure 8.15) each of radius $r$ forming a circuit.

Let the charges in coulomb per meter be $Q_A$, $Q_B$, $Q_C$,..., $Q_n$. The spacing between the conductors are denoted by $d_{AB}$, $d_{BC}$, $d_{CD}$, ..., etc. The spacing are assumed to be very large in comparison to their radii that the distribution of charge is uniform around the periphery of each conductor.

The principle of superposition will be applied here to find out the potential difference between any two conductors. According to this principle, the difference of potential between two charged conductors is equal to the potential difference due to charge on first conductor alone, plus the potential difference due to the charge on second conductor alone, plus the potential difference due to other charged conductors in the field.

Using the result obtained in Equation 8.24, the potential difference between two conductors A and B is given by

$V_{AB}$ = Potential difference between A and B due to charge $Q_A$ on A

  +Potential difference between A and B due to charge $Q_B$ on B

  +$\cdots$ + Potential difference between A and B due to charge $Q_N$ on N

$$V_{AB} = \frac{Q_A}{2\pi x \varepsilon_0} \log_e \frac{d_{AB}}{r_A} + \frac{Q_B}{2\pi x \varepsilon_0} \log_e \frac{r_B}{d_{BA}} + \frac{Q_C}{2\pi x \varepsilon_0} \log_e \frac{d_{CB}}{d_{CA}} + \cdots$$
$$+ \frac{Q_N}{2\pi x \varepsilon_0} \log_e \frac{d_{NB}}{d_{NA}}$$

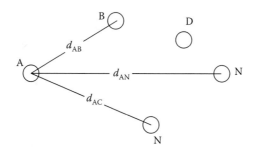

FIGURE 8.15   System of $n$ conductors.

If $r_A, r_B, r_C, \ldots$ are replaced by $d_{AA}, d_{BB}, d_{CC}, \ldots$, respectively, for the sake of symmetrical result

$$V_{AB} = \frac{1}{2\pi\varepsilon_0} \left[ Q_A \log_e \frac{d_{AB}}{d_{AA}} + Q_B \log_e \frac{d_{BB}}{d_{BA}} + Q_C \log_e \frac{d_{CB}}{d_{CA}} + \cdots \right. $$
$$\left. + Q_N \log_e \frac{d_{NB}}{d_{NA}} \right] \text{Volts} \qquad (8.25)$$

$$V_{AC} = \frac{1}{2\pi\varepsilon_0} \left[ Q_A \log_e \frac{d_{AC}}{d_{AA}} + Q_B \log_e \frac{d_{BC}}{d_{BA}} + Q_C \log_e \frac{d_{CC}}{d_{CA}} \right. $$
$$\left. + \cdots + Q_N \log_e \frac{d_{NC}}{d_{NA}} \right] \text{Volts} \qquad (8.26)$$

$$V_{AN} = \frac{1}{2\pi\varepsilon_0} \left[ Q_A \log_e \frac{d_{AN}}{d_{AA}} + Q_B \log_e \frac{d_{BN}}{d_{BA}} + Q_C \log_e \frac{d_{CN}}{d_{CA}} \right. $$
$$\left. + \cdots + Q_N \log_e \frac{d_{NN}}{d_{NA}} \right] \text{Volts} \qquad (8.27)$$

$$V_{AN} = \frac{1}{2\pi\varepsilon_0} \left[ \sum_{x=A}^{x=N} Q_x \log_e \frac{d_{xN}}{d_{xA}} \right] \text{Volts} \qquad (8.28)$$

For a system working under normal conditions,

$$Q_A + Q_B + Q_C + \cdots + Q_N = 0 \qquad (8.29)$$

The equations obtained in this section will be used in calculating the capacitance per-unit length of a conductor in any system of parallel conductors constituting a complete circuit.

## 8.11 CAPACITANCE OF A SINGLE-PHASE TWO-WIRE LINE

Figure 8.16 shows a line consisting of two conductors A and B, each of radius $r$; the distance between conductors being D.

Therefore, the potential difference between A and B is

$$V_{AB} = \frac{1}{2\pi\varepsilon_0} \left[ Q_A \log_e \frac{d_{AB}}{d_{AA}} + Q_B \log_e \frac{d_{BB}}{d_{BA}} \right]$$

FIGURE 8.16   Single-phase two-wire line.

Here, $Q_A + Q_B = 0$

$$Q_A = -Q_B$$

$$d_{AB} = d_{BA} = d$$

$$d_{AA} = d_{BB} = r$$

Substituting these values in Equation 8.29, we get

$$V_{AB} = \frac{1}{2\pi\varepsilon_0}\left[Q_A\log_e\frac{d}{r} - Q_A\log_e\frac{r}{d}\right] = \frac{1}{2\pi\varepsilon_0}Q_A\log_e\left(\frac{d}{r}\right)^2 = \frac{1}{\pi\varepsilon_0}Q_A\log_e\frac{d}{r}$$

The capacitance between the conductors is

$$C_{AB} = \frac{Q_A}{V_{AB}} = \frac{Q_A}{(2Q/2\pi\varepsilon_0)\log_e(d/r)}\,\text{F/m}$$

$$C_{AB} = \frac{\pi\varepsilon_0}{\log_e(d/r)}\,\text{F/m} \qquad (8.30)$$

$C_{AB}$ is referred to as line-to-line capacitance. It is shown in Figure 8.17.

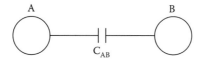

FIGURE 8.17   Line-to-line capacitance.

### 8.11.1 Capacitance to Neutral

Since the two conductors A and B are oppositely charged, the potential of the points midway between the conductors is zero, that is, zero potential plane is midway between A and B. The potential of each conductor is therefore $(1/2)V_{AB}$ with respect to neutral.

$$C_N = \frac{Q}{(1/2)V_{AB}} = \frac{2\pi\varepsilon_0}{\log_e(d/r)} \, \text{F/m} \qquad (8.31)$$

$C_N$ is called the capacitance to neutral or capacitance to ground. The term capacitance to neutral is more common in transmission calculations. Also $C_N = C_{AN} = C_{BN}$ (Figure 8.18).

Thus, the capacitance to neutral is twice the capacitance between conductors, that is,

$$C_N = 2C_{AB}$$

## 8.12 CAPACITANCE OF A THREE-PHASE OVERHEAD LINE

In a three-phase transmission line, the capacitance of each conductor is considered instead of capacitance from conductor to conductor. Here, again two conditions arise viz., symmetrical spacing and unsymmetrical spacing.

### 8.12.1 Symmetrical Spacing

Figure 8.19 shows the three conductors A, B, and C of the three-phase overhead transmission line having charges $Q_A$, $Q_B$, and $Q_C$ per meter length, respectively. Let the conductors be equidistant ($d$ meters) from each other.

Potential difference between conductor A and conductor B is given by

$$V_{AB} = \frac{1}{2\pi\varepsilon_0}\left[Q_A \log_e \frac{d}{r} + Q_B \log_e \frac{r}{d} + Q_C \log_e \frac{d}{d}\right] \text{Volts} \qquad (8.32)$$

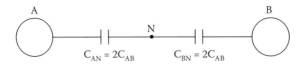

A                  B

$C_{AN} = 2C_{AB}$     $C_{BN} = 2C_{AB}$

FIGURE 8.18   Line-to-neutral capacitances.

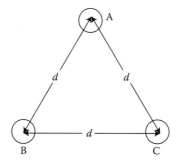

FIGURE 8.19   Three-phase line with equilateral spacing.

Similarly potential difference between A and C is

$$V_{AC} = \frac{1}{2\pi\varepsilon_0}\left[Q_A \log_e \frac{d}{r} + Q_B \log_e \frac{d}{d} + Q_C \log_e \frac{r}{d}\right] \text{Volts} \quad (8.33)$$

Adding Equations 8.32 and 8.33, we get

$$V_{AB} + V_{AC} = \frac{1}{2\pi\varepsilon_0}\left[2Q_A \log_e \frac{d}{r} + (Q_B + Q_C)\log_e \frac{r}{d}\right] \text{Volts} \quad (8.34)$$

Assuming balanced supply,   $Q_A + Q_B + Q_C = 0$ \quad (8.35)

$$\therefore Q_B + Q_C = -Q_A \quad (8.36)$$

Combining Equations 8.34 and 8.36,

$$V_{AB} + V_{AC} = \frac{1}{2\pi\varepsilon_0}\left[2Q_A \log_e \frac{d}{r} - Q_A \log_e \frac{r}{d}\right] = \frac{3Q_A}{2\pi\varepsilon_0}\log_e \frac{d}{r} \quad (8.37)$$

With balanced three-phase voltage applied to the line, it follows from the phasor diagram of Figure 8.20 that

$$V_{AB} + V_{AC} = 3V_{AN} \quad (8.38)$$

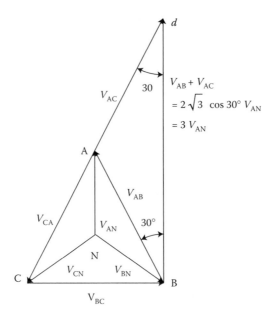

FIGURE 8.20   Phasor diagram of balanced three-phase voltage.

Substituting for $(V_{AB} + V_{AC})$ from Equation 8.38 in Equation 8.37 we get,

$$V_{AN} = \frac{Q_A}{2\pi\varepsilon_0} \log_e \frac{d}{r}$$

(8.39)

The capacitance of line-to-neutral immediately follows as

$$C_N = \frac{Q_A}{V_{AN}} = \frac{2\pi\varepsilon_0}{\log_e(d/r)}$$

(8.40)

Note that this equation is identical to capacitance to neutral for two-wire line. Derived in a similar manner, the expressions for capacitance are the same for conductors B and C.

### 8.12.2  Unsymmetrical Spacing

Figure 8.21 shows the three identical conductors of radius $r$ of a three-phase line with unsymmetrical spacing. The line is believed to be fully transposed. As the conductors are rotated cyclically in the three sections of the transposition cycle, correspondingly three expressions can be written for $V_{AB}$. These expressions are

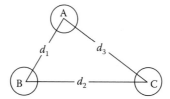

FIGURE 8.21   Cross-sectional view of a three-phase overhead line with unsymmetrical spacing (fully transposed).

For the first section of the transposition cycle,

$$V_{AB} = \frac{1}{2\pi\varepsilon_0} \left[ Q_{A1} \log_e \frac{d_1}{r} + Q_{B1} \log_e \frac{r}{d_1} + Q_{C1} \log_e \frac{d_2}{d_3} \right] \qquad (8.41)$$

For the second section of the transposition cycle,

$$V_{AB} = \frac{1}{2\pi\varepsilon_0} \left[ Q_{A2} \log_e \frac{d_2}{r} + Q_{B2} \log_e \frac{r}{d_2} + Q_{C2} \log_e \frac{d_3}{d_1} \right] \qquad (8.42)$$

For the third section of the transposition cycle,

$$V_{AB} = \frac{1}{2\pi\varepsilon_0} \left[ Q_{A3} \log_e \frac{d_3}{r} + Q_{B3} \log_e \frac{r}{d_3} + Q_{C3} \log_e \frac{d_1}{d_2} \right] \qquad (8.43)$$

If the voltage drop along the line is neglected, $V_{AB}$ is the same in each transposition cycle. On similar lines, three such equations can be written for $V_{BC} = V_{AB} \angle -120°$. Three more equations can be written equating to zero the summation of all line charges in each section of the transposition cycle. From these nine (independent) equations, it is possible to determine the nine unknown charges. The rigorous solution though possible is too involved.

With the usual spacing of conductors, sufficient accuracy is obtained by assuming:

$$Q_{A1} = Q_{A2} = Q_{A3} = Q_A, \quad Q_{B1} = Q_{B2} = Q_{B3} = Q_B, \quad Q_{C1} = Q_{C2} = Q_{C3} = Q_C \qquad (8.44)$$

This assumption of equal charge/unit length of a line in the three sections of the transposition cycle requires, on the other hand, three different values of $V_{AB}$ designated as $V_{AB1}$, $V_{AB2}$, and $V_{AB3}$, in the three sections. The solution can be considerably simplified by taking $V_{AB}$ as the average of these three voltage, that is,

$$V_{AB}(\text{avg}) = \frac{1}{3}(V_{AB1} + V_{AB2} + V_{AB3})$$

$$V_{AB} = \frac{1}{3 \times 2\pi\varepsilon_0}\left[ Q_A \log_e\left(\frac{d_1 d_2 d_3}{r^3}\right) + Q_B \log_e\left(\frac{r^3}{d_1 d_2 d_3}\right) + Q_C \log_e\left(\frac{d_1 d_2 d_3}{d_1 d_2 d_3}\right) \right]$$

$$V_{AB} = \frac{1}{2\pi\varepsilon_0}\left[ Q_A \log_e \frac{D_{eq}}{r} + Q_B \log_e \frac{r}{D_{eq}} \right] \tag{8.45}$$

where $D_{eq} = \sqrt[3]{d_1 d_2 d_3}$.
   Similarly,

$$V_{AC} = \frac{1}{2\pi\varepsilon_0}\left[ Q_A \log_e \frac{D_{eq}}{r} + Q_C \log_e \frac{r}{D_{eq}} \right] \tag{8.46}$$

Adding Equations 8.45 and 8.46, we get

$$V_{AB} + V_{AC} = \frac{1}{2\pi\varepsilon_0}\left[ 2Q_A \log_e \frac{D_{eq}}{r} + (Q_B + Q_C)\log_e \frac{r}{D_{eq}} \right] \tag{8.47}$$

And also for balanced three-phase voltages,

$$V_{AB} + V_{AC} = 3V_{AN}$$

and

$$Q_B + Q_C = -Q_A$$

Use of these relationships in Equation 8.47 leads to

$$3V_{AN} = \frac{3Q_A}{2\pi\varepsilon_0}\log_e \frac{D_{eq}}{r}$$

$$V_{AN} = \frac{Q_A}{2\pi\mu_0} \log_e \frac{D_{eq}}{r} \tag{8.48}$$

The capacitance of line-to-neutral of the transposed line is then given by

$$C_N = \frac{Q_A}{V_{AN}} = \frac{2\pi\varepsilon_0}{\log_e(D_{eq}/r)} \text{F/m to neutral} \tag{8.49}$$

It is obvious that for equilateral spacing $D_{eq} = D$, the above (approximate) formula gives the exact result presented earlier.

## 8.13 EFFECT OF EARTH ON THE TRANSMISSION LINE CAPACITANCE

While calculating the capacitance of transmission lines, the presence of earth was ignored, so far. The effect of earth on capacitance can be conveniently taken into account by the method of images.

### 8.13.1 Method of Images

The electric field of transmission line conductors must conform to the presence of the earth below. The earth for this purpose may be assumed to be a perfectly conducting horizontal sheet of infinite extent. Here it is assumed to be an equipotential surface.

The electric field of two long, parallel conductors charged $+Q$ and $-Q$ per unit is such that it has a zero potential plane midway between the conductors. If a conducting sheet of infinite dimensions is placed at the zero potential plane, the electric field remains unaltered. Further, if the conductor carrying charge $-Q$ is now eliminated, the electric field above the conducting sheet is kept intact, while electric field below it vanishes. Using these well-known results in reverse, we may equivalently replace the presence of ground below a charged conductor by a fictitious conductor having equal and opposite charge and located as far below the surface of ground as the overhead conductor above it—such a fictitious conductor is the mirror image of the overhead conductor. This method of creating the same electric field as in the presence of earth is known as the method of images originally suggested by Lord Kelvin.

### 8.13.2 Capacitance of a Single-Phase Overhead Line

Considering the case of single-phase overhead line, assume conductors A′ and B′ as image conductors of conductors A and B, respectively, as shown in Figure 8.22.

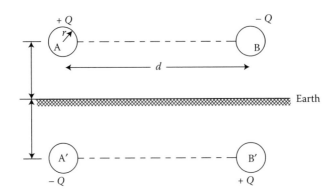

FIGURE 8.22   Single-phase overhead line with images.

Let the height of conductor be $h$ meters and $+Q$ and $-Q$ coulombs per meter length be the charges on conductors A and B, respectively.

The earth is assumed to be at zero potential and it can be done only if there is an image conductor A′ having a charge of $-Q$ coulombs per meter length at a depth of $h$ meters below the earth. Similarly, there is a conductor B′ having a charge of $+Q$ coulombs per meter length at a depth of $h$ meters below the earth.

The equation for the voltage drop is

$$V_{AB} = \frac{1}{2\pi\varepsilon_0}\left[ Q_A \log_e \frac{d}{r} + Q_B \log_e \frac{r}{d} + Q_{A'} \log_e \frac{\sqrt{4h^2 + d^2}}{2h} \right.$$

$$\left. + Q_{B'} \log_e \frac{2h}{\sqrt{4h^2 + d^2}} \right] \tag{8.50}$$

But $Q_A = Q$, $Q_B = -Q$, $Q_{A'} = -Q$, and $Q_{B'} = Q$.

Substituting the values of different charges in Equation 8.50, we get

$$V_{AB} = \frac{1}{2\pi\varepsilon_0}\left[ Q_A \log_e \frac{d}{r} - Q_B \log_e \frac{r}{d} - Q_{A'} \log_e \frac{\sqrt{4h^2 + d^2}}{2h} \right.$$

$$\left. + Q_{B'} \log_e \frac{2h}{\sqrt{4h^2 + d^2}} \right]$$

or

$$V_{AB} = \frac{Q}{\pi\varepsilon_0} \left[ \log_e \frac{2hd}{r\sqrt{4h^2 + d^2}} \right] \tag{8.51}$$

It immediately follows that

$$C_{AB} = \frac{Q}{V_{AB}} = \frac{Q}{(Q/\pi\varepsilon_0)\left[\log_e\left(d/\left(r\sqrt{1 + (d^2/4h^2)}\right)\right)\right]}$$

$$= \frac{\pi\varepsilon_0}{\left[\log_e\left(d/\left(r\sqrt{1 + (d^2/4h^2)}\right)\right)\right]} \quad \text{F/m line to line} \tag{8.52}$$

and

$$C_N = \frac{2\pi\varepsilon_0}{\left[\log_e\left(d/r\sqrt{1 + (d^2/4h^2)}\right)\right]} \quad \text{F/m to neutral} \tag{8.53}$$

It is observed from the above equation that the presence of earth modifies the radius $r$ to $r(1 + (d^2/4h^2))^{1/2}$. When $h$ is large compared to $d$ (this is the case normally), the effect of earth on line capacitance is of negligible order.

### 8.13.3 Capacitance of a Three-Phase Overhead Line

The method of images can similarly be applied for the calculation of capacitance of a three-phase line, shown in Figure 8.23. The line is considered to be fully transposed. The conductors A, B, and C carry the charges $Q_A$, $Q_B$, and $Q_C$ and occupy positions 1, 2, and 3, respectively, in the first position of the transposition cycle. The effect of earth is simulated by image conductors with charges $-Q_A$, $-Q_B$, and $-Q_C$, respectively, as shown.

The equations for the three sections of the transposition cycle can be written for the voltage drop $V_{AB}$ as determined by the three charged conductors and their images.

For the first portion of the transposition with conductor A in position 1, B in position 2, and C in position 3 (see Figure 8.23a).

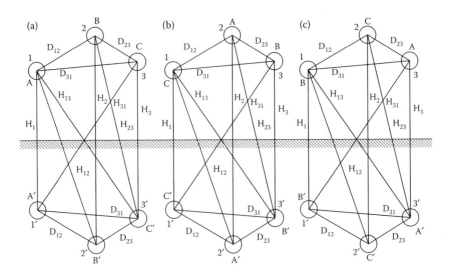

FIGURE 8.23  Three-phase line with images.

$$V_{AB1} = \frac{1}{2\pi\varepsilon_0}\left[Q_A\left(\log_e\frac{d_{12}}{r} - \log_e\frac{h_{12}}{h_1}\right) + Q_B\left(\log_e\frac{r}{d_{12}} - \log_e\frac{h_2}{h_{12}}\right)\right.$$

$$\left. + Q_C\left(\log_e\frac{d_{23}}{d_{31}} - \log_e\frac{h_{23}}{h_{31}}\right)\right] \tag{8.54}$$

For the second portion of the transposition with conductor C in position 1, A in position 2, and B in position 3 (see Figure 8.23b):

$$V_{AB2} = \frac{1}{2\pi\varepsilon_0}\left[Q_A\left(\log_e\frac{d_{23}}{r} - \log_e\frac{h_{23}}{h_2}\right) + Q_B\left(\log_e\frac{r}{d_{23}} - \log_e\frac{h_3}{h_{23}}\right)\right.$$

$$\left. + Q_C\left(\log_e\frac{d_{31}}{d_{12}} - \log_e\frac{h_{31}}{h_{12}}\right)\right] \tag{8.55}$$

For the second portion of the transposition with conductor B in position 1, C in position 2, and A in position 3 (see Figure 8.23c):

$$V_{AB3} = \frac{1}{2\pi\varepsilon_0}\left[Q_A\left(\log_e\frac{d_{31}}{r} - \log_e\frac{h_{31}}{h_3}\right) + Q_B\left(\log_e\frac{r}{d_{31}} - \log_e\frac{h_1}{h_{31}}\right)\right.$$

$$\left. + Q_C\left(\log_e\frac{d_{12}}{d_{23}} - \log_e\frac{h_{12}}{h_{23}}\right)\right] \tag{8.56}$$

If the fairly accurate assumption of constant charge per-unit length of the conductor throughout the transmission cycle is made, the average value of $V_{AB}$ for the three sections of the cycle is given by

$$V_{AB}(\text{avg}) = \frac{1}{3}(V_{AB1} + V_{AB2} + V_{AB3})$$

$$V_{AB} = \frac{1}{6\pi\varepsilon_0}\left[Q_A\left(\log_e\frac{D_{eq}}{r} - \log_e\frac{(h_{12}h_{23}h_{31})^{1/3}}{(h_1h_2h_3)^{1/3}}\right)\right. \tag{8.57}$$

$$\left. + Q_B\left(\log_e\frac{r}{D_{eq}} - \log_e\frac{(h_1h_2h_3)^{1/3}}{(h_{12}h_{23}h_{31})^{1/3}}\right)\right]$$

where $D_{eq} = \sqrt[3]{d_{12}d_{23}d_{31}}$.
Similarly,

$$V_{AC} = \frac{1}{6\pi\varepsilon_0}\left[Q_A\left(\log_e\frac{D_{eq}}{r} - \log_e\frac{(h_{12}h_{23}h_{31})^{1/3}}{(h_1h_2h_3)^{1/3}}\right)\right. $$

$$\left. + Q_C\left(\log_e\frac{r}{D_{eq}} - \log_e\frac{(h_1h_2h_3)^{1/3}}{(h_{12}h_{23}h_{31})^{1/3}}\right)\right] \tag{8.58}$$

Proceeding on the lines of Section 8.12.1 and using $V_{AB} + V_{AC} = 3V_{AN}$ and $Q_A + Q_B + Q_C = 0$, we ultimately obtained the following expression for the capacitance to neutral.

$$C_n = \frac{2\pi\varepsilon_0}{\log_e(D_{eq}/r) - \log_e\left(((h_{12}h_{23}h_{31})^{1/3})/(h_1h_2h_3)^{1/3}\right)} \text{ F/m to neutral} \tag{8.59}$$

Equation 8.59 shows that the effect of ground gives a higher value for the capacitance than that the obtained by neglecting the ground effect.

## 8.14 BUNDLED CONDUCTOR

The demand of electric power is increasing throughout the world and in many countries it doubling every 5–8 years. The power stations (hydro, thermal, or nuclear) are usually located far away from the load centers. Thus transmission of large amount of power over long distances, which can be accomplished most economically only by using extra high voltage (or simply EHV, voltage in excess of 230 kV). An increase in transmission voltage results in reduction of electrical losses, increases in transmission efficiency, improvement of voltage regulation, and reduction of conductor material requirement. At voltage above 300 kV, corona causes a significant power loss and interference with communication circuits, if a round single conductor per phase is used. Instead of going for a hollow conductor, it is preferable to use more than one conductor per phase which is called the bundling of conductors. Lines of 400 kV and higher voltage invariably use bundled conductor.

A bundled conductor is a conductor made up of two or more conductors, called the subconductors, per phase in close proximity compared with the spacing between phases (Figure 8.24). The basic difference between a composite conductor and a bundled conductor is that the subconductors of a bundled conductor are separated from each other by a constant distance varying from 0.2 to 0.6 m depending upon the length of the line with the help of spacers whereas the wires of a composite conductor touch each other. The bundled conductors have filter material or air space inside so that the overall diameter is increased.

The uses of bundled conductors per phase reduce the voltage gradient in the vicinity of the line and reduce the possibilities of corona discharge.

### 8.14.1 Bundled Conductors Have Several Advantages over Single Conductors

1. The BD lines transmit bulk power with reduced losses thereby giving increased transmission efficiency.

2. It has a higher capacitance to neutral in comparison with single conductor lines, therefore, they have higher charging current, which helps improving the power factor.

3. By bundling GMR is increased, the inductance per phase, in comparison with single conductor lines, is reduced. As a result reactance per phase is reduced.

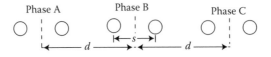

FIGURE 8.24 Cross-section of a bundled conductor three-phase transmission line.

4. Since surge impedance of a line is given by $Z_0 = \sqrt{L/C}$ and the bundled conductor lines have higher capacitance and lower inductance in comparison with single conductor lines, therefore, bundled conductor lines have comparatively lower surge impedance with a corresponding increase in the maximum power transfer capability.

GMR of a bundled conductor for

1. Two conductor (duplex arrangement):

$$D_S = \sqrt{r's}$$

2. Three conductor (triplex arrangement):

$$D_S = \sqrt[3]{r's^2}$$

3. Four conductor (quadruplex arrangement):

$$D_S = \sqrt[4]{r's^3}$$

where $r'$ is the GMR of each subconductor of bundle and $s$ is the spacing between subconductors of a bundle.

## 8.15 SKIN EFFECT

The distribution of current throughout the cross-section of a conductor is uniform only when DC is passing through it. On the contrary, when AC is flowing through a conductor, the current is nonuniformly distributed over the cross-section in a manner that the current density is higher at the surface of the conductor compared to the current density at its center (Figure 8.25). This effect becomes more pronounced, as frequency is increased. This phenomenon is called skin effect. It causes larger power loss for a given rms AC than the loss when the same value of DC is flowing through the conductor. Consequently, the effective conductor resistance is more for AC then for DC.

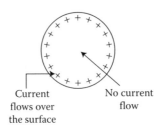

FIGURE 8.25  Cross-section of a conductor.

Imagine a solid conductor to be consisting of a large number of strands, each carrying a small part of the current. The inductance of each strand will vary according to its position. Thus, the strands near the center are surrounded by a greater magnetic flux and hence have larger inductance than that near the surface. The high reactance of inner strands causes the alternating current to flow near the surface of the conductor. This tendency of alternating current to concentrate near the surface of a conductor is known as *skin effect*.

The skin effect depends upon the following factors:

1. Nature of material

2. Diameter of wire—increase with the diameter of wire

3. Frequency—increases with increase in frequency

4. Shape of wire—less for stranded conductor than the solid conductor

It may be noted that the skin effect is negligible when the supply frequency is low (< 50 Hz), and the conductor diameter is small (< 1 cm).

## 8.16 PROXIMITY EFFECT

The inductance and, therefore, the current distribution in a conductor is also affected by the presence of other conductor in its vicinity. This effect is known as the proximity effect.

This is another electromagnetic effect which also results in the increment of the apparent resistance of the conductor due to the presence of other conductors carrying current in its vicinity. When two or more conductors are in proximity, their electromagnetic fields interact with each other with the result that the current in each of them is redistributed.

Consider a two-wire line as shown in Figure 8.26. Each line conductor can be divided into sections of equal cross-sectional area (say three

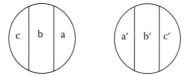

FIGURE 8.26   Two-wire line.

sections). Pairs aa′, bb′, and cc′ can form three loops in parallel. The flux linking loop aa′ (and therefore its inductance) is the least and it increases somewhat for loops bb′ and cc′. Thus the density of AC flowing through the conductors is highest at the inner edges (aa′) of the conductors and is the least at the outer edges (cc′). This type of nonuniform AC current distribution becomes more pronounced, as the distance between conductors is reduced. Like skin effect, the nonuniformity of current distribution caused by proximity effect also increases the effective conductor resistance. For normal spacing of overhead lines, this effect is always of a negligible order. However, for underground cables where conductors are located close to each other, proximity effect causes an appreciable increase in effective conductor resistance.

## WORKED EXAMPLES

### EXAMPLE 8.1

Calculate the GMR of 6/3 mm AC, 1/3 mm steel ACSR conductor (Figure 8.27).

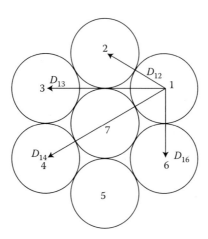

FIGURE 8.27   ACSR conductor.

### Solution

An ACSR conductor with six outer conductors of aluminum each of radius $r = 3$ mm and one central conductor of steel also of radius $r = 3$ mm.

$$d_{11} = d_{12} = d_{16} = d_{17} = 2, \quad d_{14} = 4r, \quad \text{and} \quad d_{13} = d_{15} = \sqrt{D_{14}^2 - D_{34}^2}$$
$$= \sqrt{(4r)^2 - (2r)^2} = 2\sqrt{3}r$$
$$D_{S_1} = D_{S_2} = D_{S_3} = D_{S_4} = D_{S_5} = D_{S_6}$$
$$= \sqrt[7]{0.7788r \times 2r \times 2\sqrt{3}r \times 4r \times 2\sqrt{3}r \times 2r \times 2r}$$
$$= \sqrt[7]{299r}$$

$$D_{S_7} = \sqrt[7]{2r \times 0.7788r \times 2r \times 2r \times 2r \times 2r \times 2r}$$
$$= \sqrt[7]{49.8432r}$$

Geometric mean radius is

$$D_s = \sqrt[7]{D_{S_1} \cdot D_{S_2} \cdot D_{S_3} \cdot D_{S_4} \cdot D_{S_5} \cdot D_{S_6} \cdot D_{S_7}}$$
$$= r\sqrt[49]{299^6 \times 49.8432} = 2.176r = 2.176 \times 3\,\text{mm} = 6.528\,\text{mm}$$

### EXAMPLE 8.2

A single-phase double circuit transmission line is shown in the Figure 8.28. Conductors 1 and 2 in parallel from one path where

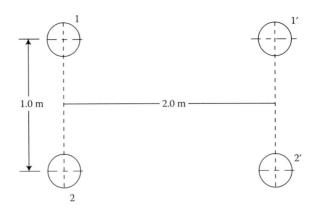

FIGURE 8.28 A single-phase double circuit transmission line.

conductors $1'$ and $2'$ in parallel from the return path. The current is equally shared by the two parallel lines. Determine the total inductance per kilometer of the line. The diameter of each conductor is 3 cm and spacing between them is 2 m.

**Solution**

Radius of each conductor = 3/2 = 1.5 cm
  GMR, $r' = 0.7788$, $r = 1.168$ cm

Spacing of conductors, $d'_{11} = 200$ cm

$$d'_{12} = \sqrt{200^2 + 100^2} = 223.60 \text{ cm}$$

$$d'_{21} = d'_{12} = 223.60 \text{ cm}$$

$$d'_{22} = 200 \text{ cm}$$

$$d_{11} = d_{22} = r' = 1.168 \text{ cm}, \quad d_{12} = d_{21} = 100 \text{ cm}$$

Mutual GMD, $D_m = \sqrt[4]{d'_{11}d'_{12}d'_{21}d'_{22}}$ cm

$\qquad = \sqrt[4]{200 \times 223.60 \times 223.60 \times 200} = 211.47$ cm

Self GMD, $D_s = \sqrt[4]{d_{11}d_{12}d_{21}d_{22}}$ cm

$\qquad = \sqrt[4]{1.168 \times 100 \times 100 \times 1.168}$ cm = 10.80 cm

Loop inductance = $0.4 \log_e \dfrac{D_m}{D_s} = 0.4 \log_e \dfrac{211.47}{10.8} = 1.189$ mH/km

**EXAMPLE 8.3**

A three-phase overhead line is designed with an equilateral spacing of 3.5 m with a conductor diameter of 1.2 cm. If the line is constructed with horizontal spacing with suitably transposed conductors, find spacing between adjacent conductors which would give the same value of inductance as in the equilateral arrangement (Figure 8.29).

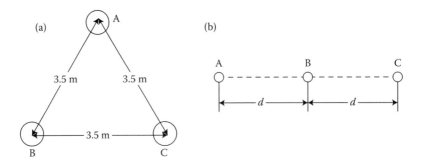

FIGURE 8.29  (a) Equilateral arrangement of line conductor. (b) Horizontally spaced line conductors.

**Solution**

Conductor radius, $r = 1.2/2 = 0.6$ cm or 6 mm

Geometric mean radius, $r' = 0.7786 \times 6 = 4.6728$ mm

In equilateral arrangement of line conductors spacing of conductors $= 3.5$ m $= 3500$ mm

Inductance per phase, $L = 2 \times 10^{-7} \log_5 \dfrac{d}{r}$, H/m

$$= 2 \times 10^{-7} \log_e \frac{3500}{4.6728} = 1.324 \times 10^{-6} \text{ H/m}$$

Let the spacing between the conductor be $d$ mm.

The effect of transposition of line is that each conductor has the same average inductance which is given as

$$L_{av} = \frac{L_A + L_B + L_C}{3}$$

Substituting the value of $L_A$, $L_B$, and $L_C$,

$$L_{av} = \frac{1}{3} \times 2 \times 10^{-7} \left[ \log_e \frac{d}{r'} + \frac{1}{2} \log_e 2 - j\,0.866 \log_e 2 \right.$$

$$\left. + \log_e \frac{d}{r'} + \log_e \frac{d}{r'} + \frac{1}{2} \log_e 2 + j\,0.866 \log_e 2 \right] \text{H/m}$$

$$= 2 \times 10^{-7} \left[ \log_e \frac{d}{r'} + \frac{1}{3} \log_e 2 \right]$$

Substituting $L_{av} = 1.324 \times 10^{-6}$ and $r' = 4.6728$ mm in above equation, we have

$$1.324 \times 10^{-6} = 2 \times 10^{-7} \left[ \log_e \frac{d}{4.6728} + \frac{1}{3} \log_e 2 \right]$$

or $d = 4.6728^{6.389} = 2781.5$ mm or 2.781 m.

**EXAMPLE 8.4**

A three-phase, 50-Hz, 132-kV overhead line conductors are placed in a horizontal plane as shown in Figure 8.30. The conductor diameter is 2 cm. If the line length is 150 km, calculate (1) capacitance per phase, (2) charging current per phase, assuming complete transposition of the line.

**Solution**

Figure shows the arrangement of conductors of the three-phase line. The equivalent equilateral spacing is

$$d = \sqrt[3]{d_1 d_2 d_3} = \sqrt[3]{1 \times 2 \times 3} = 1.81 \text{ m}$$

$$\text{Conductor radius, } r = \frac{2}{2} = 1 \text{ cm}$$

$$\text{Conductor spacing, } d = 1.81 \text{m} = 181 \text{ cm}$$

1. Line-to-neutral capacitance $= \dfrac{2\pi\varepsilon_0}{\log_e(d/r)}$ F/m

$$= \frac{2\pi \times 8.854 \times 10^{-12}}{\log_e(181/1)} \text{F/m}$$

$$= 0.0107 \times 10^{-9} \text{ F/m}$$

$$= 0.0107 \times 10^{-6} \text{ F/km}$$

$$= 0.0107 \text{ μF/km}$$

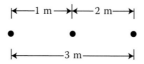

FIGURE 8.30   Three-phase overhead line conductors (placed in a horizontal plane).

∴ Line-to-neutral capacitance for 150 km line is

$$C = 0.0107 \times 150 = 1.605 \ \mu F$$

2. Charging current per phase is

$$I_C = \frac{V_{ph}}{X_C} = \frac{132,000}{\sqrt{3}} \times 2\pi f C$$

$$= \frac{132,000}{\sqrt{3}} \times 2\pi \times 50 \times 1.605 \times 10^{-6} = 38.427 \ A$$

### EXAMPLE 8.5

Find out the capacitance of a two-wire, one-phase line running at a height of $h$ meters above the earth. Calculate the capacitance to neutral in the case of single-phase line, whose conductors with radius of 0.25 cm are separated by 1.5 m and which are lying 7 m above ground. Line length is 50 km.

### Solution

Radius of each conductor, $r = 0.25$ cm
  Spacing between conductors, $d = 1.5$ m $= 150$ cm
  Height of the conductors above earth, $h = 7$ m $= 700$ cm
  Capacitance between conductor,

$$C = \frac{\pi \varepsilon_0}{\log_e \left( d / \left( \sqrt[3]{1+(d^2/4h^2)} \right) \right)} \ F/m$$

$$C = \frac{\pi \times 8.854 \times 10^{-12}}{\log_e \left( 120 / {}^{0.25}\!\sqrt{1+(150/1400)^2} \right)}$$

$$= 4.5 \times 10^{-12} \ F/m$$

Capacitance of 50 km long wire $= 4.5 \times 10^{-12} \times 50,000$

$$= 0.225 \ \mu F$$

## EXERCISES

1. What do you understand by the constants of an overhead transmission line?

2. What is skin effect? Why is it absent in the DC system?

3. Find an expression for the flux linkages:

   a. Due to a single current-carrying conductor

   b. In parallel current-carrying conductors

4. Derive an expression for the loop inductance of a single-phase line.

5. Derive an expression for the inductance per phase for a three-phase overhead transmission line when

   a. Conductors are symmetrically placed

   b. Conductors are asymmetrically placed but the line is completely transposed

6. What do you understand by electric potential? Derive an expression for electric potential

   a. At a charged single conductor

   b. At a conductor in a group of charged conductors

7. Derive an expression for the capacitance of a single-phase overhead transmission line.

8. Deduce an expression for line-to-neutral capacitance for a three-phase overhead transmission line, when the conductors are

   a. Symmetrically placed

   b. Unsymmetrically placed but transposed

# CHAPTER 9

# Performance of Transmission Lines

## 9.1 INTRODUCTION

A transmission line comprises of resistance $R$, inductance $L$, capacitance $C$, and shunt or leakage conductance $G$. All the parameters are distributed uniformly on the whole distance of the cable. These parameters along with load current and power factor determine the electrical performance of the line. The term performance includes the calculation of sending-end voltage, sending-end current and sending-end power factor, power loss in the line, efficiency of transmission, regulation, and limits of power flow during steady-state and transient conditions. The values of voltage, current, and power factor at the receiving end are usually known. Calculation of prior performance is helpful in system planning. The purpose of deriving the formulae to study the performance of a line is to know the effect of the line parameters on various loads.

## 9.2 CLASSIFICATION OF LINES

Transmission lines are classified as follows:

1. *Short transmission line.* For overhead lines up to 50 km, the capacitance $C$ is negligibly small but for cable lines, where the distance between the conductors is small, the effect of capacitance cannot be neglected. All low voltage (<20 kV) overhead lines having lengths up to 50 km are generally classified as short transmission line. While

studying the performance of a short transmission line, only resistance and inductance of the line are taken into account.

2. *Medium transmission line.* The lines ranging in length from 50 to 150 km are generally termed as medium transmission line or moderately long lines, where the voltage is moderately high ($20 < v < 100$ kV). Due to sufficient length and voltage of the line, the capacitance effects are taken into account and it is considered to be lumped at one or more points of the line.

3. *Long transmission line.* When the length of an overhead transmission line is more than 150 km and line voltage is $> 100$ kV, it is considered as a long transmission line. For the treatment of such a line, the line constants are considered uniformly distributed over the whole line length of the line.

## 9.3 PERFORMANCE OF A SINGLE-PHASE SHORT TRANSMISSION LINE

As stated earlier, the effect of the line capacitance are neglected for a short transmission line. Therefore, while studying the performance of such a line, only resistance and inductance are taken into account. The equivalent circuit of a single-phase short transmission line is shown in Figure 9.1a. Here the total line resistance and inductance are shown as concentrated or lumped instead of being distributed. The circuit is simple AC circuit, where $I$ is the load current, $R$ is the loop resistance or resistance of both conductors, $X_L$ is the loop reactance, $V_r$ is the receiving-end voltage, $\cos \phi_R$ is the receiving-end power factor, $V_s$ is the sending-end voltage, and $\cos \phi_S$ is the sending-end power factor.

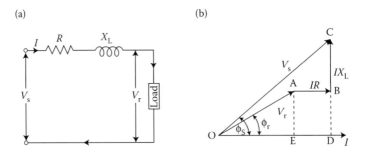

FIGURE 9.1 (a) Equivalent circuit of a single-phase short line. (b) Phasor diagram for a short line (lagging power factor).

Since the shunt capacitance and shunt conductance are neglected in short line, the load current practically remains the same at all points along the length of the lines, $I_s = I_r = I$ (say).

The phasor diagram of the line for lagging load power factor is shown in Figure 9.1b. Current $I$ is taken as the reference phasor. OA represents the receiving-end voltage $V_r$ leading $I$ by $\phi_R$. AB represents the drop $IR$ in phase with BC represents the inductive drop $IX_L$ and leads $I$ by 90°. The magnitude of $V_S$ can be found from the right angle triangle ODC.

$$(OC)^2 = (OD)^2 + (DC)^2 = (OE + ED)^2 + (DB + BC)^2$$

$$V_s^2 = (V_r \cos \phi_R + IR)^2 + (V_r \sin \phi_R + IX_L)^2$$

$$V_s = \sqrt{(V_r \cos \phi_R + IR)^2 + (V_r \sin \phi_R + IX_L)^2}$$

The power factor of the load measured at sending end is

$$\cos \phi_s = \left(\frac{OD}{OC}\right) = \frac{V_r \cos \phi_R + IR}{V_s}$$

The alternative expression for $V_s$ can be found by using complex algebra. If $V_r$ be the reference phasor,

$$\mathbf{V_r} = V_r \angle 0° = V_r + j0$$

For lagging power factor $\cos \phi_R$, $\mathbf{I} = I \angle - \phi_R = I \cos \phi_R - j I \sin \phi_R$

For leading power factor $\cos \phi_R$, $\mathbf{I} = I \angle + \phi_R = I \cos \phi_R + j I \sin \phi_R$

For unity power factor $\cos \phi_R$, $\mathbf{I} = I \angle 0° = I + j0$

The line impedance is given by

$$\mathbf{Z} = R + jX_L$$

The sending-end voltage is

$$\mathbf{V}_s = \mathbf{V}_r + \mathbf{ZI}$$

For lagging power factor

$$\mathbf{V}_s = (V_r + j0) + (R + jX_L)(I \cos \phi_R - j I \sin \phi_R)$$
$$= (V_r + IR \cos \phi_R + IX \sin \phi_R) + j(IX_L\cos \phi_R - IR \sin \phi_R)$$

$$V_s = \sqrt{[(V_r + IR \cos \phi_R + IX_L\sin \phi_R)^2 + (IX_L\cos \phi_R - IR \sin \phi_R)^2]}$$

## 9.4 SHORT THREE-PHASE LINE

A balance three-phase circuit may be considered as consisting of three separate identical single-phase circuits. Therefore, the calculations for a balanced three-phase line are carried out in a similar manner as explained for single-phase line, the difference being that per-phase basis is adopted. When working with balance three-phase line, it is usual to assume that all the given voltage are line to line values, that all the current are line current. Similarly the given power is the total power for all the three phases and the given reactive power volt-amperes represent the total volt-amperes for all the three phases. Thus for three-phase line calculations,

Power per phase = (1/3) × (Total power)

Reactive power per phase = (1/3) × (Total reactive power)

Also in Figure 9.1a and b, $I$ is the phase current, $R$ is the line resistance per phase, $X_L$ is the line reactance per phase, $Z$ is the line impedance per phase, $V_s$ is the sending-end-phase voltage, and $V_r$ is the receiving-end-phase voltage.

For balance three-phase star-connected line,

$$\text{Phase voltage} = \left(\frac{1}{\sqrt{3}}\right) \times \text{Line voltage}$$

## 9.5 TRANSMISSION LINE AS TWO-PORT NETWORK

A transmission line may be viewed as a two-port network, as shown in Figure 9.2. The voltage and current at input and output terminals are expressed in the form of general equation is given by

$$\mathbf{V_s} = \mathbf{AV_r} + \mathbf{BI_r} \tag{9.1}$$

$$\mathbf{I_s} = \mathbf{CV_r} + \mathbf{DI_r} \tag{9.2}$$

where $\mathbf{V_s}$ is the sending-end voltage, $\mathbf{I_s}$ is the sending-end current, $\mathbf{V_r}$ is the receiving-end voltage, and $\mathbf{I_r}$ is the receiving-end current.

The **A, B, C, D** constants are called *general network constants*. They depend on the line parameters and in general are complex. Equations 9.1 and 9.2 can be put in the form of matrix, where

$$\begin{bmatrix} \mathbf{V_s} \\ \mathbf{I_s} \end{bmatrix} = \begin{bmatrix} \mathbf{A} & \mathbf{B} \\ \mathbf{C} & \mathbf{D} \end{bmatrix} \begin{bmatrix} \mathbf{V_r} \\ \mathbf{I_r} \end{bmatrix}$$

The validity of Equations 9.1 and 9.2 is based on the fact that a transmission line can be represented by a linear, passive, and bilateral network. By virtue of reciprocity, the generalized constants are related to each other by the following equation.

$$\mathbf{AD} - \mathbf{BC} = 1 \tag{9.3}$$

### 9.5.1 ABCD Constants of a Short Line

The sending-end voltage and current can be written in the form of a short-line equivalent network, as shown in Figure 9.1a:

$$\mathbf{V_s} = \mathbf{V_r} + \mathbf{ZI_r} \tag{9.4}$$

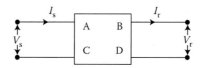

FIGURE 9.2  A transmission line as a two-port network.

$$I_s = I_r \tag{9.5}$$

Comparing the coefficient of Equations 9.4 and 9.5 with the general Equations 9.1 and 9.2, ABCD constants for short line are

$$A = 1 = D, \quad B = Z, \quad C = 0$$

## 9.6 LINE REGULATION

Voltage regulation of a line is defined as the change in voltage at the receiving end when full load at a given power factor is removed, the voltage at the sending end being kept constant.

It is expressed as a percentage of the receiving-end voltage keep the sending-end voltage constant. It can be written as

$$\text{Per-unit regulation} \triangleq \frac{|V_{rnl}| - |V_{rfl}|}{|V_{rfl}|}$$

$$\text{Percent regulation} \triangleq \frac{|V_{rnl}| - |V_{rfl}|}{|V_{rfl}|} \times 100$$

where $|V_{rnl}|$ is the magnitude of receiving-end voltage at no-load, and $|V_{rfl}|$ is the magnitude of sending-end voltage at full load.

The voltage $V_s$ at the sending end is kept constant. It can be written as

$$V_s = AV_r + BI_r$$

When the load is removed,

$$I_r = 0, \quad V_r = V_{r0}$$

Therefore, $V_s = AV_{r0}$, $V_{r0} = V_s/A$, where $V_{r0}$ is the receiving-end voltage at no-load.

$$\text{Line regulation} = \frac{|V_s|/|A| - |V_{rfl}|}{|V_{rfl}|} \text{pu}$$

### 9.6.1 Line Regulation for Short Line

In case of a short line, when the load is removed, the voltage at the receiving end is equal to the voltage at the sending end. At full load,

$$|V_{rfl}| = |V_r|$$

At no-load, $\quad |V_{rnl}| = |V_s|$

Therefore, for a short line

$$\text{Line regulation} = \frac{|V_{rnl}| - |V_{rfl}|}{|V_{rfl}|} = \frac{|V_s| - |V_r|}{|V_r|} \text{pu}$$

## 9.7  LINE EFFICIENCY

The line efficiency or efficiency of transmission ($\eta_T$) is the ratio of receiving-end power to the sending-end power of a transmission line.

$$\eta_T = \frac{\text{Power output}}{\text{Power input}} \text{pu} = \frac{\text{Power delivered at the receiving end}}{\text{Power delivered to the receiving end}} \text{pu}$$

$$= \frac{\text{Power delivered at the receiving end}}{\text{Power delivered to the receiving end} + \text{Losses}} \text{pu}$$

## 9.8  PERFORMANCE OF MEDIUM TRANSMISSION LINE

It has been mentioned already that the capacitance of medium length lines is significant. When the effect of capacitance is not negligible, it may be assumed to be concentrated at one or more definite points along the line. A number of localized capacitance models have been used to make approximate line performance calculations. The following models are commonly used:

1. Nominal T model

2. Nominal $\pi$ model

It should be noted that nominal T and $\pi$ models are not equivalent representations. They are different representations for actual lines.

### 9.8.1 Nominal T Model

In this method, the whole line capacitance is assumed to be concentrated at the middle point of the line, and half the line resistance and reactance are lumped on its either side as shown in Figure 9.3. Therefore, in this arrangement, full charging current flows over half the line. In Figure 9.3, one phase of three-phase transmission line is shown, as it is advantageous to work in phase instead of line-to-line values:

$$\text{Series impedance of the line} = \mathbf{Z} = R + jX_L$$
$$\text{Shunt admittance of the line} = \mathbf{Y} = j\omega C$$

Based on the assumption that $V_r$ and $I_r$ are known, the corresponding sending-end quantities can be obtained by application of Kirchhoff's voltage law (KVL) and Kirchhoff's current law (KCL) to the circuit shown in Figure 9.3.

By KVL,

$$\mathbf{V}_1 = \mathbf{V}_r + \frac{\mathbf{Z}}{2}\mathbf{I}_r$$

$$\text{Current in the capacitor,} \quad \mathbf{I}_C = \frac{\mathbf{V}_1}{\mathbf{Z}_1} = \mathbf{YV}_1$$

Sending-end current,

$$\mathbf{I}_s = \mathbf{I}_r + \mathbf{I}_C = \mathbf{I}_r + \mathbf{YV}_1 = \mathbf{I}_r + \mathbf{Y}\left(\mathbf{V}_r + \frac{\mathbf{Z}}{2}\mathbf{I}_r\right)$$

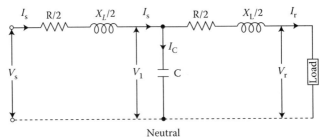

FIGURE 9.3 Nominal T model of a medium line.

$$I_s = YV_r + \left(1 + \frac{ZY}{2}\right)I_r \tag{9.6}$$

By KVL,

$$V_s = V_1 + \frac{Z}{2}I_s = V_r + \frac{Z}{2}I_r + \frac{Z}{2}\left[YV_r + \left(1 + \frac{ZY}{2}\right)I_r\right]$$

$$V_s = \left(1 + \frac{ZY}{2}\right)V_r + Z\left(1 + \frac{ZY}{4}\right)I_r \tag{9.7}$$

Equations 9.6 and 9.7 give the sending-end current and sending-end voltage, respectively. These equations can be written in the matrix form as

$$\begin{bmatrix} V_s \\ I_s \end{bmatrix} = \begin{bmatrix} 1 + \dfrac{ZY}{2} & Z\left(1 + \dfrac{ZY}{4}\right) \\ Y & 1 + \dfrac{ZY}{2} \end{bmatrix} \begin{bmatrix} V_r \\ I_r \end{bmatrix} \tag{9.8}$$

Also,

$$\begin{bmatrix} V_s \\ I_s \end{bmatrix} \begin{bmatrix} A & B \\ C & D \end{bmatrix} \begin{bmatrix} V_r \\ I_r \end{bmatrix}$$

Hence, ABCD constants for nominal T-circuit model of medium line are

$$A = D = 1 + \frac{ZY}{2}, \quad B = Z\left(1 + \frac{ZY}{4}\right), \quad C = Y$$

### 9.8.1.1 Phasor Diagram

The phasor diagram of the nominal T circuit of Figure 9.3 as shown in Figure 9.4. It is drawn for a lagging power factor $\cos \phi_R$. $V_r$ is taken as the reference phasor represented by OA. The load current $I_r$ lags behind $V_r$ by $\phi_R$. The drop AB = $I_r R/2$ is in phase with $I_r$. BC = $I_r X_L/2$ leads $I_r$ by 90°. The phasor OC represents the voltage $V_1$ across capacitor C. The capacitor

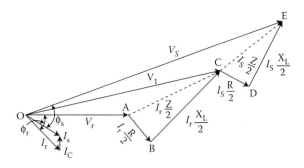

FIGURE 9.4　Phasor diagram of the nominal T circuit.

current $I_C$ leads $V_1$ by 90° as shown. The phasor sum of $I_r$ and $I_C$ gives $I_s$. Now $CD = I_s R/2$ is in phase with $I_s$, while $DE = I_s X_L/2$ leads $I_s$ by 90°. Then, OE represents the sending-end voltage $V_s$.

### 9.8.2　Nominal π Model

In this method, capacitance of each conductor (i.e., line-to-neutral) is divided into two halves; one half being lumped at the sending end and the other half at the receiving end as shown in Figure 9.5. It is obvious that capacitance at the sending end has no effect on the line drop. However, its charging current must be added to line current in order to obtain the total sending-end current.

$$\mathbf{V}_1 = \mathbf{V}_r$$

$$\mathbf{Z}_1 = \frac{1}{\mathbf{Y}_1}$$

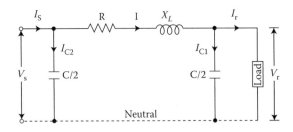

FIGURE 9.5　Nominal π model of a medium line.

By Ohm's law, charging current at load end is

$$\mathbf{I}_{C1} = \frac{\mathbf{V}_1}{\mathbf{Z}_1} = \frac{Y}{2}\mathbf{V}_r$$

Line current:

$$\mathbf{II}_R + \mathbf{I}_{C1} = \mathbf{I}_r + \frac{Y}{2}\mathbf{V}_r$$

Voltage at the sending end:

$$\mathbf{V}_s = \mathbf{V}_2 = \mathbf{V}_1 + \mathbf{ZI} = \mathbf{V}_r + \mathbf{Z}\left(\mathbf{I}_r + \frac{Y}{2}\mathbf{V}_r\right)$$

or

$$\mathbf{V}_s = \left(1 + \frac{ZY}{2}\right)\mathbf{V}_r + \mathbf{ZI}_r \tag{9.9}$$

By Ohm's law:

$$\mathbf{I}_{C2} = \frac{\mathbf{V}_2}{\mathbf{Z}_2} = \frac{Y}{2}\mathbf{V}_s = \frac{Y}{2}\left[\left(1 + \frac{ZY}{2}\right)\mathbf{V}_r + \mathbf{ZI}_r\right]$$

Sending-end current,

$$\mathbf{I}_s = \mathbf{I} + \mathbf{I}_{C2} = \mathbf{I}_r + \frac{Y}{2}\mathbf{V}_r + \frac{Y}{2}\left[\left(1 + \frac{ZY}{2}\right)\mathbf{V}_r + \mathbf{ZI}_r\right]$$

$$\mathbf{I}_s = Y\left(1 + \frac{ZY}{4}\right)\mathbf{V}_r + \left(1 + \frac{ZY}{2}\right)\mathbf{I}_r \tag{9.10}$$

Equations 9.9 and 9.10 can be written in matrix form as,

$$\begin{bmatrix} \mathbf{V}_s \\ \mathbf{I}_s \end{bmatrix} = \begin{bmatrix} \left(1 + \dfrac{ZY}{2}\right) & Z \\ Y\left(1 + \dfrac{ZY}{4}\right) & \left(1 + \dfrac{ZY}{2}\right) \end{bmatrix} \begin{bmatrix} \mathbf{V}_r \\ \mathbf{I}_r \end{bmatrix}$$

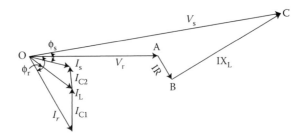

FIGURE 9.6    Phasor diagram of the nominal π circuit.

Also,

$$\begin{bmatrix} \mathbf{V}_s \\ \mathbf{I}_s \end{bmatrix} = \begin{bmatrix} \mathbf{A} & \mathbf{B} \\ \mathbf{C} & \mathbf{D} \end{bmatrix} \begin{bmatrix} \mathbf{V}_r \\ \mathbf{I}_r \end{bmatrix} \tag{9.11}$$

Hence, ABCD constants for nominal π-circuit model of medium line are

$$\mathbf{A} = \mathbf{D} = 1 + \frac{\mathbf{ZY}}{2}, \quad \mathbf{B} = \mathbf{Z}, \quad \mathbf{C} = \mathbf{Y}\left(1 + \frac{\mathbf{ZY}}{4}\right) \tag{9.12}$$

### 9.8.2.1 Phasor Diagram

The phasor diagram of the nominal π circuit is shown in Figure 9.6. It is also drawn for a lagging power factor of the load. $V_r$ is taken as the reference phasor represented by OA. The load current $I_r$ lags behind $V_r$ by $\phi_R$. The charging current $I_{C1}$ leads $V_r$ by 90°. The phasor sum of $I_r$ and $I_{C1}$ gives **I**. The drop AB = $I_L R$ is in phase with **I**, whereas BC = $I_L X_L$ leads **I** by 90°. Then, OC represents the sending-end voltage $V_s$. The charging current $I_{C2}$ leads $V_r$ by 90°. The phasor OC represents the voltage $V_1$ across capacitor C. The capacitor current $I_C$ leads $V_s$ by 90°. The phasor sum of **I** and $I_{C2}$ gives $I_s$. The angle $\phi_s$ between sending-end voltage $V_s$ and sending-end current $I_s$ determines the sending-end pf cos $\phi_s$.

## 9.9  CALCULATION OF TRANSMISSION EFFICIENCY AND REGULATION OF MEDIUM LINE

If $V_R$ denotes the line-to-neutral (phase) voltage at the receiving end in volts, $S_s$ denotes the sending-end volt-amperes, and $\mathbf{I}_s^*$ is the complex conjugate of $\mathbf{I}_s$, then

$$\mathbf{S}_s = 3\,\mathbf{V}_{SP}\,\mathbf{I}_{SP}^* = P_s + jQ_s$$

where $P_s$ is the active power in watts at the sending end, and $Q_s$ is the reactive volt-ampere at the sending end.

The transmission efficiency can be calculated as follows:

$$\eta_T = \frac{P_R}{P_s}$$

In order to calculate regulation, we have to calculate the receiving-end voltage at no-load $\mathbf{V}_{rnl}$.

The sending-end voltage can be written as

$$\mathbf{V}_s = \mathbf{A}\mathbf{V}_r + \mathbf{B}\mathbf{I}_r$$

When the load is removed,

$$\mathbf{I}_r = 0, \quad \mathbf{V}_r = \mathbf{V}_{rnl}$$

where $\mathbf{V}_{rnl}$ is the receiving-end voltage at no-load.

$$\mathbf{V}_s = \mathbf{A}\mathbf{V}_{rnl}$$

$$\mathbf{V}_{rnl} = \frac{\mathbf{V}_s}{\mathbf{A}}$$

$$|\mathbf{V}_{rnl}| = \left|\frac{\mathbf{V}_s}{\mathbf{A}}\right|$$

Per-unit line regulation is

$$\frac{|\mathbf{V}_{rnl}| - |\mathbf{V}_{rfl}|}{|\mathbf{V}_{rfl}|} = \frac{|\mathbf{V}_s/\mathbf{A}| - |\mathbf{V}_{rfl}|}{|\mathbf{V}_{rfl}|}$$

For nominal T and $\pi$ model of the line,

$$\mathbf{A} = 1 + \frac{\mathbf{Z}\mathbf{Y}}{2}$$

## 9.10 LONG TRANSMISSION LINE

Line constants of the transmission line are uniformly distributed over the entire length of the line. Figure 9.7 shows the equivalent circuit of a three-phase long transmission line on a phase-neutral basis. The whole line length is divided into $n$ sections, each section having line constants 1/$n$th of those for the whole line. The line constants are uniformly distributed over the entire length of line as is actually the case. The resistance and inductive reactance are the series elements, whereas the leakage susceptance (B) and leakage conductance (G) are shunt elements. The leakage susceptance is due to the fact that capacitance exists between line and neutral. The leakage conductance takes into account the energy losses occurring through leakage over the insulators or due to corona effect between conductors. The leakage current through shunt admittance is maximum at the sending end of the line and decreases continuously as the receiving end of the circuit is approached at which point its value is zero.

### 9.10.1 Analysis of Long Transmission Line (Rigorous Method)

Figure 9.8 shows one phase and neutral connection of a three-phase line with impedance and shunt admittance of the line uniformly distributed. Consider a small element in the line of length d$x$ situated at a distance $x$ from the receiving end of the line.

Let $r$ be the resistance per-unit length of line, $x_1$, the reactance per-unit length of line, $b$, the suseptance per-unit length of line, $g$, the conductance per-unit length of line, $z$, the series impedance of the line per-unit length $= \sqrt{r^2 + x_1^2}$, $y$, the shunt admittance of the line per-unit length $= \sqrt{g^2 + b^2}$, $V$, the voltage at a distance $x$ from the receiving end, $V + dV$,

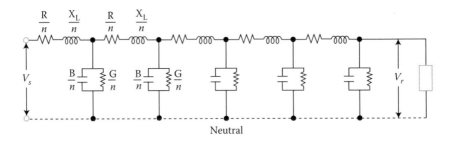

FIGURE 9.7 Representation of a transmission line showing the distributed nature of parameters.

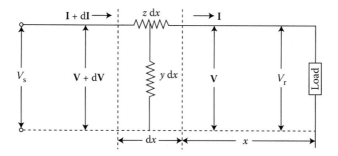

FIGURE 9.8   Incremental length of the transmission line.

the voltage at distance $(x + dx)$ from the receiving end, $\mathbf{I} + d\mathbf{I}$, the current at distance $(x + dx)$ from the receiving end, and $\mathbf{I}$ is at a distance $x$ from the receiving end.

Then for the small element $dx$,

$$\mathbf{z}\, dx = \text{Series impedance}$$
$$\mathbf{y}\, dx = \text{Shunt admittance}$$

The rise in voltage over the element length in the direction of increasing $x$. Obviously, $d\mathbf{V} = \mathbf{I}\,\mathbf{z}\, dx$:

$$\frac{d\mathbf{V}}{dx} = \mathbf{Iz} \tag{9.13}$$

The difference $d\mathbf{I}$ of the current between the two ends of the section due to leakage current through the shunt admittance $\mathbf{y}\, dx$ of the section is given by;

$$d\mathbf{I} = \mathbf{V}\,\mathbf{y}\, dx$$

$$\frac{d\mathbf{I}}{dx} = \mathbf{yV} \tag{9.14}$$

Differentiating Equation 9.13 wrt $x$, we get

$$\frac{d^2\mathbf{V}}{dx^2} = \mathbf{z}\frac{d\mathbf{I}}{dx}$$

The value of $d\mathbf{I}/dx$ is substituted from Equation 9.14 to give

$$\frac{d^2\mathbf{V}}{dx^2} = \mathbf{z}\mathbf{y}\mathbf{V} \tag{9.15}$$

The solution of the above differential equation is

$$\mathbf{V} = \mathbf{A}_1 e^{\sqrt{yz}x} + \mathbf{A}_2 e^{-\sqrt{yz}x} \tag{9.16}$$

where $\mathbf{A}_1$ and $\mathbf{A}_2$ are unknown constants, differentiating Equation 9.16 wrt $x$, we get

$$\frac{d\mathbf{V}}{dx} = \sqrt{yz}\left[\mathbf{A}_1 e^{\sqrt{yz}x} - \mathbf{A}_2 e^{-\sqrt{yz}x}\right] \tag{9.17}$$

From Equation 9.13 we have,

$$\frac{d\mathbf{V}}{dx} = \mathbf{I}\mathbf{z}$$

$$\begin{aligned}\mathbf{I} &= \frac{1}{\mathbf{z}}\sqrt{yz}\left[\mathbf{A}_1 e^{\sqrt{yz}x} - \mathbf{A}_2 e^{-\sqrt{yz}x}\right] \\ &= \sqrt{\frac{y}{z}}\left[\mathbf{A}_1 e^{\sqrt{yz}x} - \mathbf{A}_2 e^{-\sqrt{yz}x}\right]\end{aligned} \tag{9.18}$$

Equations 9.16 and 9.17 thus give the expressions for $\mathbf{V}$ and $\mathbf{I}$ in the form of unknown constants $\mathbf{A}_1$ and $\mathbf{A}_2$. The values of $\mathbf{A}_1$ and $\mathbf{A}_2$ can be determined by receiving-end condition as under

At receiving end $= 0$,   $\mathbf{V} = \mathbf{V}_r$   and   $\mathbf{I} = \mathbf{I}_r$

Substituting these values in Equations 9.16 and 9.17, we get

$$\mathbf{V}_r = \mathbf{A}_1 + \mathbf{A}_2 \tag{9.19}$$

and

$$\mathbf{I}_r = \sqrt{\frac{\mathbf{Y}}{\mathbf{Z}}}[\mathbf{A}_1 - \mathbf{A}_2] \tag{9.20}$$

For transmission line $\sqrt{z/y}$ is constant, called the characteristic constant $Z_c$, and $\sqrt{yz}$ is called propagation constant $\delta$. Both are complex quantities.

From Equations 9.19 and 9.20, we have

$$A_1 = \frac{1}{2}\left(V_r + I_r\sqrt{\frac{z}{y}}\right) = \frac{1}{2}[V_r + Z_cI_r]$$

$$A_2 = \frac{1}{2}\left(V_r - I_r\sqrt{\frac{z}{y}}\right) = \frac{1}{2}[V_r - Z_cI_r]$$

Thus, the expression for $V$ and $I$ becomes

$$V = \frac{1}{2}[V_r + Z_cI_r]e^{\delta x} + \frac{1}{2}[V_r - Z_cI_r]e^{-\delta x} \tag{9.21}$$

$$V = V_r\left(\frac{e^{\delta x} + e^{-\delta x}}{2}\right) + Z_cI_r\left(\frac{e^{\delta x} - e^{-\delta x}}{2}\right)$$

or

$$V = V_r\cosh \delta x + Z_cI_r\sinh \delta x \tag{9.22}$$

$$I = \frac{1}{2}\left[\frac{V_r}{Z_c} + I_r\right]e^{\delta x} - \frac{1}{2}\left[\frac{V_r}{Z_c} - I_r\right]e^{-\delta x} \tag{9.23}$$

$$I = I_r\left(\frac{e^{\delta x} + e^{-\delta x}}{2}\right) + \frac{V_r}{Z_c}\left(\frac{e^{\delta x} - e^{-\delta x}}{2}\right)$$

or

$$I = I_r\cosh \delta x + \frac{V_r}{Z_c}\sinh \delta x \tag{9.24}$$

The sending-end voltage $\mathbf{V_s}$ and sending-end current $\mathbf{I_s}$ can be obtained by substituting $x = l$ in the above Equations 9.23 and 9.24:

$$\mathbf{V_s} = \mathbf{V_r}\cosh \delta l + \mathbf{I_r Z_c}\sinh \delta l \tag{9.25}$$

$$\mathbf{I_s} = \mathbf{I_r}\cosh \delta l + \frac{\mathbf{V_r}}{\mathbf{Z_c}}\sinh \delta l \tag{9.26}$$

Now $\delta l = \sqrt{yzl} = \sqrt{yl \times zl} = \sqrt{\mathbf{YZ}}$, where $\mathbf{Z}$ is the total impedance of the line and $\mathbf{Y}$ is the total admittance of the line.

The expression for sending-end voltage and sending-end current is

$$\mathbf{V_s} = \mathbf{V_r} \cosh \sqrt{\mathbf{YZ}} + \mathbf{I_r Z_c} \sinh \sqrt{\mathbf{YZ}} \tag{9.27}$$

$$\mathbf{I_s} = \mathbf{I_r} \cosh \sqrt{\mathbf{YZ}} + \frac{\mathbf{V_r}}{\mathbf{Z_c}} \sinh \sqrt{\mathbf{YZ}} \tag{9.28}$$

Comparing above equations with the general voltage and current equations of the line, we define the ABCD parameters:

$$\mathbf{A} = \mathbf{D} = \cosh \sqrt{\mathbf{YZ}}$$

$$\mathbf{B} = \mathbf{Z_c} \sinh \sqrt{\mathbf{YZ}}$$

$$\mathbf{C} = \frac{1}{\mathbf{Z_c}} \sinh \sqrt{\mathbf{YZ}}$$

The relation $\mathbf{AD} - \mathbf{BC} = 1$ holds for $\mathbf{AD} - \mathbf{BC} = \left[ \cosh \sqrt{\mathbf{YZ}}^2 + \sinh \sqrt{\mathbf{YZ}}^2 \right] = 1$.

Equations 9.25 and 9.26 can be written in matrix form as

$$\begin{bmatrix} \mathbf{V_s} \\ \mathbf{I_s} \end{bmatrix} = \begin{bmatrix} \cosh \delta l & \mathbf{Z_c} \sinh \delta l \\ \dfrac{1}{\mathbf{Z_c}}\sinh \delta l & \cosh \delta l \end{bmatrix} \begin{bmatrix} \mathbf{V_r} \\ \mathbf{I_r} \end{bmatrix}$$

## 9.11 EVALUATION OF ABCD CONSTANTS

The hyperbolic functions involved in the transmission line equations are not easily evaluated with the help of ordinary tables of hyperbolic functions. This is due to the fact that they are functions of complex arguments. The following methods are used to calculate the hyperbolic functions $\cosh \delta l$ and $\sinh \delta l$ for determining the ABCD constants for a long transmission line.

We have $\delta = \alpha + j\beta$,

1. Use of complex exponential:

$$\cosh \delta l = \cosh(\alpha l + j\beta l) = \frac{e^{\alpha l}e^{j\beta l} + e^{-\alpha l}e^{-j\beta l}}{2} = \frac{1}{2}(e^{\alpha l}\angle\beta l + e^{-\alpha l}\angle - \beta l) \quad (9.29)$$

$$\sinh \delta l = \sinh(\alpha l + j\beta l) = \frac{e^{\alpha l}e^{j\beta l} - e^{-\alpha l}e^{-j\beta l}}{2} = \frac{1}{2}(e^{\alpha l}\angle\beta l + e^{-\alpha l}\angle - \beta l) \quad (9.30)$$

2. Use of identities: In this method, the hyperbolic sines and cosines of the complex argument $\delta l$ can be separated into real and imaginary parts of the use of following identities. Then

$$\cosh \delta l = \cosh(\alpha l + j\beta l) = \cosh \alpha l \cos\beta l + j\sinh \alpha l \sin\beta l \quad (9.31)$$

and

$$\sinh \delta l = \sinh(\alpha l + j\beta l) = \sinh \alpha l \cos\beta l + j\cosh \alpha l \sin\beta l \quad (9.32)$$

It should be noted that the unit of $\beta l$ is the radian.

3. Use of power series: In this method the hyperbolic sine and cosine are expressed in terms of their power series. The expressions are

$$\cosh \delta l = 1 + \frac{\delta^2 l^2}{2!} + \frac{\delta^4 l^4}{4!} + \cdots \quad (9.33)$$

$$\sinh \delta l = \delta l + \frac{\delta^3 l^3}{3!} + \frac{\delta^5 l^5}{5!} + \cdots \quad (9.34)$$

The above series converge rapidly for the values of $\delta l$ usually found for power lines. Sufficient accuracy can be obtained by taking only the first two terms. Thus,

$$\cosh \delta l = 1 + \frac{\delta^2 l^2}{2!} = 1 + \frac{YZ}{2} \tag{9.35}$$

$$\sinh \delta l = \delta l + \frac{\delta^3 l^3}{3!} = \sqrt{YZ}\left(1 + \frac{1}{2}YZ\right) \tag{9.36}$$

Usually above approximation are satisfactory for overhead lines up to 500 km.

$$A = D = \cosh \delta l = 1 + \frac{YZ}{2} \tag{9.37}$$

$$B = Z_c \sinh \delta l = \sqrt{\frac{Z}{Y}} \times \sqrt{YZ}\left(1 + \frac{1}{2}YZ\right) = Z\left(1 + \frac{1}{2}YZ\right) \tag{9.38}$$

$$C = \frac{1}{Z_c} \sinh \delta l = \sqrt{\frac{Y}{Z}} \times \sqrt{YZ}\left(1 + \frac{1}{2}YZ\right) = Y\left(1 + \frac{1}{2}YZ\right) \tag{9.39}$$

## 9.12 FERRANTI EFFECT

A long-transmission line has a large capacitance. When a long line is operating under no-load condition, the receiving-end voltage is greater than the sending-end voltage. This is known as Ferranti effect. This phenomenon can be explained with the following reasoning. It was first noticed by Ferranti on overhead lines supplying a lightly loaded network. Ferranti effect is due to charging current of the line. The value of current at the sending end at no-load and normal operating voltage applied at the sending end is called the charging current.

A simple explanation of Ferranti effect can be given by approximating the distributed parameters of the line by lumped impedance as shown in Figure 9.9a. Since usually the capacitive reactance of the line is quite large as compared to the inductive reactance, under no-load or lightly loaded condition, the line current is of leading pf. The phasor diagram

(a)                                                         (b)

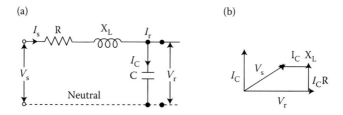

FIGURE 9.9   (a) Line representation (lumped) under no-load condition. (b) Its phasor diagram.

(Figure 9.9b) is given below for this operating condition. The charging current produces drop in the reactance of the line which is in phase opposition to the receiving-end voltage and hence the sending-end voltage becomes smaller than the receiving-end voltage.

## 9.13  ABCD CONSTANTS

We know that the sending-end quantities, that is, $V_s$ and $I_s$ are given by

$$V_s = AV_r + BI_r \tag{9.40}$$

$$I_s = CV_r + DI_r \tag{9.41}$$

Similar expressions for $V_r$ and $I_r$ can be found from these equations as follows.

Multiply Equation 9.40 by C and Equation 9.41 by A, we get

$$CV_s = CAV_r + CBI_r \tag{9.42}$$

$$AI_s = ACV_r + ADI_r \tag{9.43}$$

Subtracting Equation 9.42 from Equation 9.43,

$$AI_s - CV_s = (AD - BC)I_r$$

Since $AD - BC = 1$ and $A = D$,

$$I_r = -CV_s + DI_s \tag{9.44}$$

Next to eliminate $I_r$ from Equations 9.40 and 9.41, multiply Equation 9.40 by D and Equation 9.41 by B,

$$DV_s = ADV_r + BDI_r \qquad (9.45)$$

$$BI_s = BCV_r + BDI_r \qquad (9.46)$$

Subtracting Equation 9.45 from Equation 9.46,

$$DV_s - BI_s = (AD - BC)V_r$$

Again,

$$AD - BC = 1 \quad \text{and} \quad V_r = DV_s - BI_s \qquad (9.47)$$

### 9.13.1 Proof for the Relation AD − BC = 1

Consider Figure 9.10a, where a two-terminal pair network with parameters A, B, C, and D is connected to an ideal voltage source with zero internal impedance at one end and the other end being short circuited.
The equation representing this condition is

$$V_s = E = 0 + BI_2$$

$$I_2 = \frac{E}{B} \qquad (9.48)$$

Now we short circuit the sending end and connect the generator at the receiving end as shown in Figure 9.10b. The positive directions of flow of current are shown in the figures.

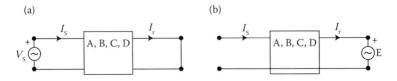

FIGURE 9.10   (a) Two terminal pair network. (b) Source at the receiving end.

$$\therefore 0 = AE + BI_r \tag{9.49}$$

Since transmission line is a linear passive bilateral network, therefore

$$I_s = -I_2 = CE + DI_r \tag{9.50}$$

Eliminating $I_r$ from Equations 9.49 and 9.50, we obtain

$$-I_r = CE - D\frac{-AE}{B} \tag{9.51}$$

Since from Equation 9.48, $I_2 = (E/B)$, Equation 9.51 becomes

$$-\frac{E}{B} = CE + D\frac{-AE}{B}$$

or

$$-\frac{1}{B} = C - D\frac{A}{B}$$

or

$$-1 = BC - AD$$

or

$$AD - BC = 1 \tag{9.52}$$

As is said earlier that if A, B, C, and D parameters are calculated independently, Equation 9.52 gives a check on the accuracy of the values calculated.

### 9.13.2 Constants for Two Networks in Tandem

Two networks are said to be connected in tandem when the output of one network is connected to the input of the other network. Let the constants of these networks be $A_1$, $B_1$, $C_1$, $D_1$ and $A_2$, $B_2$, $C_2$, $D_2$ which are connected in tandem as shown in Figure 9.11. These two networks could be two transmission connected to a transmission line, etc.

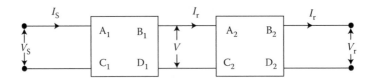

FIGURE 9.11    Two networks in tandem.

The net constants of the system relating the terminal conditions can be found as follows:

$$V = D_1 V_s - B_1 I_s \tag{9.53}$$

$$I = -C_1 V_s + A_1 I_s \tag{9.54}$$

$$V = A_2 \cdot V_r + B_2 I_r \tag{9.55}$$

$$I = C_2 V_r + D_2 I_r \tag{9.56}$$

From Equations 9.53 and 9.55 and Equations 9.54 and 9.56, respectively,

$$D_1 V_s - B_1 I_s = A_2 \cdot V_r + B_2 I_r \tag{9.57}$$

$$-C_1 V_s + A_1 I_s = C_2 V_r + D_2 I_r \tag{9.58}$$

Multiplying Equation 9.57 by $A_1$ and Equation 9.58 by $B_1$ and adding the resulting equation.

$$(A_1 D_1 - B_1 C_1) V_s = (A_1 A_2 + B_1 C_2) V_r + (A_1 B_2 + B_1 D_2) I_r \tag{9.59}$$

Multiplying Equation 9.57 by $C_1$ and Equation 9.58 by $D_1$ and adding the resulting equation.

$$(A_1 D_1 - B_1 C_1) I_s = (A_2 C_1 + C_2 D_1) V_r (B_2 C_1 + D_1 D_2) I_r \tag{9.60}$$

Since $A_1 D_1 - B_1 C_1 = 1$, the constants for the two networks in tandem are

$$A = A_1 A_2 + B_1 C_2$$

$$B = A_1B_2 + B_1D_2$$

$$C = A_2C_1 + C_2D_1 \tag{9.61}$$

$$D = B_2C_1 + D_1D_2$$

The relation is given in matrix form.

If network 2 is at the sending end and 1 is at the receiving end the overall constants for the two networks in tandem can be obtained by interchanging the subscripts in equation.

### 9.13.3 Constants for Two Networks in Parallel

In case two networks are connected in parallel as shown in Figure 9.12, the constants for the overall network can be determined as follows.

For this derivation, it is assumed that the transmission line is a reciprocal network (symmetrical network) and we know when two reciprocal networks are connected in parallel, the resulting network is also reciprocal. (The resulting network is not reciprocal in case the two networks are connected in tandem.)

Writing the equation for the terminal conditions,

$$V_s = A_1V_r + B_1I_{r_1} \tag{9.62}$$

$$V_s = A_2V_r + B_2I_{r_2} \tag{9.63}$$

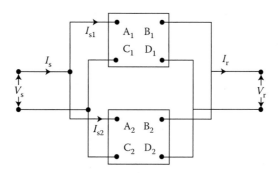

FIGURE 9.12   Two networks in parallel.

Since the overall expression required is

$$V_s = AV_r + BI_r \tag{9.64}$$

where $I_r = I_{r_1} + I_{r_2}$.

Therefore, multiplying Equations 9.62 and 9.63 by $B_2$ and $B_1$, respectively, and adding, we get

$$(B_1 + B_2)V_s = (A_1B_2 + A_2B_1)V_r + B_1B_2(I_{r_1} + I_{r_2})$$

or

$$V_s = \frac{A_1B_2 + A_2B_1}{B_1 + B_2}V_r + \frac{B_1B_2}{B_1 + B_2}I_r \tag{9.65}$$

Comparing the coefficients of Equations 9.64 and 9.65, we get

$$A = \frac{A_1B_2 + A_2B_1}{B_1 + B_2}$$

and

$$B = \frac{B_1B_2}{B_1 + B_2} \tag{9.66}$$

Since transmission line is a symmetrical line is a symmetrical network,

$$A = D = \frac{A_1B_2 + A_2B_1}{B_1 + B_2} = \frac{D_1B_2 + D_2B_1}{B_1 + B_2} \tag{9.67}$$

Also since transmission line is a two-terminal pair network,

$$AD - BC = 1 \tag{9.68}$$

Using relations (9.66) through (9.68), we obtain

$$C = C_1 + C_2 + \frac{(A_1 - A_2)(D_2 - D_1)}{B_1 + B_2} \tag{9.69}$$

### 9.13.4 Measurement of A, B, C, and D Constants

If a transmission line is already erected, the constants can be obtained by conducting the open- and short-circuit tests at two ends of the line as follows.

In Figure 9.13a and b, the connection diagrams for conducting OC and SC test on the sending end are shown, respectively. Similar connections are made on the receiving end for performing these tests. Before proceeding further to determine the constants, the following impedances are defined as

$Z_{so}$ = Sending-end impedance with receiving-end open circuited

$Z_{ss}$ = Sending-end impedance with receiving-end short circuited

$Z_{ro}$ = Receiving-end impedance with sending-end open circuited

$Z_{rs}$ = Receiving-end impedance with sending-end short circuited

Using equations

$$V_s = AV_r + BI_r \tag{9.70}$$

$$I_s = CV_r + DI_r \tag{9.71}$$

For making impedance measurements on the sending-end side, we get

$$Z_{so} = \frac{V_s}{I_s} = \frac{A}{C} \quad \text{for } I_r = 0 \quad \text{(open-circuit test)} \tag{9.72}$$

(a)

(b)

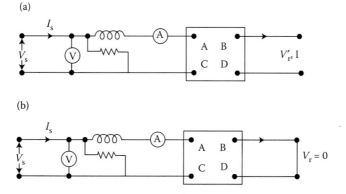

FIGURE 9.13   (a) Open-circuit tests. (b) Short-circuit tests.

$$Z_{ss} = \frac{V_s}{I_s} = \frac{B}{D} \quad \text{for } V_r = 0 \, (\text{short-circuit test}) \qquad (9.73)$$

It is to be noted here that the impedances are complex quantities, the magnitudes are obtained by the ratio of the voltage and current and the angle is obtained with the help of Watt-meter reading.

To determine the impedances on the receiving end the following equation are made use of

$$V_r = DV_s - BI_s \qquad (9.74)$$

$$I_r = -CV_s + AI_s \qquad (9.75)$$

Since the direction of sending-end current according to the above equation enters the network, while performing the tests on receiving-end side, the direction of the current will be leaving the network, therefore, these equations become

$$V_r = DV_s + BI_s \qquad (9.76)$$

$$-I_r = -CV_s - AI_s$$

or

$$I_r = CV_s + AI_s \qquad (9.77)$$

Therefore, for sending-end open-circuits $I_s = 0$,

$$Z_{ro} = \frac{V_r}{I_r} = \frac{D}{C} \qquad (9.78)$$

And for short-circuit $V_s = 0$,

$$Z_{rs} = \frac{V_r}{I_r} = \frac{B}{A} \qquad (9.79)$$

From Equations 9.78 and 9.79, we obtain

$$Z_{ro} - Z_{rs} = \frac{D}{C} - \frac{B}{A} = \frac{AD - BC}{AC} = \frac{1}{AC}$$

$$\frac{Z_{ro} - Z_{rs}}{Z_{so}} = \frac{1}{AC}\frac{C}{A} = \frac{1}{A^2}$$

$$A = \sqrt{\frac{Z_{so}}{Z_{ro} - Z_{rs}}} \qquad (9.80)$$

$$Z_{rs} = \frac{B}{A}$$

$$B = AZ_{rs} = Z_{rs}\sqrt{\frac{Z_{so}}{Z_{ro} - Z_{rs}}} \qquad (9.81)$$

$$Z_{so} = \frac{A}{C}$$

$$C = \frac{A}{Z_{so}} = \frac{1}{Z_{so}}\sqrt{\frac{Z_{so}}{Z_{ro} - Z_{rs}}} \qquad (9.82)$$

$$Z_{ro} = \frac{D}{C}$$

$$D = CZ_{ro} = \frac{Z_{ro}}{Z_{so}}\sqrt{\frac{Z_{so}}{Z_{ro} - Z_{rs}}}$$

$$= Z_{ro}\sqrt{\frac{1}{Z_{so}(Z_{ro} - Z_{rs})}}$$

Since for a symmetric network $Z_{ro} = Z_{so}$,

$$D = A = \sqrt{\frac{Z_{so}}{Z_{ro} - Z_{rs}}} \qquad (9.83)$$

## WORKED EXAMPLES

### EXAMPLE 9.1

ABCD constants of a 220-kV line are $A = D = 0.94\angle1°$, $B = 130\angle73°$, and $C = 0.001\angle90°$. If the sending-end voltage of a line for a given load delivered at nominal voltage is 240 kV, determine the voltage regulation of the line.

**Solution**

$$\text{Voltage regulation} = \frac{|\mathbf{V}_{R0}| - |\mathbf{V}_R|}{|\mathbf{V}_R|} \times 100\%$$

The voltage $\mathbf{V}_S$ at the sending end is

$$\mathbf{V}_S = \mathbf{A}\mathbf{V}_R + \mathbf{B}\mathbf{I}_R$$

When the load is removed,

$$\mathbf{I}_R = 0, \quad \mathbf{V}_R = \mathbf{V}_{R0}$$

Therefore, $\mathbf{V}_S = \mathbf{A}\mathbf{V}_{R0}$, $\mathbf{V}_{R0} = \mathbf{V}_S/\mathbf{A}$, where $\mathbf{V}_{R0}$ is the receiving-end voltage at no-load.

$$\text{Line regulation} = \frac{|\mathbf{V}_S|/|\mathbf{A}| - |\mathbf{V}_R|}{|\mathbf{V}_R|} \times 100\%$$

$$= \frac{|240/0.94| - |220|}{|220|} \times 100\% = 16.05\%$$

### EXAMPLE 9.2

A 220-kV transmission line represented as T model. Parameters are $A = D = 0.8\angle1°$, $B = 170\angle85°$, and $C = 0.2 \times 10^{-3}\angle90.4°$. Sending-end voltage = 400 kV. Determine:

1. The receiving-end voltage under no-load condition
2. Increase in receiving-end voltage

**Solution**

Receiving-end voltage under no-load condition is

$$\mathbf{V}_{R0} = \frac{\mathbf{V}_s}{\mathbf{A}} \quad (\because \mathbf{I}_R = 0)$$

$$= \frac{400}{0.8}$$

$$= 500 \text{ kV}$$

The increase in receiving-end voltage $= \mathbf{V}_{R0} - \mathbf{V}_R$

$$= (500 - 220) \text{ kV}$$

$$= 280 \text{ kV}$$

### EXAMPLE 9.3

Generalized circuit constants of a three-phase, 220-kV rated voltage medium length are A = D = 0.936 ∠0.98°, B = 142 ∠76.4°. If the load at receiving end is 50 MW at 220 kV with a pf 0.9 (lagging), determine the magnitude of line to line sending-end voltage.

**Solution**

Sending-end line to neutral voltage $= \mathbf{A}\mathbf{V}_R + \mathbf{B}\mathbf{I}_R$

$$= (0.936\angle 0.98°)\left(\frac{220 \times 10^3}{\sqrt{3}}\right)$$

$$+ 142\angle 76.4° \times \mathbf{I}_R \qquad (9.84)$$

Power at the receiving end,

$$P_R = \sqrt{3} \times \text{Line voltage at the receiving end} \times \text{Line current}$$
$$\text{at the receiving end} \times \text{pf}$$

$$P_R = \sqrt{3} \times V_R \times I_R \times \cos\phi$$

$$50 \times 10^6 = \sqrt{3} \times 220 \times 10^3 \times I_R \times 0.9$$

$$I_R = 145.8\angle -\cos^{-1}(0.9)$$
$$= 145.8\angle -25.8° \, \text{A}$$

$$\text{Sending-end line-to-neutral} = (0.936\angle 0.98°)\left(\frac{220 \times 10^3}{\sqrt{3}}\right)$$
$$+ 142\angle 76.4° \times 145.8\angle -25.8°$$
$$= 133\angle 7.77° \, \text{V}$$

$$\therefore \text{Sending-end line to line} = \sqrt{3} \times 133\angle 7.77° \, \text{V}$$
$$= 230.78 \, \text{kV}$$

### EXAMPLE 9.4

Three-phase, 50 Hz transmission line of length 100 km has a capacitance $(0.03/\pi)\mu\text{F/km}$. It is represented as $\pi$ model. Determine the shunt admittance at each end of transmission line.

### Solution

Capacitance of each end of a $\pi$ model = $C/2$

$$\therefore \text{Admittance} = \frac{Y}{2} = \frac{1}{2}j\left(\frac{1}{X_C}\right)$$
$$= \frac{1}{2}j\left(\frac{1}{X_C}\right)$$
$$= \frac{1}{2}j\omega C = \frac{1}{2}j2\pi fC$$
$$= \frac{1}{2}j(2\pi \times 50)\left(\frac{0.03}{\pi} \times 100\right)$$
$$= j150 \times 10^{-6} \, \text{Siemens}$$
$$= 150 \times 10^{-6} \angle 90° \, \text{Siemens}$$

Therefore, the shunt admittance is $150 \times 10^{-6}\angle 90°$ Siemens.

### EXAMPLE 9.5

Estimate the distance over which a load of 10,000 kW at a pf 0.8 lagging can be delivered by a three-phase transmission line having conductors each of resistance 1.5 Ω/km. The voltage at the receiving end is to be 66 kV and the loss in the transmission is to be 5%.

**Solution**

$$\text{Line current,} \quad I = \frac{\text{Power delivered}}{\sqrt{3} \times \text{Line voltage} \times \text{Power factor}}$$

$$= \frac{10000 \times 10^3}{\sqrt{3} \times 66 \times 10^3 \times 0.8} = 109.346 \text{ A}$$

Line losses = 5% of power delivered = $0.05 \times 10,000 = 500$ kW
Let $R \ \Omega$ be the resistance of one conductor.
Line losses = $3I^2R$

$$500 \times 10^3 = 3 \times (109.346)^2 \times R$$

$$R = \frac{500 \times 10^3}{3 \times (109.346)^2} = 13.94 \ \Omega$$

Resistance of each conductor per kilometer is 1.5 Ω (given).

$$\therefore \text{Length of line} = \frac{13.94}{1.5} \text{km} = 9.2928 \text{ km}$$

## EXERCISES

1. What is the purpose of an overhead transmission line? How are these lines classified?

2. Discuss the terms voltage regulation and transmission efficiency as applied to transmission line.

3. Deduce an expression for voltage regulation of a short transmission line, giving the vector diagram.

4. What is the effect of load power factor on regulation and efficiency of a transmission line?

5. What do you understand by medium transmission lines? How capacitance effects are taken into account in such lines?

6. Show how regulation and transmission efficiency are determined for medium lines using

   a. Nominal T method

   b. Nominal $\pi$ method

   Illustrate your answer with suitable vector diagrams.

7. What do you understand by long transmission lines? How capacitance effects are taken into account in such lines?

8. Using rigorous method, derive expressions for sending-end voltage and current for a long transmission line.

9. What do you understand by generalized circuit constants of a transmission line? What is their importance?

10. Evaluate the generalized circuit constants for

   a. Short transmission line

   b. Medium line—nominal T method

   c. Medium line—nominal $\pi$ method

# Underground Cables

## 10.1 INTRODUCTION

Cables are used for transmission and distribution of electrical energy in highly populated areas of towns and cities. Cables form the artery system for the transmission and distribution of electrical energy. A cable is basically an insulated conductor. External protection against mechanical injury, moisture entry, and chemical reaction is provided on the cable. The conductor is usually aluminum or annealed copper, while the insulation is mostly polyvinyl chloride (PVC) or other chemical compounds.

Transmission line is more expensive than overhead lines, especially at high potentials. Besides increase in temperature is high in cables. However, there is limitation of raising the operating voltage. In low and medium voltage distribution in urban areas, cables are more widespread.

An underground cable essentially consists of one or more conductors covered with suitable insulation and surrounded by a protecting cover.

In general, a cable must fulfill the following necessary requirements:

- The conductor used in cables should be tinned stranded copper or aluminum of high conductivity. Stranding is done so that conductor may become flexible and carry more current.

- The conductor size should be such that the cable carries the desired load current without overheating and causes voltage drop within permissible limit.

- The cable must have proper thickness of insulation in order to give high degree of safety and reliability at the voltage for which it is designed.

- The cable must be provided with suitable mechanical protection so that it may withstand the rough use in laying it.

- The material used in the manufacture of cables should be such that there is complete chemical and physical stability throughout.

## 10.2 INSULATING MATERIALS FOR CABLES

The satisfactory operation of a cable depends to a great extent upon the characteristics of insulation used. Therefore, the proper choice of insulating material for cables is of considerable importance. In general, the insulating materials used in cables should have the following properties:

1. High insulation resistance to avoid leakage current.

2. High dielectric strength to avoid electrical breakdown of the cable.

3. High mechanical strength to withstand the mechanical handling of cables.

4. Nonhygroscopic, that is, it should not absorb moisture from air or soil. The moisture tends to decrease the insulation resistance and hastens the breakdown of the cable. In case, the insulating material is hygroscopic, it must be enclosed in a waterproof covering like lead sheath.

5. Noninflammable.

6. Low cost so as to make the underground system a viable proposition.

7. Unaffected by acids and alkalies to avoid any chemical action.

None insulating material possesses all the above-mentioned properties. Therefore, the type of insulating material to be used depends upon the purpose for which the cable is required and the quality of insulation to be aimed at. The principal insulating materials used in cables are rubber, vulcanized India rubber (VIR), impregnated paper, varnished cambric, and PVC.

1. *Rubber*: Rubber may be obtained from milky sap of tropical trees or it may be produced from oil products. It has relative permittivity varying between 2 and 3, dielectric strength is about 30 kV/mm, and resistivity of insulation is $10^{17}$ Ω cm. Although pure rubber has reasonably high insulating properties, it suffers from some major

drawbacks viz., readily absorbs moisture, maximum safe temperature is low (about 38°C), soft and liable to damage due to rough handling, and ages when exposed to light. Therefore, pure rubber cannot be used as an insulating material.

2. *VIR*: It is prepared by mixing pure rubber with mineral matter such as zinc oxide, red lead, etc., and 3%–5% of sulfur. The compound so formed is rolled into thin sheets and cut into strips. The rubber compound is then applied to the conductor and is heated to a temperature of about 150°C. The whole process is called vulcanization and the product obtained is known as VIR. VIR has greater mechanical strength, durability, and wear resistant property than pure rubber. Its main drawback is that sulfur reacts very quickly with copper and for this reason, cables using VIR insulation have tinned copper conductor. The VIR insulation is generally used for low and moderate voltage cables.

3. *Impregnated paper*: It consists of chemically pulped paper made from wood chippings and impregnated with some compound such as paraffinic or naphthenic material. This type of insulation has almost superseded the rubber insulation. It is because it has the advantages of low cost, low capacitance, high dielectric strength, and high insulation resistance. The only disadvantage is that paper is hygroscopic and even if it is impregnated with suitable compound, it absorbs moisture and thus lowers the insulation resistance of the cable. For this reason, paper-insulated cables are always provided with some protective covering and are never left unsealed. If it is required to be left unused on the site during laying, its ends are temporarily covered with wax or tar. Since the paper insulated cables have the tendency to absorb moisture, they are used where the cable route has a few joints. For instance, they can be profitably used for distribution at low voltages in congested areas where the joints are generally provided only at the terminal apparatus. However, for smaller installations, where the lengths are small and joints are required at a number of places, VIR cables will be cheaper and durable than paper-insulated cables.

4. *Varnished cambric*: It is a cotton cloth impregnated and coated with varnish. This type of insulation is also known as empire tape. The cambric is lapped on to the conductor in the form of a tape and its surfaces are coated with petroleum jelly compound to allow for the sliding of one turn over another as the cable is bent. As the varnished

cambric is hygroscopic, such cables are always provided with metallic sheath. Its dielectric strength is about 4 kV/mm and permittivity is 2.5–3.8.

5. *PVC*: This insulating material is a synthetic compound. It is obtained from the polymerization of acetylene and is in the form of white powder. For obtaining this material as a cable insulation, it is compounded with certain materials known as plasticizers which are liquids with high boiling point. The plasticizer forms a gel and renders the material plastic over the desired range of temperature. PVC has high insulation resistance, good dielectric strength, and mechanical toughness over a wide range of temperatures. It is inert to oxygen and almost inert to many alkalies and acids. Therefore, this type of insulation is preferred over VIR in extreme environmental conditions such as in cement factory or chemical factory. As the mechanical properties (i.e., elasticity, etc.) of PVC are not as good as those of rubber, PVC-insulated cables are generally used for low and medium domestic lights and power installations.

## 10.3 CONSTRUCTION OF CABLES

Figure 10.1 shows the general construction of a three-conductor cable. The various parts are as follows.

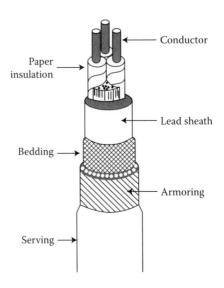

FIGURE 10.1   General construction of a three-conductor cable.

1. *Cores or conductor*: A cable may have one or more than one core (conductor) depending upon the type of service for which it is intended. For instance, the three-conductor cable shown in Figure 10.1 is used for three-phase service. The conductors are made of tinned copper or aluminum and are usually stranded in order to provide flexibility to the cable.

2. *Insulation*: Each core or conductor is provided with a suitable thickness of layer depending upon the voltage to be withstood by the cable. The commonly used materials for insulation are impregnated paper, varnished cambric, or rubber mineral compound.

3. *Metallic sheath*: In order to protect the cable from moisture, gases or other damaging liquids (acids or alcohols) in the soil and atmosphere, a metallic sheath of lead or aluminum is provided over the insulation.

4. *Bedding*: Over the metallic sheath is applied a layer of bedding which consists of a fibrous material like jute or hessian type. The purpose of bedding is to protect the metallic sheath against corrosion and from mechanical injury due to armoring.

5. *Armoring*: Over the bedding, armoring is provided which consists of one or two layers of galvanized steel wire or steel tape. Its purpose is to protect the cable from mechanical injury while laying it and during the course of handling.

6. *Serving*: In order to protect armoring from atmospheric conditions, a layer of fibrous material (like jute) similar to bedding is provided over the armoring. This is known as serving.

## 10.4 CLASSIFICATION OF CABLE

Cables for underground service may be classified into two ways:

1. The type of insulating material used in their manufacture.

2. The voltage for which they are manufactured.

However, the latter method of classification is generally preferred, according to which cables can be divided into the following groups:

1. Low-tension (LT) cables—up to 1000 V

2. High-tension (HT) cables—up to 1100 V

3. Super-tension (ST) cables—from 22 to 33 kV

4. Extra high-tension (EHV) cables—from 33 to 66 kV

5. Extra super-voltage cables—beyond 132 kV

A cable may have one or more than one core depending upon the type of service for which it is intended. It may be (a) single core, (b) two cores, (c) three cores, and (d) four cores.

For a three-phase service, either three single core cables or three cores can be used depending upon the operating voltage and load demand. For three-phase, four-wire system, four-core cables may be used.

The constructional details of a single-core, low-tension cable is shown in Figure 10.2. The cable has ordinary construction because the stress developed in the cable for low voltage (up to 6600 V) is generally small. It consists of one circular core of tinned stranded copper (or aluminum) insulated by layers of impregnated paper. The insulation is surrounded by a lead sheath that prevents the entry of moisture into the inner parts. In order to project the lead sheath from corrosion, an overall serving of compounded fibrous material (jute, etc.) is provided. Single core cables are not usually armored in order to avoid excessive sheath losses. The principal advantages of single core cables are simple construction and availability of larger copper section.

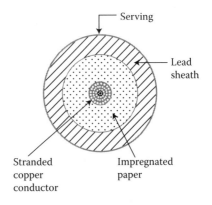

FIGURE 10.2 Constructional details of a single-core, low-tension cable.

## 10.5 CABLES FOR THREE-PHASE SERVICE

In practice, underground cables are generally required to deliver three-phase power. For the purpose, either three-core cable or three single-core cables may be used. For voltages up to 66 kV, three-core cable (i.e., multi-core construction) is preferred due to economic reasons. However, for voltages beyond 66 kV, three-core cables become too large and unwieldy and, therefore, single-core cables are used. The following types of cables are generally used for three-phase service:

1. Belted cables—up to 11 kV

2. Screened cables—from 22 to 66 kV

3. Pressure cables—beyond 66 kV

### 10.5.1 Belted Cables

These cables are used for voltages up to 11 kV but in extraordinary cases, their use may be extended up to 22 kV. Figure 10.3 shows the constructional details of a three-core belted cable. The cores are insulated from each other by layers of impregnated paper. Another layer of impregnated paper tape, called paper belt, is wound round the grouped insulated cores. The gap between the insulated cores is filled with fibrous insulating material (jute, etc.) so as to give circular cross-section to the cable. The cores are generally stranded and may be of noncircular shape to make better use of available space. The belt is covered with lead sheath to protect the cable against ingress of moisture and mechanical injury. The lead sheath

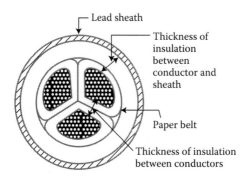

FIGURE 10.3   Three-core belted cable.

is covered with one or more layers of armoring with an outer serving (not shown in the figure).

The belted type construction is suitable only for low and medium voltage, as the electrostatic stresses developed in the cables for these voltages are more or less radial, that is, across the insulation. However, for high voltages (beyond 22 kV), the tangential stresses also become important. These stresses act along the layers of paper insulation. As the insulation resistance of paper is quite small along the layers, therefore, tangential stresses setup leakage current along the layers of paper insulation. The leakage current causes local heating, resulting in the risk of breakdown of insulation at any moment. In order to overcome this difficulty, screened cables are used where leakage currents are conducted to earth through metallic screens.

### 10.5.2 Screened Cables

These cables are meant for use up to 33 kV, but in particular cases, their use may be extended to operating voltages up to 66 kV. Two principal types of screened cables are H-type cables and separate lead (SL)-type cables.

1. *H-type cables*: This type of cable was first designed by H. Hochstadter and hence the name. Figure 10.4 shows the constructional details of a typical three-core, H-type cable. Each core is insulated by layers of impregnated paper. The insulation on each core is covered with a metallic screen which usually consists of a perforated aluminum foil. The cores are laid in such a way that metallic screens make contact with one another. An additional conducting belt (copper woven fabric tape) is wrapped round the three cores. The cable has

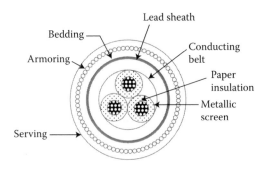

FIGURE 10.4   Typical three-core, H-type cable.

no insulating belt but lead sheath, bedding, armoring, and serving follow as usual. It is easy to see that each core screen is in electrical contact with the conducting belt and the lead sheath. As all the four screens (three-core screens and one conducting belt) and the lead sheath are at earth potential, the electrical stresses are purely radial and consequently dielectric losses are reduced. Two principal advantages are claimed for H-type cables. Firstly, the perforations in the metallic screens assist in the complete impregnation of the cable with the compound and thus the possibility of air pockets or voids (vacuous spaces) in the dielectric is eliminated. The voids if present tend to reduce the breakdown strength of the cable and may cause considerable damage to the paper insulation. Secondly, the metallic screens increase the heat dissipating power of the cable.

2. *SL-type cables*: Figure 10.5 shows the constructional details of a three-core, SL-type cable. It is basically H-type cable but the screen round each core insulation is covered by its own lead sheath. There is no overall lead sheath but only armoring and serving are provided. The SL-type cables have two main advantages over H-type cables. Firstly, the separate sheaths minimize the possibility of core-to-core breakdown. Secondly, bending of cables becomes easy due to the elimination of overall lead sheath. However, the disadvantage is that the three lead sheaths of SL cable are much thinner than the single sheath of H-cable and, therefore, call for greater care in manufacture.

### 10.5.2.1 Limitations of Solid-Type Cables

All the cables of above construction are referred to as solid-type cables because solid insulation is used and no gas or oil circulates in the cable

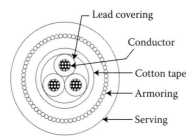

FIGURE 10.5  Three-core SL type cable.

sheath. The voltage limit for solid type cables is 66 kV due to the following reasons:

1. As a solid cable carries the load, its conductor temperature increases and the cable compound (i.e., insulating compound over paper) expands. This action stretches the lead sheath which may be damaged.

2. When the load on the cable decreases, the conductor cools and a partial vacuum is formed within the cable sheath. If the pinholes are present in the lead sheath, moist air may be drawn into the cable. The moisture reduces the dielectric strength of insulation and may eventually cause the breakdown of the cable.

3. In practice, voids are always present in the insulation of a cable. Modern techniques of manufacturing have resulted in void free cables. However, under operating conditions, the voids are formed as a result of the differential expansion and contraction of the sheath and impregnated compound. The breakdown strength of voids is considerably less than that of the insulation. If the void is small enough, the electrostatic stress across it may cause its breakdown. The voids nearest to the conductor are the first to break down, the chemical and thermal effects of ionization causing permanent damage to the paper insulation.

### 10.5.3 Pressure Cables

For voltages beyond 66 kV, solid-type cables are unreliable because there is a danger of breakdown of insulation due to the presence of voids. When the operating voltages are greater than 66 kV, pressure cables are used. In such cables, voids are eliminated by increasing the pressure of compound and for this reason they are called pressure cables. Two types of pressure cables viz. oil-filled cables and gas pressure cables are commonly used.

#### 10.5.3.1 Oil-Filled Cables

In such types of cables, channels or ducts are provided in the cable for oil circulation. The oil under pressure (it is the same oil used for impregnation) is kept constantly supplied to the channel by means of external reservoirs placed at suitable distances (say 500 m) along the route of the cable. Oil under pressure compresses the layers of paper insulation and is forced into any voids that may have formed between the layers. Due to the elimination of voids, oil-filled cables can be used for higher voltages, the range being from 66 kV to 230 kV. Oil-filled cables are of three types viz., single-core conductor channel, single-core sheath channel, and three-core filler-space channels.

Figure 10.6 shows the constructional details of a single-core conductor channel, oil-filled cable. The oil channel is formed at the center by stranding the conductor wire around a hollow cylindrical steel spiral tape. The oil under pressure is supplied to the channel by means of external reservoir. As the channel is made of spiral steel tape, it allows the oil to percolate between copper strands to the wrapped insulation. The oil pressure compresses the layers of paper insulation and prevents the possibility of void formation. The system is so designed that when the oil gets expanded due to increase in cable temperature, the extra oil collects in the reservoir. However, when the cable temperature falls during light load conditions, the oil from the reservoir flows to the channel. The disadvantage of this type of cable is that the channel is at the middle of the cable and is at full voltage wrt earth, so that a very complicated system of joints is necessary.

Figure 10.7 shows the constructional details of a single-core sheath channel oil-filled cable. In this type of cable, the conductor is solid similar to that of solid cable and is paper insulated. However, oil ducts are provided in the metallic sheath as shown. In the three-core, oil-filled cable shown in Figure 10.8, the oil ducts are located in the filler spaces. These channels are composed of perforated metal-ribbon tubing and are at earth potential. The oil-filled cables have three principal advantages. Firstly, formation of voids and ionization are avoided. Secondly, allowable temperature range and dielectric strength are increased. Thirdly, if there is leakage, the defect

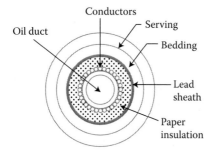

FIGURE 10.6   Single-core conductor channel, oil-filled cable.

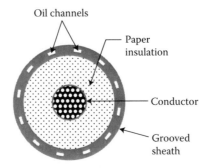

FIGURE 10.7   Single-core sheath channel oil-filled cable.

in the lead sheath is at once indicated and the possibility of earth faults is decreased. However, their major disadvantages are the high initial cost and complicated system of laying.

### 10.5.3.2 Gas Pressure Cables

The voltage required to set up ionization inside a void increases as the pressure is increased. Therefore, if ordinary cable is subjected to a sufficiently high pressure, the ionization can be altogether eliminated. At the same time, the increased pressure produces radial compression which tends to close any voids. This is the underlying principle of gas pressure cables.

Figure 10.9 shows the section of external pressure cable designed by Hochstadter, Vogal, and Bowden. The construction of the cable is similar to that of an ordinary solid type except that it is of triangular shape and thickness of lead sheath is 75% that of solid cable. The triangular section reduces the weight and gives low thermal resistance but the main reason

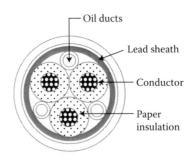

FIGURE 10.8   Three-core oil-filled cable.

FIGURE 10.9    External pressure cable.

for triangular shape is that the lead sheath acts as a pressure membrane. The sheath is protected by a thin metal tape. The cable is laid in a gas-tight steel pipe. The pipe is filled with dry nitrogen gas at 12–15 atmospheres. The gas pressure produces radial compression and closes the voids that may have formed between the layers of paper insulation. Such cables can carry more load current and operate at higher voltages than a normal cable. Moreover, maintenance cost is small and the nitrogen gas helps in quenching any flame. However, it has the disadvantage that the overall cost is very high.

## 10.6  LAYING OF UNDERGROUND CABLES

The reliability of underground cable network depends to a considerable extent upon the proper laying and attachment of fittings, that is, cable end boxes, joints, branch connectors, etc. There are three main methods of laying underground cables viz., direct laying, draw-in system, and the solid system.

### 10.6.1  Direct Laying

Direct laying system is shown in Figure 10.10. This method of laying underground cables is simple and cheap and is much favored in modern practice. In this method, a trench of about 1.5 m deep and 45 cm wide is dug. The trench is covered with a layer of fine sand (of about 10 cm thickness) and the cable is laid over this sand bed. The sand prevents the entry of moisture from the ground and thus protects the cable from decay. After the cable has been laid in the trench, it is covered with another layer of sand of about 10 cm thickness.

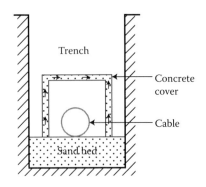

FIGURE 10.10   Direct laying system.

The trench is then covered with bricks and other materials in order to protect the cable from mechanical injury. When more than one cable is to be laid in the same trench, a horizontal or vertical inter-axial spacing of at least 30 cm is provided in order to reduce the effect of mutual heating and also to ensure that a fault occurring on one cable does not damage the adjacent cable. Cables to be laid in this way must have serving of bituminized paper and hessian tape so as to provide protection against corrosion and electrolysis.

*Advantages:*

1. It is a simple and less costly method.

2. It gives the best conditions for dissipating the heat generated in the cables.

3. It is a clean and safe method as the cable is invisible and free from external disturbances.

*Disadvantages:*

1. The extension of load is possible only by a completely new excavation which may cost as much as the original work.

2. The alterations in the cable network cannot be made easily.

3. The maintenance cost is very high.

4. Localization of fault is difficult.

5. It cannot be used in congested areas where excavation is expensive and inconvenient.

This method of laying cables is used in open areas where excavation can be done conveniently and at low cost.

## 10.6.2 Draw-in System

In this method, conduit or duct of glazed stone or cast iron or concrete is laid in the ground with manholes at suitable positions along the cable route. The cables are then pulled into position from manholes. Figure 10.11 shows section through four-way underground duct line. Three of the ducts carry transmission cables and the fourth duct carries relay protection connection, pilot wires. Care must be taken that where the duct line changes direction; depths, dips, and offsets be made with a very long radius or it will be difficult to pull a large cable between the manholes. The distance between the manholes should not be too long so as to simplify the pulling in of the cables. The cables to be laid in this way need not be armored but must be provided with serving of hessian and jute in order to protect them when being pulled into the ducts.

*Advantages:*

1. Repairs, alterations, or additions to the cable network can be made without opening the ground.

2. As the cables are not armored, joints become simpler and maintenance cost is reduced considerably.

3. There are very less chances of fault occurrence due to strong mechanical protection provided by the system.

*Disadvantages:*

1. The initial cost is very high.

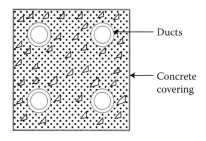

FIGURE 10.11   Four-way underground duct line.

2. The current carrying capacity of the cables is reduced due to the close grouping of cables and unfavorable conditions for dissipation of heat.

This method of cable laying is suitable for congested areas where excavation is expensive and inconvenient, for once the conduits have been laid, repairs or alterations can be made without open undergrounding the ground. This method is generally used for short length cable routes such as in workshops, road crossings where frequent digging is costlier or impossible.

### 10.6.3 Solid System

In this method of laying, the cable is laid in open pipes or troughs dug out in earth along the cable route. The toughing is of cast iron, stoneware, asphalt, or treated wood. After the cable is laid in position, the troughing is filled with a bituminous or asphaltic compound and covered over. Cables laid in this manner are usually plain lead covered because troughing affords good mechanical protection.

*Disadvantages:*

1. It is more expensive than direct laid system.

2. It requires skilled labor and favorable weather conditions.

3. Due to poor heat dissipation facilities, the current carrying capacity of the cable is reduced.

In view of these disadvantages, this method of laying underground cables is rarely used nowadays.

## 10.7 INSULATION RESISTANCE OF A SINGLE-CORE CABLE

The cable conductor is provided with a suitable thickness of insulating material in order to prevent leakage current. The path for leakage current is radial through the insulation. The opposition offered by insulation to leakage current is known as insulation resistance of the cable. For satisfactory operation, the insulation resistance of the cable should be very high.

Consider a single-core cable of conductor radius $r_1$ and internal sheath radius $r_2$ as shown in Figure 10.12. Let $l$ be the length of the cable and $\rho$ be the resistivity of the insulation.

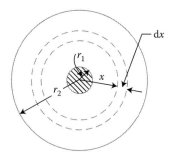

FIGURE 10.12   Cross-sectional view of a single-core cable for measurement of insulation resistance.

Consider a very small layer of insulation of thickness $dx$ at a radius $x$. The length through which leakage current tends to flow is $dx$ and the area of cross-section offered to this flow is $2\pi x l$.

Therefore, insulation resistance of considered layer is

$$\rho \frac{dx}{2\pi x l}$$

Insulation resistance of the whole cable is

$$R = \int_{r_1}^{r_2} \rho \frac{dx}{2\pi x l} = \frac{\rho}{2\pi l} \int_{r_1}^{r_2} \frac{dx}{x}$$

$$R = \frac{\rho}{2\pi l} \log_e \frac{r_2}{r_1}$$

This shows that insulation resistance of a cable is inversely proportional to its length. In other words, if the cable length increases, its insulation resistance decreases and vice versa.

## 10.8  CAPACITANCE OF A SINGLE-CORE CABLE

The single-core cable can be considered to be equivalent to two co-axial cylinders. The conductor of the cable is the inner cylinder while the outer cylinder is represented by lead sheath which is at the earth potential. Consider a single core cable with conductor diameter $d$ and inner sheath diameter $D$ (Figure 10.13). Let the charge per meter axial length of the cable be $Q$ coulombs and $\varepsilon$ be the permittivity of the insulation material between core and lead sheath. Obviously $\varepsilon = \varepsilon_0 \, \varepsilon_r$, where $\varepsilon_r$ is the relative permittivity of the insulation.

Consider a cylinder of radius $x$ meters and axial length $l$ meter. The surface area of this cylinder is $= 2\pi x \times 1 = 2\pi x$ m².

Therefore, electric flux density at point P on the considered cylinder is $D_x = (Q/2\pi x)$ c/m².

Electric intensity at point P, $E_x = (D_x/\varepsilon) = Q/(2\pi x \varepsilon_0 \varepsilon_r)$ V/m.

Potential difference between the capacitor plates (between core and sheath):

$$V = \int_{d/2}^{D/2} E_x dx = \int_{d/2}^{D/2} \frac{Q}{2\pi x \varepsilon_0 \varepsilon_r} dx = \frac{Q}{2\pi \varepsilon_0 \varepsilon_r} \log_e \frac{D}{d}$$

Capacitance of the cable is

$$C = \frac{Q}{V} = \frac{Q}{(Q/2\pi \varepsilon_0 \varepsilon_r) \log_e (D/d)} \, \text{F/m}$$

or

$$c = \frac{2\pi \varepsilon_0 \varepsilon_r}{\log_e (D/d)} \, \text{F/m} = \frac{2\pi \varepsilon_r \times 8.854 \times 10^{-12}}{2.303 \log_{10} (D/d)} \, \text{F/m}$$

$$= \frac{\varepsilon_r}{41.4 \log_{10} (D/d)} \times 10^{-9} \, \text{F/m}$$

If the cable has a length of $l$ meters, then capacitance of the cable is

$$= \frac{\varepsilon_r l}{41.4 \log_{10} (D/d)} \times 10^{-9} \, \text{F/m}$$

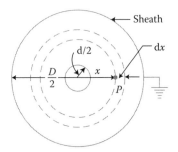

FIGURE 10.13 Cross-sectional view of a single-core cable for measurement of capacitance.

## 10.9 DIELECTRIC STRESS IN A SINGLE-CORE CABLE

Under the operating conditions, the insulation of a cable is subjected to electrostatic forces. This is known as dielectric stress. The dielectric stress at any point in a cable is in fact the potential gradient (or electric intensity) at that point.

Consider a single core cable with diameter $d$ and internal sheath diameter $D$.

Electric intensity at a point $x$ meters from the center of the cable is

$$E_x = \frac{Q}{2\pi\varepsilon_0\varepsilon_r x} \, \text{V/m}$$

By definition, electric intensity is equal to potential gradient. Therefore, potential gradient $s$ at a point $x$ meters from the center of cable is

$$g = E_x$$

or

$$g = \frac{Q}{2\pi\varepsilon_0\varepsilon_r x} \, \text{V/m} \qquad (10.1)$$

Now potential difference $V$ between conductor and sheath is

$$V = \frac{Q}{2\pi\varepsilon_0\varepsilon_r} \log_e \frac{D}{d} \, \text{V}$$

or

$$Q = \frac{2\pi\varepsilon_0\varepsilon_r V}{\log_e(D/d)} \qquad (10.2)$$

Substituting the value of $Q$ from Equation 10.2 in Equation 10.1,

$$g = \frac{(2\pi\varepsilon_0\varepsilon_r V/\log_e(D/d))}{2\pi\varepsilon_0\varepsilon_r x} = \frac{V}{x\log_e(D/d)} \, \text{V/m} \qquad (10.3)$$

It is clear from Equation 10.3 that potential gradient varies inversely as the distance $x$. Therefore, potential gradient will be maximum when $x$

is minimum that is, when $x = d/2$ or at the surface of the conductor. On the other hand, potential gradient will be minimum at $x = D/2$ at sheath surface.

Putting $x = d/2$ in Equation 10.3, maximum potential gradient is

$$g_{max} = \frac{2V}{d \log_e (D/d)} \text{ V/m}$$

Putting $x = D/2$ in Equation 10.3, minimum potential gradient is

$$g_{min} = \frac{2V}{D \log_e (D/d)} \text{ V/m}$$

$$\therefore \frac{g_{max}}{g_{min}} = \frac{2V/(d \log_e (D/d))}{2V/(D \log_e (D/d))} = \frac{D}{d}$$

The variation of stress in the dielectric is shown in Figure 10.14. It is clear that the dielectric stress is maximum at the conductor surface and its value goes on decreasing as we move away from the conductor. It may be noted that maximum stress is an important consideration in the design of a cable. For instance, if a cable is to be operated at such a voltage that maximum stress (at the conductor surface) is 5 kV/mm, then the insulation used must have a dielectric strength of at least 5 kV/mm, otherwise breakdown of the cable will become inevitable.

## 10.10 MOST ECONOMICAL CONDUCTOR SIZE IN A CABLE

It has already been shown that maximum stress in a cable occurs at the surface of the conductor. For safe working of the cable, dielectric strength of the insulation should be more than the maximum stress. Rewriting the expression for maximum stress, we get

$$g_{max} = \frac{2V}{d \log_e (D/d)} \text{ V/m} \tag{10.4}$$

The values of working voltage $V$ and internal sheath diameter $D$ have to be kept fixed at certain values due to design considerations. This

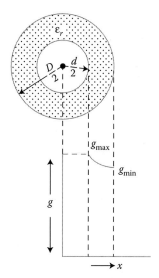

FIGURE 10.14   Cross-sectional view of a single-core cable for measurement of dielectric stress.

leaves conductor diameter $d$ to be the only variable in Equation 10.4. For given values of $V$ and $D$, the most economical conductor diameter will be one for which $g_{max}$ will be minimum when $d \log_e D/d$ is maximum, that is,

$$\frac{d}{dd}\left[d\log_e \frac{D}{d}\right] = 0$$

or

$$\log_e \frac{D}{d} + d \cdot \frac{d}{D} \cdot \frac{-D}{d^2} = 0$$

or

$$\log_e \frac{D}{d} - 1 = 0$$

or

$$\log_e \frac{D}{d} = 1 \quad \frac{D}{d} = e = 2.718$$

Therefore, most economical conductor diameter is $d = D/2.718$ and as putting $\log_e(D/d) = 1$ in Equation 10.4, the value of $g_{max}$ under this condition is

$$g_{max} = \frac{2V}{d} \, V/m$$

For low- and medium-voltage cables, the value of conductor diameter arrived at by this method (i.e., $d = 2V/g_{max}$) is often too small from the point of view of current density. Therefore, the conductor diameter of such cable is determined from the consideration of safe current density. For high-voltage cables, design based on this theory give a very high value of $d$, much too large from the point of view of current carrying capacity and it is, therefore, advantageous to increase the conductor diameter to this value. There are three ways of doing this without using excessive copper:

1. Using aluminum instead of copper because for the same current, diameter of aluminum will be more than that of copper.

2. Using copper wires stranded round a central core of hemp.

3. Using a central lead tube instead of hemp.

## 10.11 GRADING OF CABLES

The process of achieving uniform electrostatic stress in the dielectric of cables is known as grading of cables.

It has already been shown that electrostatic stress in a single core cable has a maximum value $(g_{max})$ at the conductor surface and goes on decreasing as we move toward the sheath. The maximum voltage that can be safely applied to a cable depends upon $g_{max}$, that is, electrostatic stress at the conductor surface. For safe working of a cable homogeneous dielectric, the strength of dielectric must be more than $g_{max}$. If a dielectric of high strength is used for a cable, it is useful only near the conductor where it is maximum. But as we move away from the conductor, the electrostatic stress decreases. So the dielectric will be unnecessarily overstrong.

The unequal stresses distribution in a cable is undesirable for two reasons. Firstly, insulation of greater thickness is required which increases the cable size. Secondly, it may lead to the breakdown of insulation. In order to overcome above disadvantages, it is necessary to have a uniform stress distribution in cables. This can be achieved by distributing the stress in such a way that its value is increased in the outer layers of dielectric. This is known as grading of cables. The following are the two main methods of grading of cables:

1. Capacitance grading

2. Intersheath grading

### 10.11.1 Capacitance Grading

The process of achieving uniformity in the dielectric stress by using layer of different dielectrics is known as capacitance grading.

In capacitance grading, the homogeneous dielectric is replaced by a composite dielectric. The composite dielectric consists of various layers of different dielectrics in such a manner that relative permittivity $\varepsilon_r$ of any layer is inversely proportional to its distance from the center is shown in Figure 10.15. Under such conditions, the value of potential gradient at any point in the dielectric is constant and is independent of its distance from the center. In other words, the dielectric stress in the cable is same everywhere and the grading is ideal one. However, ideal grading requires the use of an infinite number of dielectric which is an impossible task. In practice, two or three dielectrics are used in the decreasing order of permittivity. The dielectric of highest permittivity being used near the core.

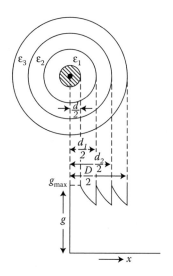

FIGURE 10.15  Capacitance grading.

There are three dielectrics of outer diameter $d_1$, $d_2$, and $D$ and relative permittivity $\varepsilon_1$, $\varepsilon_2$, and $\varepsilon_3$, respectively. If the permitivities such that $\varepsilon_1 > \varepsilon_2 > \varepsilon_3$ and three dielectric are worked at the same maximum stress, then

$$\varepsilon_r \propto \frac{1}{x}, \quad \therefore \varepsilon_r \propto \frac{k}{x}$$

where $k$ is a constant.

$$\frac{Q}{2\pi\varepsilon_0\varepsilon_r x} = \frac{Q}{2\pi\varepsilon_0(k/x)x} = \frac{Q}{2\pi\varepsilon_0 k} = \text{constant}$$

This shows that if the condition $\varepsilon_r \propto (1/x)$ is fulfilled, potential gradient will be content throughout the dielectric cable.

$$g_{1max} = \frac{Q}{2\pi\varepsilon_0\varepsilon_1 d}, \quad g_{2max} = \frac{Q}{\pi\varepsilon_0\varepsilon_2 d_1}, \quad g_{3max} = \frac{Q}{\pi\varepsilon_0\varepsilon_3 d_2}$$

If $g_{1max} = g_{1max} = g_{2max} = g_{3max} = g_{max}$ (say), then

$$\frac{1}{\varepsilon_1 d} = \frac{1}{\varepsilon_2 d_1} = \frac{1}{\varepsilon_3 d_2}$$

$$\therefore \varepsilon_1 d = \varepsilon_2 d_1 = \varepsilon_3 d_2$$

Potential difference across the linear layer is

$$V_1 = \int_{d/2}^{d_1/2} g\,dx = \int_{d/2}^{d_1/2} \frac{Q}{2\pi\varepsilon_0\varepsilon_1 x}\,dx = \frac{Q}{2\pi\varepsilon_0\varepsilon_1}\log_e\frac{d_1}{d} = \frac{g_{max}}{2}d\log_e\frac{d_1}{d}$$

Similarly, potential across second layer ($V_2$) and third layer ($V_3$) is given by

$$V_2 = \frac{g_{max}}{2}d_1\log_e\frac{d_2}{d_1}, \quad V_3 = \frac{g_{max}}{2}d_2\log_e\frac{D}{d_2}$$

Total potential difference between core and earthed sheath is

$$V = V_1 + V_2 + V_3 = \frac{g_{max}}{2}\left[d\log_e\frac{d_1}{d} + d_1\log_e\frac{d_2}{d_1} + d_2\log_e\frac{D}{d_2}\right]$$

If the cable had homogeneous dielectric, then for the same values of $d$, $D$, and $g_{max}$, the permissible potential difference between core earthed sheath would have been

$$V' = \frac{g_{max}}{2}d\log_e\frac{D}{d}$$

Obviously $V > V'$, that is, for given dimensions of the cable, a graded cable can be worked at a greater potential than nongraded cable. Alternatively, for the same safe potential, the size of graded cable will be less than that of nongraded cable. The following points may be noted:

1. As the permissible values of $g_{max}$ are peak values, therefore all the voltages in above expressions should be taken as peak values and not the rms values.

2. If the maximum stress in the three dielectrics is not the same, then,

$$V = \frac{g_{1max}}{2}d\log_e\frac{d_1}{d} + \frac{g_{2max}}{2}d_1\log_e\frac{d_2}{d_1} + \frac{g_{max}}{2}d_2\log_e\frac{D}{d_2}$$

The principal disadvantage of this method is that there are a few high-grade dielectrics of reasonable cost whose permitivities vary over the required range.

## 10.11.2 Intersheath Grading

In this method of cable grading, a homogeneous dielectric is used, but it is divided into various layers by placing metallic intersheaths between the core and lead sheath. The intersheaths are held at suitable potentials which are in between the core potential and the earth potential. This arrangement improves voltage distribution in the dielectric of the cable and consequently more uniform potential gradient is obtained.

Consider a cable of core diameter $d$ and outer lead sheath of diameter $D$ shown in Figure 10.16. Suppose that two intersheaths of diameters $d_1$ and $d_2$ are inserted into the homogeneous dielectric and maintained at some fixed potentials. Let $V_1$, $V_2$, and $V_3$, respectively, be the voltage between core and intersheaths 1, between intersheaths 1 and 2 and between intersheath 2 and outer lead sheath. As there is a definite potential difference between the inner and outer layers of each intersheath, therefore each sheath can be treated like a homogeneous single core cable.

Therefore, maximum stress between core and intersheath 1 is

$$g_{1\max} = \frac{V_1}{(d/2)\log_e(d_1/d)}$$

Similarly,

$$g_{2\max} = \frac{V_2}{(d_1/2)\log_e(d_2/d_1)}$$

$$g_{3\max} = \frac{V_3}{(d_2/2)\log_e(D/d_2)}$$

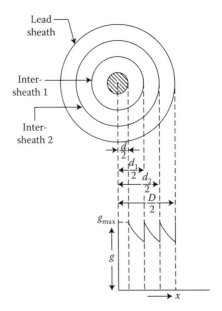

FIGURE 10.16   Intersheath grading.

Since the dielectric is homogeneous, the maximum stress in each layer is the same, that is, $g_{1max} = g_{2max} = g_{3max} = g_{max}$ (say)

$$\therefore \frac{V_1}{(d/2)\log_e(d_1/d)} = \frac{V_2}{(d_1/2)\log_e(d_2/d_1)} = \frac{V_3}{(d_2/2)\log_e(D/d_2)}$$

As the cable behaves like three capacitors in series, all the potential are in phase, that is, voltage between conductor and earthed lead sheath is

$$V = V_1 + V_2 + V_3$$

*Disadvantages*

1. There are complications in fixing the sheath potentials.

2. The intersheaths are likely to be damaged during transportation and insulation which might result in local concentration of potential gradient.

3. There are considerable losses in the intersheaths due to charging.

For this reasons, intersheath grading is rarely used.

## 10.12 CAPACITANCE IN A THREE-CORE-BELTED CABLE

The conductors in a cable are separated from each other by the dielectric. Similarly, there is dielectric between the conductors and the sheath. When a potential difference is applied between the conductors the cable, in effect, is a combination of six capacitances. The capacitances between the conductors are represented by $C_c$. While those between conductors and the sheath by $C_s$ (or $C_e$). Thus a three-phase, belted cable may be represented by a system of capacitances connected in star and delta as shown in Figure 10.17.

The delta-connected capacitances $C_c$ may be replaced by equivalent star connected capacitances $C_1$ (Figure 10.18). The capacitances between pairs of terminals will be same in the two systems.

Capacitance between A and B in the delta system = $C_c + 0.5\ C_c = 1.5\ C_c$ and the capacitance between A and B in the star system = $0.5C_1$.

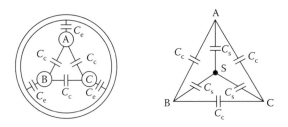

FIGURE 10.17   Capacitance in a three-core belted cable.

For the two systems to be equivalent,

$$1.5C_c = 0.5C_1$$

or

$$C_1 = 3C_c$$

The cable may, therefore, be represented by Figure 10.19. If the neutral point N of the system be earthed, and the sheath be also at zero potential, N and S will become equipotential, and Figure 10.19 then becomes equivalent to that shown in Figure 10.20.

Since $C_1$ and $C_s$ are in parallel, they are combined into a single capacitance $(C_1 + C_s)$.

Finally, we see that the system of capacitances in a cable is reduced to a system of three star-connected capacitances (Figure 10.21). The capacitances of each conductor to neutral or equivalent capacitance is given by

$$C_0 = C_1 + C_s = 3C_c + C_s$$

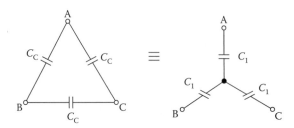

FIGURE 10.18   Equivalent star connect capacitances.

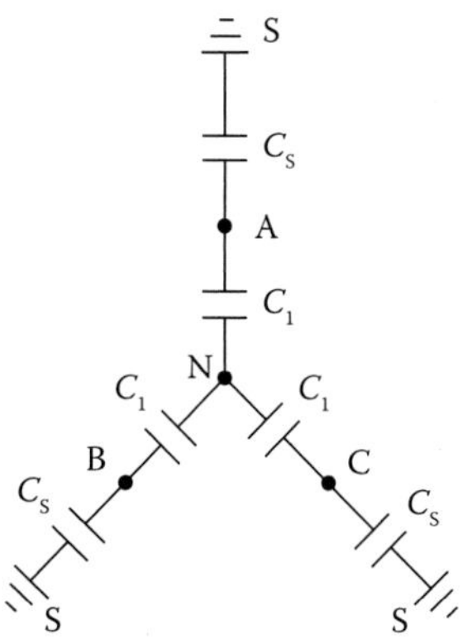

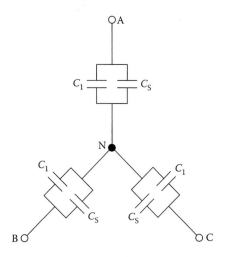

FIGURE 10.19 Equivalent circuit capacitances of a three-core belted cable.

If $V_L$ = line voltage, $V_p$ = phase voltage, the charging current per phase is

$$I_c = V_p \omega C_0 = \frac{V_L}{\sqrt{3}} \omega (3C_c + C_s) A$$

The charging kVA/phase = $V_p I_c \times 10^{-3}$.

Total charging kVA = $3 V_p I_c \times 10^{-3} = \sqrt{3} V_L I_c \times 10^{-3}$ kVA.

It is to be noted that $C_0$ is the capacitance between any conductor and screen for a three-core screened cable.

FIGURE 10.20 Equivalent circuit with $C_1$ and $C_S$ in parallel.

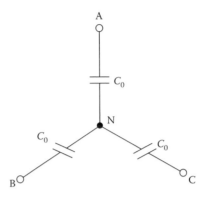

FIGURE 10.21    Final equivalent circuit with three capacitances in star connection.

## 10.13  MEASUREMENT OF $C_C$ AND $C_S$

Cable capacitance is determined from actual measurements instead of relying on the results obtained from the geometry of the cable. The nonuniformity of the insulation material, the variation in the shape of conductors, and the use of fillers make it difficult to estimate the capacitance of a cable from its diameter. The following tests are generally performed:

1. One conductor, say $C$, is connected to the sheath or insulated and the capacitance is measured between the remaining two conductors A and B. Figure 10.22a then reduced to Figure 10.22b.

   The total capacitance $C_L$ measured between the cores A and B is

$$C_L = C_c + \frac{C_c + C_s}{2} = \frac{1}{2}(3C_c + C_s) = \frac{1}{2}C_0$$

   The single measurement is sufficient for calculating the charging current per conductor.

2. The three conductors are connected or bunched together (Figure 10.23) and the capacitance is measured between this bunch and the sheath. Let it be denoted by $C_b$. Here $C_c$ becomes zero and $C_b = 3C_s$.

   Two conductors, say A and B, are joined together and the capacitance is measured between them and the remaining conductor. The arrangement then becomes as shown in Figure 10.24.

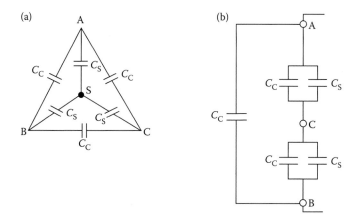

FIGURE 10.22   (a) Conductor C is connected to the sheath. (b) Equivalent circuit of (a).

The capacitance between B and $C = C_c + C_c + (2/3)C_c = (2/3)$ $(3C_c + C_s) = (2/3)C_0$

3. Two conductors say B and C are connected to sheath and the capacitance measured between these and the third conductor A (Figure 10.25).

The capacitance measured in this case $= C_s + C_c + C_c = 2C_c + C_s$.
From the above tests the value of $C_c$ and $C_s$ can also be determined separately.

## 10.14  CURRENT-CARRYING CAPACITY

The safe current carrying capacity of an underground cable is determined by the maximum permissible temperature rise. The cause of temperature rise is the losses that occur in a cable which appear as heat. These losses are

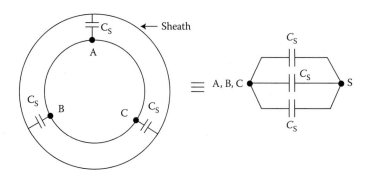

FIGURE 10.23   Conductors A, B, and C are bunched together.

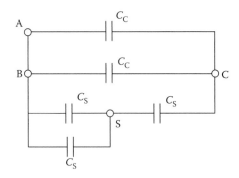

FIGURE 10.24   Two conductors A and B are joined together.

1. Copper losses in the conductor

2. Hysteresis losses in the dielectric

3. Eddy current losses in the sheath

The safe working temperature is 65°C for armored cables and 50°C for lead-sheathed cables laid in ducts. The maximum steady temperature conditions prevail when the heat generated in the cable is equal to the heat dissipated. The heat dissipation of the conductor losses is by conduction through the insulation to the sheath from which the total loss (including

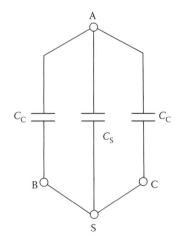

FIGURE 10.25   Two conductors B and C are connected to the sheath.

dielectric and sheath losses) may be conducted to the earth. Therefore, in order to find permissible current loading, the thermal resistivities of the insulation, the protective covering, and the soil must be known.

## 10.15 THERMAL RESISTANCE

The thermal resistance between two points in a medium (e.g., insulation) is equal to temperature difference between these points divided by the heat flowing between them in a unit time. That is, thermal resistance is

$$S = \frac{\text{Temperature difference between point D}}{\text{Heat flowing in a unit time}}$$

In SI units, heat flowing in a unit time is measured in watts. Therefore, thermal resistance is

$$S = \frac{\text{Temperature rise } (t)}{\text{Watts dissipated } (P)}$$

or

$$S = \frac{t}{P}$$

The SI unit of thermal resistance is °C per watt. This is called thermal ohm.

Like electric resistance, thermal resistance is directly proportional to length $L$ in the direction of transmission of heat and inversely proportional to the cross-section area $a$ at right angles to that direction

$$\therefore S \propto \frac{L}{a}$$

or

$$S = k\frac{L}{a}$$

where $k$ is the constant of proportionality and known as thermal resistivity $k = (Sa/L)$.

Unit of $k$ = (Thermal ohm × m²/m) = Thermal ohm meter.

## 10.16 THERMAL RESISTANCE OF DIELECTRIC OF A SINGLE-CORE CABLE

Let us now find the thermal resistance of the dielectric of a single-core cable.

Let $r$ be the radius of the core in meter, $r_1$, the inside radius of the sheath in meter, and $k$, the thermal resistivity of the insulation (i.e., dielectric).

Consider 1 m length of the cable. The thermal resistance of small element of thickness d$x$ at radius $x$ is in Figure 10.26.

$$dS = k \times \frac{dx}{2\pi x}$$

Therefore, thermal resistance of the dielectric is

$$S = \int_{r}^{r1} k \times \frac{dx}{2\pi x}$$

$$= \frac{k}{2\pi} \int_{r}^{r1} \frac{dx}{x}$$

Therefore, $S = (k/2\pi)\log_e(r1/r)$ thermal ohms per meter length of the cable.

The thermal resistance of lead sheath is small and is generally neglected in calculations.

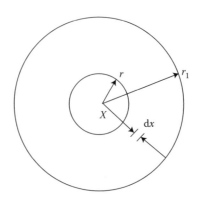

FIGURE 10.26   Cross-sectional view of single-core cable.

## 10.17 TYPES OF CABLE FAULTS

Cables are generally laid directly in the ground or in ducts in the underground distribution system. For this reason, there are little chances of faults in underground cables. However, if a fault does occur, it is difficult to locate and repair the fault because conductors are not visible. Nevertheless, the following are the faults most likely to occur in underground cables.

1. *Open-circuit fault.* When there is a break in the conductor of a cable, it is called open-circuit fault. The open-circuit fault can be checked by a megger. For this purpose, the three conductors of the three-core cable at the far end are shorted and earthed. Then resistance between each conductor and earth is measured by a megger. The megger will indicate zero resistance in the circuit of the conductor that is not broken. However, if the conductor is broken, the megger will indicate infinite resistance in its circuit.

2. *Short-circuit fault.* When two conductors of a multi-core cable come in electrical contact with each other due to insulation failure, it is called a short-circuit fault. Again, we can seek the help of a megger to check this fault. For this purpose, the two terminals of the megger are connected to any two conductors. If the megger gives zero reading, it indicates short-circuit fault between these conductors. The same step is repeated for other conductors taking two at a time.

3. *Earth fault.* When the conductor of a cable comes in contact with earth, it is called earth fault or ground fault. To identify this fault, one terminal of the megger is connected to the conductor and the other terminal connected to earth. If the megger indicates zero reading, it means the conductor is earthed. The same procedure is repeated for other conductors of the cable.

## 10.18 LOOP TESTS FOR LOCATION OF FAULTS IN UNDERGROUND CABLES

There are several methods for locating the faults in underground cables. However, two popular methods known as loop tests are

1. Murray loop test

2. Varley loop test

These simple tests can be used to locate the earth fault or short-circuit fault in underground cables provided that a sound cable runs along the faulty cable. Both these tests employ the principle of Wheatstone bridge for fault location.

### 10.18.1 Murray Loop Test

The Murray loop test is the most common and accurate method of locating earth fault or short-circuit fault in underground cables.

#### 10.18.1.1 Earth Fault

Figure 10.27 shows the circuit diagram for locating the earth fault by Murray loop test. Here AB is the sound cable and CD is the faulty cable; the earth fault occurring at point F. The far end D of the faulty cable is joined to the far end B of the sound cable through a low resistance link. Two variable resistances $P$ and $Q$ are joined to ends A and C, respectively, and serve as the ratio arms of the Wheatstone bridge.

Let $R$ be resistance of the conductor loop up to the fault from the test end, and $X$, resistance of the other length of the loop.

Note that $P$, $Q$, $R$, and $X$ are the four arms of the Wheatstone bridge. The resistances $P$ and $Q$ are varied till the galvanometer indicates zero deflection.

In the balanced position of the bridge, we have

$$\frac{P}{Q} = \frac{R}{X}$$

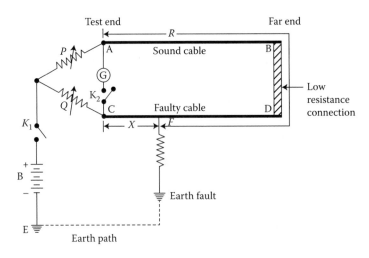

FIGURE 10.27   Circuit diagram for locating the earth fault by Murray loop test.

or

$$\frac{P}{Q} + 1 = \frac{R}{X} + 1$$

or

$$\frac{P + Q}{Q} = \frac{R + X}{X}$$

If $r$ is the resistance of each cable, then $R + X = 2$

$$\therefore \frac{P + Q}{Q} = \frac{2r}{X}$$

or

$$X = \frac{Q}{P + Q} \times 2r$$

If $l$ is the length of each cable in meters, then resistance per meter length of cable = $r/1$.

Therefore, distance of fault point from test end is

$$d = \frac{X}{(r/l)} = \frac{Q}{P + Q} \times 2r \times \frac{1}{r} = \frac{Q}{P + Q} \times 2l$$

or

$$d = \frac{Q}{P + Q} \times (\text{Loop length}) \text{ meters}$$

Thus the position of the fault is located. Note that resistance of the fault is in the battery circuit and not in the bridge circuit. Therefore, fault resistance does not affect the balancing of the bridge. However, if the fault resistance is high, the sensitivity of the bridge is reduced.

### 10.18.1.2 Short-Circuit Fault

Figure 10.28 shows the circuit diagram for locating the short-circuit fault by Murray loop test. Again $P$, $Q$, $R$, and $X$ are the four arms of the bridge. Note that fault resistance is in the battery circuit and not in the bridge circuit. The bridge in balanced by adjusting the resistances $P$ and $Q$. In the balanced position of the bridge:

$$\frac{P}{Q} = \frac{R}{X}$$

or

$$\frac{P+Q}{Q} = \frac{R+X}{X} = \frac{2r}{X}$$

or

$$\therefore X = \frac{Q}{P+Q} \times 2r$$

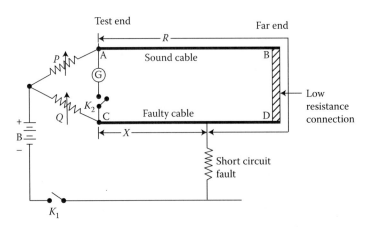

FIGURE 10.28 Circuit diagram for locating the short-circuit fault by Murray loop test.

$$X = \frac{Q}{P + Q} \times (\text{Loop length}) \text{ meters}$$

Thus, the position of the fault is located.

### 10.18.2 Varley Loop Test

The Varley loop test is also used to locate earth fault or short-circuit fault in underground cables. This test also employs Wheatstone bridge principle. It differs from Murray loop test in that here the ratio arms $P$ and $Q$ are fixed resistances. Balance is obtained by adjusting the variable resistance $S$ connected to the test end of the faulty cable. The connection diagrams for locating the earth fault and short-circuit fault by Varley loop test are shown in Figure 10.29 and Figure 10.30, respectively.

For earth fault or short-circuit fault, the key $S_1$ is first thrown to position 1. The variable resistance $S$ is varied till the bridge is balanced for resistance value of $S_1$. Then,

$$\frac{P}{Q} = \frac{R}{X + S_1}$$

or

$$\frac{P + Q}{Q} = \frac{R + X + S_1}{X + S_1}$$

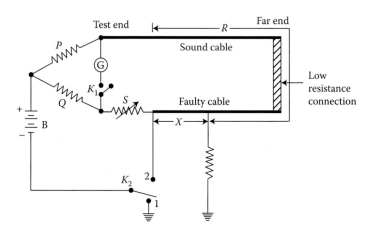

FIGURE 10.29   Circuit diagram for locating the earth fault by Varley loop test.

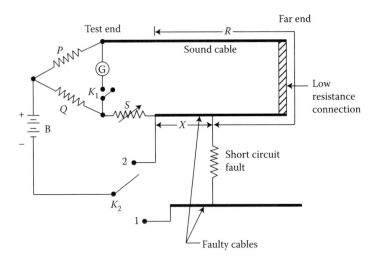

FIGURE 10.30   Circuit diagram for locating the short-circuit fault by Varley loop test.

or

$$X = \frac{Q(R + X) - PS_1}{P + Q} \qquad (10.5)$$

Now, key $K_2$ is thrown to position 2 (for earth fault or short-circuit fault) and bridge is balanced with new value of resistance $S_2$. Then,

$$\frac{P}{Q} = \frac{R + X}{S_2}$$

or

$$(R + X)Q = PS_2 \qquad (10.6)$$

From Equations 10.5 and 10.6, we get

$$X = \frac{P(S_2 - S_1)}{P + Q}$$

Since the values of $P$, $Q$, $S_1$, and $S_2$ are known, the value of $X$ can be determined.

$$\text{Loop resistance} = R + X = \frac{P}{Q}S_2$$

If $r$ is the resistance of the cable per meter length, then distance of fault from the test end is

$$d = \frac{X}{r}\,\text{m}$$

## WORKED EXAMPLES

### EXAMPLE 10.1

The insulation resistance of a single-core cable is 500 MΩ/km. If the core diameter is 3 cm and resistivity of insulation is $3.5 \times 10^{14}$ Ω cm, find the insulation thickness.

**Solution**

Length of cable, $l = 1$ km $= 1000$ m
　　Cable insulation resistance, $R = 500$ MΩ $= 500 \times 10^6$ Ω
　　Conductor radius, $r_1 = (3/2) = 1.5$ cm
　　Resistivity of insulation, $\rho = 3.5 \times 10^{14}$ Ω cm $= 3.5 \times 10^{12}$ Ω m
　　Let $r_2$ centimeter be the internal sheath radius.

Now, $R = \dfrac{\rho}{2\pi l}\log_e\dfrac{r_2}{r_1}$

or

$$\log_e\frac{r_2}{r_1} = \frac{2\pi lR}{\rho} = \frac{2\pi \times 1000 \times 500 \times 10^6}{3.5 \times 10^{12}} = 0.897$$

or

$$2.3\log_{10}\frac{r_2}{r_1} = 0.897$$

or

$$\frac{r_2}{r_1} = \text{Antilog}\,\frac{0.897}{2.3} = 2.45$$

or

$$r_2 = 2.45r_1 = 2.45 \times 1.5 = 3.68 \text{ cm}$$

$\therefore$ Insulation thickness $= R_2 - r_1 = 3.68{-}1.5 = 2.18$ cm.

## EXAMPLE 10.2

A single core cable for use on 11 kV, 50 Hz system has conductor area of 0.5 cm² and internal diameter of sheath is 3 cm. The permittivity of the dielectric used in the cable is 3. Find (1) the maximum electrostatic stress in the cable, (2) minimum electrostatic stress in the cable, (3) capacitance of the cable per kilometer length, and (4) charging current.

**Solution**

Area of cross-section of conductor, $a = 0.5$ cm²
   Diameter of the conductor,

$$d = \sqrt{\frac{4a}{\pi}} = \sqrt{\frac{4 \times 0.5}{\pi}} = 0.798 \text{ cm}$$

Internal diameter of sheath, $D = 3$ cm.

1. Maximum electrostatic stress in the cable is

$$g_{max} = \frac{2V}{d\log_e(D/d)} = \frac{2 \times 11}{0.798\log_e(3/0.798)} = 20.82 \text{ kV/cm rms}$$

2. Minimum electrostatic stress in the cable is

$$g_{min} = \frac{2V}{D\log_e(D/d)} = \frac{2 \times 11}{3\log_e(3/0.798)} = 5.53 \text{ kV/cm rms}$$

3. Capacitance of cable, $C = \dfrac{\varepsilon_r l}{41.4\log_{10}(D/d)} \times 10^{-9}$ F
   Here $\varepsilon_r = 3$, $l = 1$ km $= 1000$ m

$$C = \frac{3 \times 1000}{41.4\log_{10}(3/0.798)} \times 10^{-9} = 0.125 \times 10^{-6} \text{ F} = 0.125 \text{ μF}$$

4. Charging current, $I_c = (V/X_C) = 2\pi fCV = 2\pi \times 50 \times 0.125$
$\times 10^{-6} \times 11{,}000 = 0.431$ A

## EXAMPLE 10.3

Find the most economical size of a single-core cable working on a 66 kV, three-phase system, if a dielectric stress of 40 kV/cm can be allowed.

### Solution

Phase voltage of cable $= (66/\sqrt{3}) = 38.10$ kV

Peak value of phase voltage, $V = 38.10 \times \sqrt{2} = 53.88$ kV
Maximum permissible stress, $g_{max} = 40$ kV/cm
Therefore, most economical conductor diameter is

$$d = \frac{2V}{g_{max}} = \frac{2 \times 53.88}{40} = 2.694 \text{ cm}$$

Internal diameter of sheath, $D = 2.718\,d = 2.718 \times 2.694 = 7.32$ cm.
Therefore, the cable should have a conductor diameter of 2.694 cm and internal sheath diameter of 7.32 cm.

## EXAMPLE 10.4

A single core cable employing three layers of insulation with dielectric constant $E_{r_1} = 8$, $E_{r_2} = 6$, and $E_{r_3} = 4$, respectively, has conductor of radius 3 cm. Assuming that all the three insulating materials are worked at a same maximum potential gradient, work out the potential difference in kV between core and earthed sheath. The inner radius of the sheath is 3 cm and the maximum potential gradient is 50 kV/cm.

### Solution

Diameter of conductor $= 2 \times 1.5 = 3$ cm
The diameter over the insulation, $D = 2 \times 3.0 = 6$ cm
Maximum potential gradient, $g_{max} = 50$ kV/cm
Let the diameters over the insulation of relative permittivities 8 and 6 be $d_1$ and $d_2$, respectively.
As the maximum stress in three dielectric is same

$$E_{r_1}d = E_{r_2}d_1 = E_{r_3}d_2$$

or

$$d_1 = \frac{E_{r_1}}{E_{r_2}} d = \frac{8}{6} \times 3 = 4 \text{ cm}$$

and

$$d_2 = \frac{E_{r_1}}{E_{r_3}} d = \frac{8}{4} \times 3 = 6 \text{ cm}$$

Permissible peak voltage for the cable is

$$\frac{g_{\max}}{2} \left[ d \log_e^{(d_1/d)} + d_1 \log_e^{(d_2/d_1)} + d_2 \log_e^{(D/d_2)} \right]$$

$$\frac{50}{2} \left[ 3 \log_e^{(4/3)} + 4 \log_e^{(6/4)} + 6 \log_e^{(6/6)} \right] = 62.12 \text{ kV}$$

So working voltage (rms) for the cable = $(62.12/\sqrt{2})$ = 43.92 kV.

### EXAMPLE 10.5

A 33 kV, single core metal sheathed cable is to be guarded by means of a metallic inter sheath. Calculate the diameter of inter sheath and the voltage at which it must be maintained in order to obtain minimum overall cable diameter. The maximum voltage gradient at which the insulating material can be worked is 40 kV/cm.

### Solution

RMS value of cable voltage, $V_{\max}$ = 33 kV

Peak value of cable voltage = $33 \times \sqrt{2}$ = 46.67 kV

Maximum permissible potential gradient of dielectric, $g_{\max}$ = 40 kV/cm

Let $d$ be the conductor diameter, $d_1$, the outer diameter of the insulating layer, and $D$, the outer diameter of outer insulating layer.

If $V_1$ = potential difference between core and sheath and $V_2$ = potential difference between inter sheath and the outer sheath, then

$$g_{max} = \frac{V_1}{(d/2)\log_e(d_1/d)} = \frac{V_2}{(d_1/2)\log_e(D/d_1)}$$

For minimum overall diameter of the cable, conductor diameter, $d$, is

$$d = \frac{2V_{max}}{e \times g_{max}} = \frac{2 \times 46.67}{2.71828 \times 40} = 0.858 \text{ cm}$$

Diameter of inter sheath $= e \times d = 2.71828 \times 0.858 = 2.33$ cm. Voltage between conductor and inter sheath is

$$V_1 = \frac{V}{e} = \frac{46.67}{2.71828} = 17.17 \text{ kV}$$

Voltage between the inter sheath and sheath overall is

$$V - V_1 = 46.67 - 17.17 = 29.50 \text{ kV}$$

## EXERCISES

1. Compare the merits and demerits of underground system versus overhead system.

2. With a neat diagram, show the various parts of a high voltage single-core cable.

3. What should be the desirable characteristics of insulating materials used in cables?

4. Describe briefly some commonly used insulating materials for cables.

5. What is the most general criterion for the classification of cables? Draw the sketch of a single-core, low-tension cable and label the various parts.

6. Draw a neat sketch of the cross-section of the following:

   a. Three-core belted cable

b. H-type cable

c. SL-type cable

7. What are the limitations of solid-type cables? How are these overcome in pressure cables?

8. Write a brief note on oil-filled cables.

9. Describe the various methods of laying underground cables. What are the relative advantages and disadvantages of each method?

10. Derive an expression for the insulation resistance of a single-core cable.

11. Deduce an expression for the capacitance of a single-core cable.

12. Show that maximum stress in a single-core cable is

$$\frac{2V}{d\log_e(D/d)}$$

where $V$ is the operating voltage, and $d$ and $D$ are the conductor and sheath diameter.

13. Prove that $g_{max}/g_{min}$ in a single-core cable is equal to $D/d$.

14. Find an expression for the most economical conductor size of a single core cable.

15. Explain the following methods of cable grading:

a. Capacitance grading

b. Intersheath grading

16. Write short notes on the following:

a. Laying of 11 kV underground power cable

b. Capacitance grading in cables

c. Capacitance of three-core belted cables

17. Derive an expression for the thermal resistance of dielectric of a single-core cable.

18. What do you mean by permissible current loading of an underground cable?

19. With a neat diagram, describe Murray loop test for the location of (i) earth fault and (ii) short-circuit fault in an underground cable.

20. Describe Varley loop test for the location of earth fault and short-circuit fault in an underground cable.

# CHAPTER 11

# Distribution Systems

## 11.1 INTRODUCTION

The electrical energy produced at the generating station is conveyed to the consumers through a network of transmission and distribution systems. It is often difficult to draw a line between the transmission and distribution systems of a large power system. The transmission and distribution systems are similar to a human circulatory system. The transmitting system may be compared to arteries in the human body and distribution system with capillaries. They serve the same purpose of supplying the ultimate consumer in the city with the life giving blood of civilization—electricity.

That part of power system which distributes electric power for local use is known as distribution system.

In the general distribution system is the electric system between the substation fed by the transmitting system and the consumers' meters. It mostly consists of feeders, distributors, and the service main. Figure 11.1 shows the single-line diagram of a typical low-tension distribution system. Good voltage regulation of a distribution network is probably the most significant factor responsible for delivering expert service to the consumers. For this purpose, design of feeders and distributors requires careful consideration.

1. *Feeders.* A feeder is a conductor which connects the substation to the area where power is to be distributed. Generally no tapings are taken from the feeder so that current in it remains the same throughout. A feeder is designed from the point of view of its current

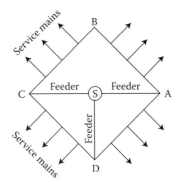

FIGURE 11.1    Typical distribution system.

carrying capacity while the voltage drop consideration is relatively unimportant. It is because voltage drop in a feeder can be compensated by means of voltage regulating equipment at the substation.

2. *Distributor.* A distributor is a conductor from which tapings are taken for supply to the consumers. In Figure 11.1, AB, BC, CD, and DA are the distributors. The current through a distributor is not constant because tapings are taken from various places along its length. A distributor is designed from the point of view of the voltage drop in it. It is because a distributor supplies power to the consumers and there is a statutory limit of voltage variations at the consumer's terminals (±6% of rated value). The size and length of the distributor should be such that voltage at the consumer's terminals is within the permissible limits.

3. *Service mains.* A service main is generally a small cable which connects the distributor to the consumer's terminal.

## 11.2  CLASSIFICATION OF DISTRIBUTION SYSTEMS

A distribution system may be classified in various ways:

1. According to nature of current, the distribution systems may be classified as

    a.  DC distribution systems

    b.  AC distribution systems

2. According to the type of construction, the distribution systems may be classified as

   a. Overhead systems

   b. Underground systems

3. According to the scheme of connection, the distribution systems may be classified as

   a. Radial systems

   b. Ring main systems

   c. Interconnected systems

4. According to the number of wires, the distribution systems may be classified as

   a. Two-wire systems

   b. Three-wire systems

   c. Four-wire systems

5. According to the character of service, the distribution systems may be classified as

   a. General light and power

   b. Industrial power

   c. Railway

   d. Street lighting

## 11.3 DC DISTRIBUTION

It is a common knowledge that electric power is almost exclusively generated, transmitted, and distributed as AC. However, for certain applications, DC supply is absolutely necessary. For instance, DC supply is required for the operation of variable speed machinery (i.e., DC motors), for electrochemical work and for congested areas where storage battery reserves are necessary. For this purpose, AC power is converted into DC power at the substation by using converting machinery, for example, mercury arc rectifiers, rotary converters, and motor-generator sets. The DC supply from the substation may be obtained in the form of (a) two wires or (b) three wires for distribution.

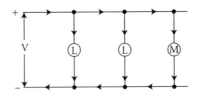

FIGURE 11.2   Two-wire DC distribution system.

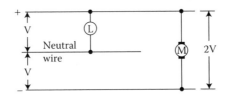

FIGURE 11.3   Three-wire DC distribution system.

### 11.3.1  Two-Wire DC System

This system of distribution consists of two wires. One is the outgoing or positive wire and the other is the return or negative wire. The loads such as lamps, motors, etc., are connected in parallel between the two wires as shown in Figure 11.2. This system is never used for transmission purposes due to low efficiency but may be employed for distribution of DC power.

### 11.3.2  Three-Wire DC System

It consists of two outer and a middle or neutral wire which is earthed at the substation. The voltage of the outer is twice the voltage between outer and neutral wire, as shown in Figure 11.3. The principal advantage of this system is that it makes available two voltage at the consumer terminals viz., $V$ between any outer and the neutral and $2V$ at the outer. Loads requiring high voltage (e.g., motors) are connected across the outer, whereas lamps and heating circuits requiring less voltage are connected between outer and neutral.

## 11.4  CONNECTION SCHEMES OF DISTRIBUTION SYSTEM

All distribution of electrical energy is done by constant voltage system. In practice, the following distribution circuits are generally used.

### 11.4.1  Radial System

In this system, separate feeders radiate from a single substation and feed the distributors at one end only. Figure 11.4a shows a single-line diagram of a

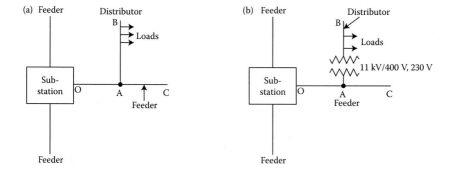

FIGURE 11.4 (a) Radial DC distribution system. (b) Radial AC distribution system.

radial system for DC distribution, where a feeder OC supplies a distributor AB at point A. Obviously, the distributor is fed at one end only, that is, point A is this case. Figure 11.4b shows a single-line diagram of radial system for AC distribution. The radial system is employed only when power is generated at low voltage, and the substation is located at the center of the load.

This is the simplest distribution circuit and has the lowest initial cost. However, it suffers from the following drawbacks:

1. The end of the distributor nearest to the feeding point will be heavily loaded.

2. The consumers are dependent on a single feeder and single distributor. Therefore, any fault on the feeder or distributor cuts off supply to the consumers who are on the side of the fault away from the substation.

3. The consumers at the distant end of the distributor would be subjected to serious voltage fluctuations when the load on the distributor changes.

## 11.4.2 Ring Main System

In this system, the primaries of distribution transformers form a loop. The loop circuit starts from the substation bus bars, makes a loop through the area to be served, and returns to the substation. Figure 11.5 shows the single-line diagram of ring main system for AC distribution, where substation supplies to the closed feeder LMNOPQRS. The distributors are tapped from different points M, O, and Q of the feeder through distribution transformers.

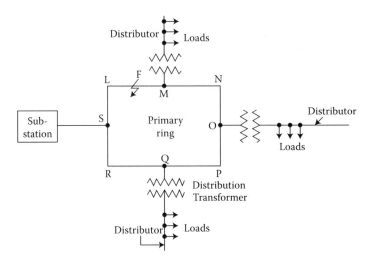

FIGURE 11.5    Ring main AC distribution system.

*Advantages*:

1. There are less voltage fluctuations at consumer's terminals.

2. The system is very reliable as each distributor is fed via two feeders. In the event of fault on any section of the feeder, the continuity of supply is maintained. For example, suppose that fault occurs at any point F of section SLM of the feeder. Then section SLM of the feeder can be isolated for repairs and at the same time continuity of supply is maintained to all the consumers via the feeder SRQPONM.

### 11.4.3  Interconnected System

When the feeder ring is energized by two or more than two generating stations or substations, it is called interconnected system. Figure 11.6 shows the single-line diagram of interconnected system where the closed feeder ring ABCD is supplied by two substations S1 and S2 at points D and C, respectively. Distributors are connected to points O, P, Q, and R of the feeder ring through distribution transformers.

*Advantages*:

1. It increases the service reliability.

2. Any area fed from one generating station during peak load hours can be fed from the other generating station. This reduces reserve power capacity and increases efficiency of the system.

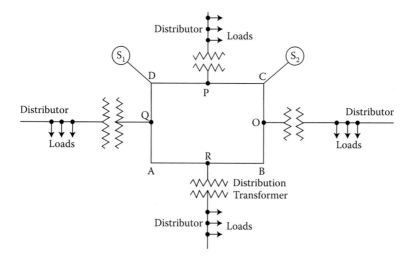

FIGURE 11.6   Interconnected system.

## 11.5  TYPES OF DC DISTRIBUTORS

The most general method of classifying DC distributors is the way they are fed by the feeders. On this basis, DC distributors are classified as

1. Distributor fed at one end

2. Distributor fed at both ends

3. Distributor fed at the center

4. Ring distributor

### 11.5.1  Distributor Fed at One End

In this type of feeding, the distributor is connected to the supply at one end, and loads are taken at different points along the length of the distributor. Figure 11.7 shows the single-line diagram of a DC distributor AB fed at the end A (also known as singly fed distributor) and loads $I_1$, $I_2$, and $I_3$ tapped off at points C, D, and E, respectively.

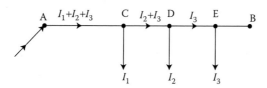

FIGURE 11.7   Distributor fed at one end.

The following points are worth noting in a singly fed distributor:

1. The current in the various sections of the distributor away from feeding point goes on decreasing. Thus current in section AC is more than the current in section CD and current in section CD is more than the current in section DE.

2. The voltage across the loads away from the feeding point goes on decreasing. Thus in Figure 11.7, the minimum voltage occurs at the load point E.

3. In case a fault occurs on any section of the distributor, the whole distributor will have to be disconnected from the supply mains.

### 11.5.2 Distributor Fed at Both Ends

In this type of feeding, the distributor is connected to the supply mains at both ends and loads are tapped off at different points along the length of the distributor. The voltage at the feeding points may or may not be equal. Figure 11.8 shows a distributor AB fed at the ends A and B and loads of $I_1$, $I_2$, and $I_3$ tapped off at points C, D, and E, respectively. Here, the load voltage goes on decreasing as we move away from one feeding point say A, reaches minimum value, and then again starts rising and reaches maximum value when we reach the other feeding point B. The minimum voltage occurs at some load point and is never fixed. It is shifted with the variation of load on different sections of the distributor.

*Advantages*:

1. If a fault occurs on any feeding point of the distributor, the continuity of supply is maintained from the other feeding point.

2. In case of fault on any section of the distributor, the continuity of supply is maintained from the other feeding point.

3. The area of cross-section required for a doubly fed distributor is much less than that of a singly fed distributor.

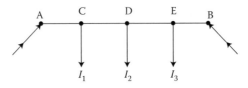

FIGURE 11.8   Distributor fed at both ends.

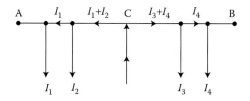

FIGURE 11.9   Distributor fed at the center.

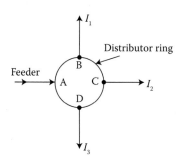

FIGURE 11.10   Ring mains distributor.

### 11.5.3 Distributor Fed at the Center

In this type of feeding, the center of the distributor is connected to the supply mains as shown in Figure 11.9. It is equivalent to two singly fed distributors, each distributor having a common feeding point and length equal to half of the total length.

### 11.5.4 Ring Mains

In this type, the distributor is in the form of a closed ring as shown in Figure 11.10. It is equivalent to a straight distributor fed at both ends with equal voltage, the two ends being brought together to form a closed ring. The distributor ring may be fed at one or more than one point.

## 11.6 DC DISTRIBUTION CALCULATIONS

In addition to the methods of feeding discussed above, a distributor may have

1. Concentrated loading

2. Uniform loading

3. Both concentrated and uniform loadings

The concentrated loads are those which act on particular points of the distributor. A common example of such loads is that tapped off for domestic use. On the other hand, distributed loads are those which act uniformly on all points of the distributor. Ideally, there are no distributed loads. However, a nearest example of distributed load is a large number of loads of same wattage connected to the distributor at equal distances.

In DC distribution calculations, one important point of interest is the determination of point of minimum potential on the distributor. The point where it occurs depends upon the loading conditions and the method of feeding the distributor. The distributor is so designed that the minimum potential on it is not less than 6% of rated voltage at the consumer's terminals.

## 11.7 DC DISTRIBUTOR FED AT ONE END— CONCENTRATED LOADING

Figure 11.11 shows the single-line diagram of a two-wire DC distributor AB fed at one end A and having concentrated loads $I_1$, $I_2$, $I_3$, and $I_4$ tapped off at points C, D, E, and F, respectively.

Let $r_1$, $r_2$, $r_3$, and $r_4$ be the resistances of both wires (go and return) of the sections AC, CD, DE, and EF of the distributor, respectively.

Current fed from point $A = I_1 + I_2 + I_3 + I_4$

Current in section $AC = I_1 + I_2 + I_3 + I_4$

Current in section $CD = I_2 + I_3 + I_4$

Current in section $DE = I_1 I_3 + I_4$

Current in section $EF = I_4$

Voltage drop in section $AC = r_1(I_1 + I_2 + I_3 + I_4)$

Voltage drop in section $CD = r_2(I_2 + I_3 + I_4)$

Voltage drop in section $DE = r_3(I_3 + I_4)$

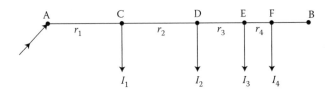

FIGURE 11.11   DC distributor fed at one end with concentrated loading.

Voltage drop in section EF $= r_4 I_4$

∴ Total voltage drop in the distributor

$$= r_1(I_1 + I_2 + I_3 + I_4) + r_2(I_2 + I_3 + I_4) + r_3(I_3 + I_4) + r_4 I_4$$

It is easy to see that the minimum potential will occur at point F, which is farthest from the feeding point A.

## 11.8 DC DISTRIBUTOR FED AT ONE END—UNIFORMLY LOADED

Figure 11.12 shows the single-line diagram of a two-wire DC distributor AB fed at one end A and loaded uniformly with $i$ amperes per meter length. It means that at every 1 m length of the distributor, the load tapped is $i$ amperes. Let $l$ meters be the length of the distributor and $r$ ohm be the resistance per meter run.

Consider a point C on the distributor at a distance $x$ meters from the feeding point A as shown in Figure 11.13. Then current at point C is

$$\text{Current} = il - ix \, \text{ampere} = i(1 - x) \, \text{ampere}$$

Now, consider a section of very small length $dx$ near point C. Its resistance is $r \, dx$ and the voltage drop over length $dx$ is,

$$dv = i(l - x)r \, dx = ir(l - x) \, dx$$

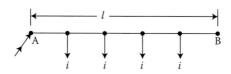

FIGURE 11.12   DC distributor fed at one end with uniform loading.

FIGURE 11.13   Point C on the distributor at a distance $x$ meters from the feeding point A.

Total voltage drop in the distributor up to point C is

$$v = \int_0^x ir(l-x)\,dx = ir\left(lx - \frac{x^2}{2}\right)$$

The voltage drop up to point B (i.e., over the whole distributor) can be obtained by putting $x = l$ in the above expression.

∴ Voltage drop over the distributor AB

$$= ir\left(l \times l - \frac{l^2}{2}\right)$$

$$= \frac{1}{2}irl^2 = \frac{1}{2}(il)(rl) = \frac{1}{2}IR$$

where $il = I$, the total current entering at point A, and $rl = R$, the total resistance of the distributor.

Thus, in a uniformly loaded distributor fed at one end, the total voltage drop is equal to that produced by the whole of the load assumed to be concentrated at the middle point.

## 11.9 DISTRIBUTOR FED AT BOTH ENDS—CONCENTRATED LOADING

The two ends of the distributor may be supplied with (1) equal voltage and (2) unequal voltage.

### 11.9.1 Two Ends Fed with Equal Voltage

Consider a distributor AB fed at both ends with equal voltage $V$ volts and having concentrated loads $I_1$, $I_2$, $I_3$, $I_4$, and $I_5$ at points C, D, E, F, and G, respectively, as shown in Figure 11.14. As we move away from one of the feeding points, say A, voltage goes on decreasing till it reaches the minimum value at some load point, say E, and then again starts rising and becomes $V$

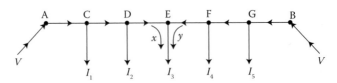

FIGURE 11.14 Distributor fed at both ends at equal voltage with concentrated loading.

volts as we reach the other feeding point B. All the current tapped off between points A and E (minimum pd point) will be supplied from the feeding point A, while those tapped off between B and E will be supplied from the feeding point B. The current tapped off at point E itself will be partly supplied from A and partly from B. If these current are $x$ and $y$, respectively, then

$$I_3 = x + y$$

Therefore, we arrive at a very important conclusion that at the point of minimum potential, current comes from both ends of the distributor.

### 11.9.1.1 Point of Minimum Potential

It is generally desired to locate the point of minimum potential. There is a simple method for it. Consider a distributor AB having three concentrated loads $I_1$, $I_2$, and $I_3$ at points C, D, and E, respectively. Suppose that current supplied by feeding end A is $I_A$. Then current distribution in the various sections of the distributor can be worked out as shown in Figure 11.15a. Thus

$$I_{AC} = I_A, \qquad I_{CD} = I_A - I_1$$
$$I_{DE} = I_A - I_1 - I_2, \qquad I_{EB} = I_A - I_1 - I_2 - I_3$$

Voltage drop between A and B = Voltage drop over AB

or

$$V - V = I_A R_{AC} + (I_A - I_1)R_{CD} + (I_A - I_1 - I_2)R_{DE}$$
$$+ (I_A - I_1 - I_2 - I_3)R_{EB}$$

From this equation, the unknown $I_A$ calculated as the values of other quantities are generally given. Suppose actual directions of current in the various sections of the distributor are indicated in Figure 11.15b. The load point where the current are coming from both sides of the distributor is the point of minimum potential, that is, point E in this case.

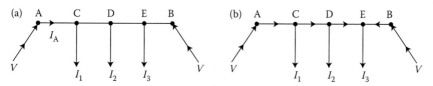

FIGURE 11.15　(a) Current distribution in the various sections of the distributor. (b) Actual directions of current in the various sections of the distributor.

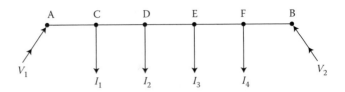

FIGURE 11.16   Distributor fed at both ends at unequal voltage with concentrated loading.

### 11.9.2 Two Ends Fed with Unequal Voltage

Figure 11.16 shows the distributor AB fed with unequal voltage; end A being fed at $V_1$ volts and end B at $V_2$ volts. The point of minimum potential can be found by following the same procedure as discussed above.

$$\text{Voltage drop between A and B} = \text{Voltage drop over AB}$$

$$\text{or } V_1 - V_2 = \text{Voltage drop over AB}$$

## 11.10  DISTRIBUTOR FED AT BOTH ENDS—UNIFORMLY LOADED

### 11.10.1  Distributor Fed at Both Ends with Equal Voltage

Consider a distributor AB of length $l$ meters, having resistance $r$ ohms per meter run and with uniform loading of $i$ amperes per meter run as shown in Figure 11.17. Let the distributor be fed at the feeding points A and B at equal voltage, say $V$ volts. The total current supplied to the distributor is $il$. As the two end voltage are equal, current supplied from each feeding point is $il/2$.

Consider a point C at a distance $x$ meters from the feeding point A, then current at point C is

$$\frac{il}{2} - ix = i\left(\frac{l}{2} - x\right)$$

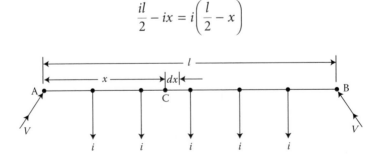

FIGURE 11.17   Distributor fed at both ends at equal voltage with uniform distributed loading.

Now, consider a small length d$x$ near point C. Its resistance is $r \, dx$ and the voltage drop over length d$x$ is

$$dv = i\left(\frac{l}{2} - x\right)rdx = ir\left(\frac{l}{2} - x\right)$$

Voltage drop up to point C is

$$\int_0^x ir\left(\frac{l}{2} - x\right)dx = ir\left(\frac{lx}{2} - \frac{x^2}{2}\right) = \frac{ir}{2}(lx - x^2)$$

Obviously, the point of minimum potential will be the midpoint. Therefore, maximum voltage drop will occur at midpoint, that is, where $x = 1/2$. Hence,

$$\text{Max. voltage drop} = \frac{ir}{2}(lx - x^2)$$

$$= \frac{ir}{2}\left(1 \times \frac{1}{2} - \frac{1^2}{4}\right) \text{[Putting } x = 1/2]$$

$$= \frac{1}{8}irl^2 = \frac{1}{8}(il)(rl) = \frac{1}{8}IR$$

where $il = I$, the total current fed to the distributor from both ends, and $rl = R$, the total resistance of the distributor.

$$\text{Minimum voltage} = V - \frac{IR}{8}\text{volts}$$

### 11.10.2 Distributor Fed at Both Ends with Unequal Voltage

Consider a distributor AB of length $l$ meters having resistance $r$ ohms per meter run and with a uniform loading of $i$ amperes per meter run as shown in Figure 11.18. Let the distributor be fed from feeding points A and B at voltage $V_A$ and $V_B$, respectively.

Suppose that the point of minimum potential C is situated at a distance $x$ meters from the feeding point A. Then current supplied by the feeding point A will be $ix$.

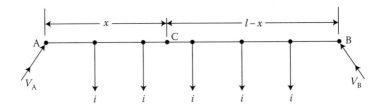

FIGURE 11.18 Distributor fed at both ends at unequal voltage with uniform distributed loading.

$$\text{Voltage drop in section AC} = \frac{irx^2}{2} \text{ volts}$$

As the distance of C from feeding point B is $(l - x)$, current fed from B is $i(l - x)$.

$$\text{Voltage drop in section BC} = \frac{ir(l - x)^2}{2} \text{ volts}$$

$$\text{Voltage at point C,} \quad V_C = V_A - \text{Drop over AC}$$

$$= V_A - \frac{irx^2}{2} \tag{11.1}$$

Also, voltage at point C,

$$V_C = V_B - \text{Drop over BC}$$

$$= V_B - \frac{ir(l - x)^2}{2} \tag{11.2}$$

From Equations 11.1 and 11.2, we get

$$V_A - \frac{irx^2}{2} = V_B - \frac{ir(l - x)^2}{2}$$

Solving the equation for $x$, we get

$$x = \frac{V_A - V_B}{irl} + \frac{1}{2}$$

As all the quantities on the right hand side of the equation are known, the point on the distributor where minimum potential occurs can be calculated.

## 11.11  RING DISTRIBUTOR

A distributor arranged to form a closed loop and fed at one or more points is called a ring distributor. Such a distributor starts from one point, makes a loop through the area to be served, and returns to the original point. For the purpose of calculating voltage distribution, the distributor can be considered as consisting of a series of open distributors fed at both ends. The principal advantage of ring distributor is that by proper choice in the number of feeding points, great economy in copper can be affected.

## 11.12  RING MAIN DISTRIBUTOR WITH INTERCONNECTOR

Sometimes a ring distributor has to serve a large area. In such a case, voltage drops in the various sections of the distributor may become excessive. In order to reduce voltage drops in various sections, distant points of the distributor are joined through a conductor called interconnector. Figure 11.19 shows the ring distributor ABCDEA. The points B and D of the ring distributor are joined through an interconnector BD. There are several methods for solving such a network. However, the solution of such a network can be readily obtained by applying Thevenin's theorem. The steps of procedure are as follows:

1. Consider the interconnector BD to be disconnected (Figure 11.20a) and find the potential difference between B and D. This gives Thevenin's equivalent circuit voltage $E_0$.

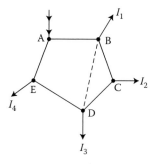

FIGURE 11.19   Ring distributor ABCDEA (with interconnector).

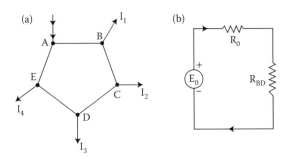

FIGURE 11.20 (a) Ring distributor ABCDEA (without interconnector). (b) Thevenin's equivalent circuit.

2. Next, calculate the resistance viewed from points B and D of the network composed of distribution lines only. This gives Thevenin's equivalent circuit series resistance $R_0$.

3. If $R_{BD}$ is the resistance of the interconnector BD, then Thevenin's equivalent circuit will be as shown in Figure 11.20b.

$$\text{Current in interconnector BD} = \frac{E_0}{R_0 + R_{BD}}$$

Therefore, current distribution in each section and the voltage of load points can be calculated.

## 11.13 AC DISTRIBUTION

One important reason for the widespread use of alternating current in preference to direct current is the fact that alternating voltage can be conveniently changed in magnitude by means of a transformer.

It can be classified as follows.

### 11.13.1 Primary Distribution System

It is that part of AC distribution system which operates at voltage somewhat higher than general utilization and handles large blocks of electrical energy than the average low-voltage consumer uses. The voltage used for primary distribution depends upon the amount of power to be conveyed and the distance of the substation required to be fed. The most commonly used primary distribution voltage are 11, 6.6, and 3.3 kV. Due to economic

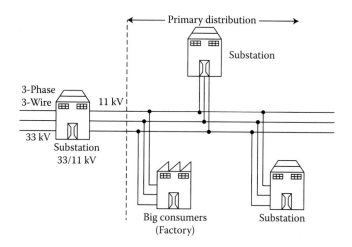

FIGURE 11.21   Three-phase, three-wire primary distribution system.

considerations, primary distribution is carried out by three-phase, three-wire system.

Figure 11.21 shows a typical primary distribution system. Electric power from the generating station is transmitted at high voltage to the substation located in or near the city. At this substation, voltage is stepped down to 11 kV with the help of step-down transformer. Power is supplied to various substations for distribution or to big consumers at this voltage.

### 11.13.2  Secondary Distribution System

It is that part of AC distribution system which includes the range of voltage at which the ultimate consumer utilizes the electrical energy delivered to consumer. The secondary distribution employs 400/230 V, three-phase, four-wire system.

Figure 11.22 shows a typical secondary distribution system. The primary distribution circuit delivers power to various substations, called distribution substations. The substations are situated near the consumers' localities and contain step down transformers. At each distribution substation, the voltage is stepped down to 400 V and power is delivered by three-phase, four-wire AC system. The voltage between any two phases is 400 V and between any phase and neutral is 230 V. The single-phase domestic loads are connected between any one phase and the neutral, whereas three-phase, 400-V motor loads are connected across three-phase lines directly.

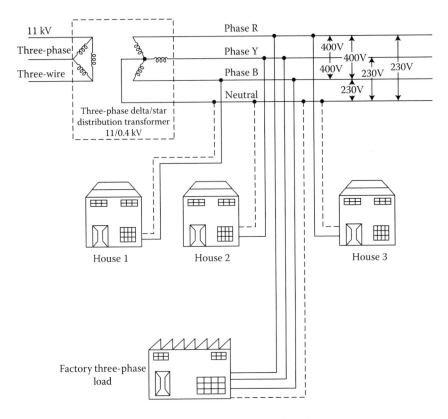

FIGURE 11.22   Three-phase, four-wire secondary distribution system.

## 11.14 AC DISTRIBUTION CALCULATION

AC distribution calculations differ from those of DC distribution in the following respects:

1. In case of DC system, the voltage drop is due to resistance alone. However, in AC system, the voltage drops are due to the combined effects of resistance, inductance, and capacitance.

2. In a DC system, additions and subtractions of current or voltage are done arithmetically but in case of AC system, these operations are done vectorially.

3. In an AC system, power factor has to be taken into account. Loads tapped off from the distributor are generally at different power factors. There are two ways of referring power factor viz.

a. It may be referred to supply or receiving end voltage which is regarded as the reference vector.

b. It may be referred to the voltage at the load point itself.

## 11.15 METHODS OF SOLVING AC DISTRIBUTION PROBLEMS

As mentioned above, there are two ways of referring power factors.

### 11.15.1 Power Factors Referred to Receiving End Voltage

Consider an AC distributor AB with concentrated loads of $I_1$ and $I_2$ tapped off at points C and B as shown in Figure 11.23. Taking the receiving end voltage $V_B$ as the reference vector, let lagging power factors at C and B be $\cos \phi_1$ and $\cos \phi_2$ wrt $V_B$. Let $R_1$, $X_1$, and $R_2$, $X_2$ be the resistance and reactance of sections AC and CB of the distributor.

$$\text{Impedance of section AC,} \quad \mathbf{Z_{AC}} = R_1 + jX_1$$

$$\text{Impedance of section CB,} \quad \mathbf{Z_{CB}} = R_2 + jX_2$$

$$\text{Load current at point C,} \quad \mathbf{I_1} = I_1(\cos\phi_1 - j\sin\phi_1)$$

$$\text{Load current at point B,} \quad \mathbf{I_2} = I_2(\cos\phi_2 - j\sin\phi_2)$$

$$\text{Current in section CB,} \quad \mathbf{I_{CB}} = \mathbf{I_2} = I_2(\cos\phi_2 - j\sin\phi_2)$$

$$\text{Current in section AC,} \quad \mathbf{I_{AC}} = \mathbf{I_1} + \mathbf{I_2}$$
$$= I_1(\cos\phi_1 - j\sin\phi_1) + I_2(\cos\phi_2 - j\sin\phi_2)$$

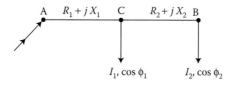

FIGURE 11.23  AC distributor.

Voltage drop in section CB, $\mathbf{V_{CB}} = \mathbf{I_{CB}Z_{CB}} = I_2(\cos\phi_2 - j\sin\phi_2)$
$$(R_2 + jx_2)$$

Voltage drop in section AC, $\mathbf{V_{AC}} = \mathbf{I_{AC}Z_{AC}} = (\mathbf{I_1} + \mathbf{I_2})\mathbf{Z_{AC}}$
$$= I_1(\cos\phi_1 - j\sin\phi_1) + I_2(\cos\phi_2$$
$$-j\sin\phi_2)(R_1 + jx_1)$$

Sending end voltage, $\quad \mathbf{V_A} = \mathbf{V_B} + \mathbf{V_{CB}} + \mathbf{V_{AC}}$

Sending end current, $\quad \mathbf{I_A} = \mathbf{I_1} + \mathbf{I_2}$

The vector diagram of the AC distributor under this condition is shown in Figure 11.24. Here, the receiving end voltage $V_B$ is taken as the reference vector. As power factors of loads are given wrt $V_B$, $I_1$ and $I_2$ lag behind $V_B$ by $\phi_1$ and $\phi_2$, respectively.

## 11.15.2 Power Factors Referred to Respective Load Voltage

Suppose the power factors of loads in the previous Figure 11.23 are referred to their respective load voltage. Then $\phi_1$ is the phase angle between $V_C$ and $I_1$ and $\phi_2$ is the phase angle between $V_B$ and $I_2$. The vector diagram of the AC distributor under this condition is shown in Figure 11.25.

Voltage drop in section CB $= \mathbf{I_2Z_{CB}} = I_2(\cos\phi_2 - j\sin\phi_2)(R_2 + jx_2)$

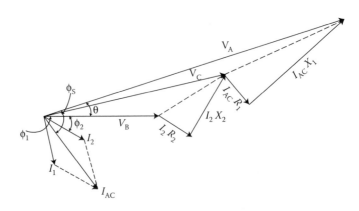

FIGURE 11.24  Phasor diagram of Figure 11.23 for pf with respect to receiving end.

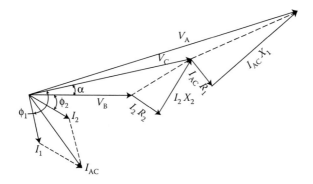

FIGURE 11.25  Phasor diagram of Figure 11.23 but with pf with respect to load points.

Voltage at point C = $\mathbf{V_B}$ + Drop in section CB = $V_C \angle \propto$ (say)

$$\mathbf{I_1} = I_1 \angle - \phi_1 \quad \text{wrt voltage } V_C$$

$$\mathbf{I_1} = I_1 \angle - (\phi_1 - \propto) \quad \text{wrt voltage } V_B$$

$$\mathbf{I_1} = I_1[\cos(\phi_1 - \propto) - j\sin\phi_1 - \propto)]$$

Now, $\mathbf{I_{AC}} = \mathbf{I_1} + \mathbf{I_2}$
$$= I_1[\cos(\phi_1 - \propto) - j\sin(\phi_1 - \propto)] + I_2(\cos\phi_2 - j\sin\phi_2)$$

Voltage drop in section AC = $\mathbf{I_{AC}}\mathbf{Z_{AC}}$

Voltage at point A = $V_B$ + Drop in CB + Drop in AC

## WORKED EXAMPLES

### EXAMPLE 11.1

The load distribution on a two-wire DC distributor is shown in Figure 11.26. The cross-sectional area of each conductor is 0.36 cm². The end A is supplied at 230 V. Resistivity of the wire is ρ = 1.97 μΩ cm. Calculate (1) the current in each section of the conductor, (2) the two-core resistance of each section, and (3) the voltage at each tapping point.

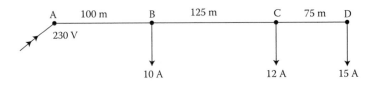

FIGURE 11.26   Load distribution on a two-wire DC distributor.

**Solution**

1. Current in the various sections are

$$\text{Section CD,} \quad I_{CD} = 15 \text{ A}$$
$$\text{Section BC,} \quad I_{BC} = 15 + 12 = 27 \text{ A}$$
$$\text{Section AB,} \quad I_{AB} = 15 + 12 + 10 = 37 \text{ A.}$$

2. Single-core resistance of the section of 100 m length is

$$\rho \frac{l}{a} = 1.97 \times 10^{-6} \times \frac{100 \times 100}{0.36} = 0.054 \text{ } \Omega$$

The resistances of the various sections are

$$R_{AB} = 0.054 \times 1 \times 2 = 0.108 \text{ } \Omega$$
$$R_{BC} = 0.054 \times 1.25 \times 2 = 0.135 \text{ } \Omega$$
$$R_{CD} = 0.054 \times 0.75 \times 2 = 0.081 \text{ } \Omega$$

3. Voltage at tapping point B is

$$V_B = V_A - I_{AB} \times R_{AB} = 230 - (37 \times 0.108) = 226 \text{ V}$$

Voltage at tapping point C is

$$V_C = V_B - I_{BC} \times R_{BC} = 226 - (27 \times 0.135) = 222.35 \text{ V}$$

Voltage at tapping point D is

$$V_D = V_C - I_{CD} \times R_{CD} = 222.35 - (15 \times 0.081) = 221.14 \text{ V}$$

**EXAMPLE 11.2**

Calculate the voltage at a distance of 250 m of a 350 m long distributor uniformly loaded at the rate of 1 A/m. The distributor is fed at one end at 230 V. The resistance of the distributor (go and return) per meter is 0.00018 Ω. Also find the power loss in the distributor.

**Solution**

Voltage drop at a distance $x$ from supply end is

$$ir\left(lx - \frac{x^2}{2}\right)$$

Here $i = 1$ A/m, $l = 350$ m, $x = 250$ m, $r = 0.00018$ Ω/m. Therefore,

$$\text{Voltage drop} = 1 \times 0.00018 \times \left(350 \times 250 - \frac{250^2}{2}\right) = 10.125 \text{ V}$$

Voltage at a distance of 250 m from supply end $= 230 - 10.125 = 219.875$ V.

Power loss in the distributor is

$$P = \frac{i^2 r l^3}{3} = \frac{1^2 \times 0.00018 \times 350^3}{3} = 2572.5 \text{ W}$$

## EXAMPLE 11.3

A ring distributor is supplied through a feeder AB and is loaded as shown below: calculate the cross-section volume of copper. Assume that the maximum voltage drop from A to the point of minimum potential is 15 V. Take $\rho = 1.73$ μΩ/cm³ (Figure 11.27).

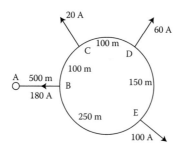

FIGURE 11.27  A ring distributor is supplied through a feeder AB.

**Solution**

Let the resistance per 100 m of ring distributor length (both conductors) be $r$ ohms.

Resistance of both conductors of various section is determined as below:

$$R_{BC} = \frac{r}{100} \times 100 = r\,\Omega$$

Similarly,

$$R_{CD} = \frac{r}{100} \times 100 = r\,\Omega$$

$$R_{DE} = \frac{r}{100} \times 150 = 1.5r\,\Omega$$

and

$$R_{EB} = \frac{r}{100} \times 250 = 2.5r\,\Omega$$

Let current $I_x$ be flowing in section BC, then current in sections CD, DE, and EB will be $(I_x - 20)$, $(I_x - 80)$, and $(I_x - 180)$ amperes, respectively.

According to Kirchoff's second law voltage drop in this closed loop will be equal to zero. So,

$$I_{BC}R_{BC} + I_{CD}R_{CD} + I_{DE}R_{DE} + I_{EB}R_{EB} = 0$$

or

$$I_x r + (I_x - 20)r + (I_x - 80)1.5r + (I_x - 180)2.5r = 0$$

or

$$6I_x r = 590 \quad \text{or} \quad I_x = 98.333\,\text{A}$$

Hence current in section BC, CD, DE, and EB will be equal to 98.333, 78.333, 18.333, and −81.667 A, respectively.

Hence, $E$ is the point of minimum potential.

Let $\alpha_F$ and $\alpha_D$ be the cross-sectional areas of the feeder and distributor, respectively, resistance of section BE for both go and return:

$$\frac{1.73 \times 10^{-6} \times 250 \times 2 \times 100}{\alpha_D} = \frac{0.0865}{\alpha_D} \, \Omega$$

Resistance of feeder AB for both go and return:

$$\frac{1.73 \times 10^{-6} \times 500 \times 2 \times 100}{\alpha_F} = \frac{0.173}{\alpha_F} \, \Omega$$

Maximum voltage drop = Voltage drop in feeder + Voltage drop in distributor

$$= v_F + v_D = 15 \text{ V}$$

and

$$v_F = \frac{180 \times 0.173}{\alpha_F} \tag{11.3}$$

and

$$v_D = \frac{81.667 \times 0.0865}{\alpha_D} \tag{11.4}$$

Volume of copper used in feeder $= 500 \times 100 \times 2 \times \alpha_F \text{ cm}^3$

$$= 500 \times 100 \times 2 \times \frac{180 \times 0.173}{v_F}$$

$$= \frac{3,114,000}{v_F} = \frac{3,114,000}{15 - v_D} \tag{11.5}$$

Volume of copper used in distributor $= 600 \times 100 \times 2 \times \alpha_D$

$$= \frac{120,000 \times 81.667 \times 0.08}{v_D}$$

$$= \frac{847,700}{v_D} \text{cm}^3 \qquad (11.6)$$

Total volume of copper required,

$$v_{OC} = \frac{3,114,000}{15 - v_D} + \frac{847,700}{v_D}$$

For volume of copper to be minimum $d(v_{OC})/dv_D$ should be zero. That is,

$$\frac{3,114,000}{(15 - v_D)^2} - \frac{847,700}{v_D^2} = 0$$

or

$$2.2663\, v_D^2 + 25.431\, v_D - 190.7325 = 0$$

or

$$v_D = \frac{-25.431 \pm \sqrt{25.431^2 - 4 \times 2.2663 \times 190.7325}}{2 \times 2.2663}$$

$$= 5.143 \text{ V}$$

$v_P = 15 - 5.143 = 9.857$ V. Neglecting negative sign giving higher value.

Substituting the value of $v_D$ and $v_F$ in expressions 11.3 and 11.4, we get

$$\alpha_F = \frac{180 \times 0.173}{v_F} = \frac{180 \times 0.173}{9.857} = 3.16 \text{ cm}^2$$

and

$$\alpha_D = \frac{81.667 \times 0.0865}{v_D} = \frac{81.667 \times 0.0865}{5.143} = 1.3736 \, \text{cm}^2$$

## EXAMPLE 11.4

A two-wire DC street mains AB, 600 m long is fed from both ends at 230 V. Loads of 20, 40, 50, and 30 A are tapped at distances of 100, 250, 400, and 500 m from the end A, respectively. If the area of cross-section of distributor conductor is 1 cm², find the minimum consumer voltage. Take $\rho = 1.7 \times 10^{-6} \, \Omega$ cm.

### Solution

Figure 11.28 shows the distributor with its tapped current. Let $I_A$ amperes be the current supplied from the feeding end A. Then current in the various sections of the distributor are as shown in Figure 11.28.

Resistance of 1 m length of distributor,

$$2 \times \frac{1.7 \times 10^{-6} \times 100}{1} = 3.4 \times 10^{-4} \Omega$$

Resistance of section AC,  $R_{AC} = (3.4 \times 10^{-4}) \times 100 = 0.034 \, \Omega$

Resistance of section CD,  $R_{CD} = (3.4 \times 10^{-4}) \times 150 = 0.051 \, \Omega$

Resistance of section DE,  $R_{DE} = (3.4 \times 10^{-4}) \times 150 = 0.051 \, \Omega$

Resistance of section EF,  $R_{EF} = (3.4 \times 10^{-4}) \times 100 = 0.034 \, \Omega$

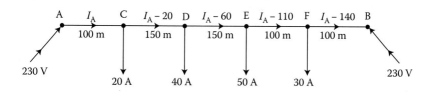

FIGURE 11.28   A distributor with its tapped currents.

Resistance of section FB, $R_{FB} = (3.4 \times 10^{-4}) \times 100 = 0.034 \, \Omega$

Voltage at B = Voltage at A − Drop over length AB

or

$$V_B = V_A - [I_A R_{AC} + (I_A - 20)R_{CD} + (I_A - 60)R_{DE}]$$
$$+(I_A - 110)R_{EF} + (I_A - 140)R_{FB}$$

or

$$230 = 230 - [0.034I_A + 0.051(I_A - 20) + 0.051(I_A - 60)]$$
$$+0.034(I_A - 110) + 0.034(I_A - 140)$$

or

$$230 = 230 - [0.204I_A - 12.58]$$

or

$$0.204I_A = 12.58$$

or

$$I_A = \frac{12.58}{0.204} = 61.7 \, A$$

The actual distribution of current in the various sections of the distributor is shown in Figure 11.28. It is clear that current are coming to load point E from both sides, that is, from point D and point F. Hence, E is the point of minimum potential.

Therefore, minimum consumer voltage is

$$V_E = V_A - [I_{AC}R_{AC} + I_{CD}R_{CD} + I_{DE}R_{DE}]$$
$$= 220 - [61.7 \times 0.034 + 41.7 \times 0.051 + 1.7 \times 0.051]$$
$$= 230 - 4.31$$
$$= 225.69 \, V$$

## EXAMPLE 11.5

A two-wire DC distributor AB is fed from both ends. At feeding point A, the voltage is maintained as at 220 V and at B 225 V. The total length of the distributor is 200 m and loads are tapped off as under: 25 A at 50 m from A, 50 A at 75 m from A, 30 A at 100 m from A, and 40 A at 150 m from A. The resistance per kilometer of one conductor is 0.3 Ω. Calculate:

a. Current in various sections of the distributor
b. Minimum voltage and the point at which it occurs

**Solution:**

Figure 11.29 shows the distributor with its tapped current. Let $I_A$ amperes be the current supplied from the feeding point A. Then current in the various sections of the distributor are as shown in Figure 11.29.

Resistance of 1000 m length of distributor (both wires):

$$2 \times 0.3 = 0.6\,\Omega$$

Resistance of section AC, $\quad R_{AC} = 0.6 \times \dfrac{50}{1000} = 0.03\,\Omega$

Resistance of section CD, $\quad R_{CD} = 0.6 \times \dfrac{5}{1000} = 0.015\,\Omega$

Resistance of section DE, $\quad R_{DE} = 0.6 \times \dfrac{5}{1000} = 0.015\,\Omega$

Resistance of section EF, $\quad R_{EF} = 0.6 \times \dfrac{50}{1000} = 0.03\,\Omega$

Resistance of section FB, $\quad R_{FB} = 0.6 \times \dfrac{50}{1000} = 0.03\,\Omega$

Voltage at B = Voltage at A − Drop over length AB

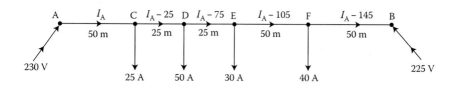

FIGURE 11.29   A two-wire DC distributor AB is fed from both ends.

or

$$V_B = V_A - [I_A R_{AC} + (I_A - 25)R_{CD} + (I_A - 75)R_{DE}]$$
$$+ (I_A - 105)R_{EF} + (I_A - 145)R_{FB}$$

or

$$225 = 220 - [0.03I_A + 0.015(I_A - 25) + 0.015(I_A - 75)]$$
$$+ 0.03(I_A - 105) + 0.03(I_A - 145)$$

or

$$225 = 220 - [0.12I_A - 9]$$

or

$$I_A = \frac{229 - 225}{0.012} = 33.34 \ \Omega$$

Therefore, current in section AC is

$$I_{AC} = I_A = 33.34 \ A$$

Current in section CD,   $I_{CD} = I_A - 25 = 33.34 - 25 = 8.34 \ A$

Current in section DE,   $I_{DE} = I_A - 75 = 33.34 - 75$
$$= -41.66 \ A \text{ from D to E}$$
$$= 41.66 \ A \text{ from E to D}$$

Current in section EF,   $I_{EF} = I_A - 105 = 33.34 - 105$
$$= -71.66 \ A \text{ from E to F}$$
$$= 71.66 \ A \text{ from F to E}$$

Current in section FB,   $I_{FB} = I_A - 145 = 33.34 - 145$
$$= -111.66 \ A \text{ from F to B}$$
$$= 111.66 \ A \text{ from B to F}$$

The current are coming to load point D from both sides of the distributor. Therefore, load point D is the point of minimum potential.

$$\text{Voltage at D,} \quad V_D = V_A - [I_A R_{AC} + I_{CD} R_{CD}]$$
$$= 220 - [33.34 \times 0.03 + 8.34 \times 0.015]$$
$$= 220 - 1.125 = 218.875 \text{ V}$$

## EXAMPLE 11.6

A two-wire DC distributor cable 900 m long is loaded with 0.6 A/m. Resistance of each conductor is 0.08 $\Omega$/km. Calculate the maximum voltage drop if the distributor is fed from both ends with equal voltage of 230 V. What is the minimum voltage and where it occurs?

### Solution

Current loading, $i = 0.6$ A/m
Resistance of distributor/m, $r = 2 \times 0.08/1000 = 0.16 \times 10^{-3}$ $\Omega$
Length of distributor, $l = 900$ m
Total current supplied by distributor, $I = il = 0.6 \times 900 = 540$ A
Total resistance of the distributor, $R = rl = 0.16 \times 10^{-3} \times 900 = 0.1442$ $\Omega$

$$\therefore \text{Maximum voltage drop} = \frac{IR}{8} = \frac{540 \times 0.1442}{8} = 9.72 \text{ V}$$

Minimum voltage will occur at the midpoint of the distributor and its value is

$$230 - 9.72 = 220.28 \text{ V}$$

## EXAMPLE 11.7

A 800-m, two-wire DC distributor AB fed from both ends is uniformly loaded at the rate of 1.25 A/m run. Calculate the voltage at the feeding points A and B if the minimum potential of 220 V occurs at point C at a distance of 450 m from the end A. Resistance of each conductor is 0.05 Ω/km.

**Solution**

Figure 11.30 shows the single-line diagram of the distributor.

$$\text{Current loading,} \quad i = 1.25\,\text{A/m}$$

$$\text{Resistance of distributor/m,} \quad r = 2 \times \frac{0.05}{1000} = 0.0001\,\Omega$$

$$\text{Voltage at C,} \quad V_C = 220\,\text{V}$$

$$\text{Length of distributer,} \quad l = 800\,\text{m}$$

$$\text{Distance of point C from A,} \quad x = 450\,\text{m}$$

$$\text{Voltage drop in section AC} = \frac{irx^2}{2} = \frac{1.25 \times 0.0001 \times (450)^2}{2}$$
$$= 12.65\,\text{V}$$

$$\therefore \text{Voltage at feeding point A,} \quad V_A = 220 + 12.65 = 232.65\,\text{V}$$

$$\text{Voltage drop in section BC} = \frac{ir(l-x)^2}{2} = \frac{1.25 \times 0.0001 \times (800-450)^2}{2}$$
$$= 7.65\,\text{V}$$

$$\therefore \text{Voltage at feeding point B,} \quad V_B = 220 + 7.65 = 227.65\,\text{V}$$

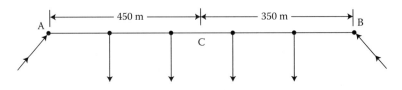

FIGURE 11.30  A two-wire DC distributor AB fed from both ends is uniformly loaded.

## EXAMPLE 11.8

A single-phase distributor 1 km long has resistance and reactance per conductor of 0.2 and 0.3 $\Omega$, respectively. At the far end, the voltage $V_B = 220$ V and the current is 150 A at a pf of 0.8 lagging. At the midpoint M of the distributor, a current of 150 A is tapped at a pf of 0.6 lagging with reference to the voltage $V_M$ at the midpoint. Calculate:

1. Voltage at midpoint
2. Sending end voltage
3. Phase angle between $V_A$ and $V_B$

**Solution**

Figure 11.31 shows the single-line diagram of the distributor AB with M as the midpoint.

Total impedance of distributor $= 2(0.2 + j0.3) = (0.4 + j0.6)\Omega$

Impedance of section AM, $\mathbf{Z_{AM}} = (0.2 + j0.3)\Omega$

Impedance of section MB, $\mathbf{Z_{MB}} = (0.2 + j0.3)\Omega$

Let the voltage $V_B$ at point B be taken as the reference vector. Then, $\mathbf{V_B} = 220 + j0$.

1. Load current at point B, $I_2 = 150 (0.8 - j\,0.6) = 120 - j\,90$

   Current in section MB, $\mathbf{I_{MB}} = \mathbf{I_2} = 120 - j90$

   Drop in section MB, $\mathbf{V_{MB}} = \mathbf{I_{MB}}\,\mathbf{Z_{MB}}$
   $$= (120 - j90) \times (0.2 + j0.3) = 51 + j18$$

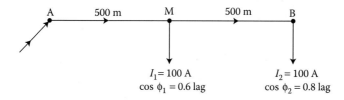

FIGURE 11.31　Single-line diagram of the distributor AB with M as the midpoint.

$\therefore$ Voltage at point M, $\mathbf{V_M} = \mathbf{V_B} + \mathbf{V_{MB}} = (220 + j0) + (51 + j18)$
$$= 271 + j18$$

Its magnitudes is

$$\sqrt{271^2 + 18^2} = 271.6 \text{ V.}$$

Phase angle between $V_M$ and $V_B$, $\alpha = \tan^{-1}\dfrac{18}{217} = \tan^{-1}0.0276 = 3.8°.$

2. The load current $\mathbf{I}_1$ has a lagging pf of 0.6 wrt $V_M$. It lags behind $V_M$ by an angle $\phi_1 = \cos^{-1}0.6 = 53.13°.$

$\therefore$ Phase angle between $\mathbf{I}_1$ and $\mathbf{V_B}$, $\phi'_1 = \phi_1 - \alpha = 53.13 - 3.8 = 49.33°$

Load current at M, $\mathbf{I}_1 = \mathbf{I}_1(\cos\phi'_1 - j\sin\phi'_1) = 100(\cos 49.33°$
$$- j\sin 49.33°)$$
$$= 97.75 - j113.77$$

Current in section AM, $\mathbf{I_{AM}} = \mathbf{I}_1 + \mathbf{I}_2 = (97.75 - j113.77) + (120 - j90)$
$$= 217.75 - j203.77$$

Drop in section AM, $\mathbf{V_{AM}} = \mathbf{I_{AM}}\,\mathbf{Z_{AM}} = (217.75 - j203.77)(0.2 + j0.3)$
$$= 104.681 + j24.571$$

Sending end voltage, $\mathbf{V_A} = \mathbf{V_M} + \mathbf{V_{AM}} = (271 + j18)$
$$+ (104.681 + j24.571)$$
$$= 375.68 + j42.571$$

Its magnitudes is

$$\sqrt{(375.681)^2 + (42.571)^2} = 378.08 \text{ V.}$$

3. The phase difference $\theta$ between $V_A$ and $V_B$ is given by:

$$\tan\theta = \frac{42.571}{375.681}$$

or

$$\theta = 6.46°$$

Hence, supply voltage is 378.08 V and leads $V_B$ by 6.46°.

## EXERCISES

1. Describe briefly the different types of DC distributors.

2. What are the advantages of a doubly fed distributor over singly fed distributor?

3. Derive an expression for the voltage drop for a uniformly loaded distributor fed at one end.

4. What is the purpose of interconnector in a DC ring main distributor?

5. Explain three-wire DC system of distribution of electrical power.

6. What are the advantages of three-wire distribution over two-wire distribution?

7. How does AC distribution differ from DC distribution?

8. What is the importance of load power factors in AC distribution?

9. Describe briefly how will you solve AC distribution problems?

# CHAPTER 12

# Fault Analysis

## 12.1 INTRODUCTION

Most faults on the power system occur when two or more conductors that normally operate with a potential difference come in contact with each other. These faults may be caused by sudden failure of a piece of equipment, accidental damage, short circuit to overhead lines, or by insulation failure resulting from lightning surges. When such a condition occurs, a large current (called short-circuit current) flows through the equipment, doing reasonable damage to the equipment and interruption of service to the consumer. The choice of apparatus and the design and arrangement of practically every equipment in the power system depends upon short-circuit current considerations.

## 12.2 CLASSIFICATION OF FAULTS

In power system, fault may be classified as follows:

1. *Unsymmetrical faults.* The faults on the power system which give rise to unsymmetrical fault current (i.e., unequal fault current in the lines with unequal phase displacement) are known as unsymmetrical faults.

   If an unsymmetrical fault occurs, the current in the three lines become unequal and so is the phase displacement among them. It may be noted that the term "unsymmetry" applies only to the fault itself and the resulting line current. However, the system impedances and the source voltage are always symmetrical through its main elements, namely generators, transmission lines, synchronous reactors, etc. There are three ways in which unsymmetrical faults may occur in a power system.

a.  Single line-to-ground fault (L–G)

b.  Line-to-line fault (L–L)

c.  Double line-to-ground fault (L–L–G)

The solution of unsymmetrical faults can be obtained by either (a) Kirchhoff's laws or (b) symmetrical component method.

Symmetrical component method is preferred because of the following reasons:

a.  It is a simple method and gives more generality to be given to fault performance studies.

b.  It provides an useful tool for the protection engineers, particularly in connection with tracing out of fault current.

2. *Symmetrical faults.* The fault on the power system which gives rise to symmetrical fault current (i.e., equal fault current in the line with 120° displacements) is called a symmetrical fault.

The symmetrical fault occurs when all the three conductors of a three-phase line are brought together simultaneously into a short-circuit condition. This type of fault gives rise to symmetrical current, that is, equal fault current with 120° displacement. Because of balanced nature of fault, only one phase need to be considered in calculations since condition in the other two phases will also be similar. The following points may be particularly noted:

a.  The symmetrical fault rarely occurs in practice as majority of the faults are of unsymmetrical nature. However, symmetrical fault calculations are being discussed in this chapter to enable the reader to understand the problems that short-circuit conditions present to the power system.

b.  The symmetrical fault is the most severe and imposes more heavy duty on the circuit breaker.

## 12.3 SYMMETRICAL COMPONENT METHOD

In 1918, Dr. C.L. Fortescue, an American scientist, showed that any unbalanced system of three-phase current (or voltage) may be regarded as being composed of three separate sets of balance vectors, that is,

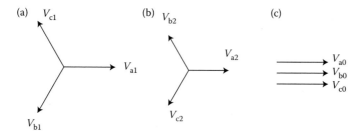

FIGURE 12.1   (a) Positive-sequence component. (b) Negative-sequence component. (c) Zero-sequence component.

1. Positive-sequence components

2. Negative-sequence components

3. Zero-sequence components

1. *Positive-sequence components.* These components are present in all kinds of fault. Positive-sequence components have three vectors of equal magnitude but displaced in phase from each other by 120° and have the same phase sequence as the original vectors.

2. *Negative-sequence components.* These components are present in all kinds of fault except three-phase fault. Negative sequence components have three vectors of equal magnitude but displaced in phase from each other by 120° and have the phase sequence opposite to the original vectors.

3. *Zero-sequence components.* These components have three vectors of equal magnitude and is also in phase with each other. Zero-sequence components present when the neutral of the system is grounded and the fault is also grounded.

The components have been shown in Figure 12.1. The voltage vectors have been designated as $V_a$, $V_b$, and $V_c$ and the phase sequence is assumed here as a, b, and c. The subscripts 1, 2, 0 are being used to represent positive, negative, and zero-sequence quantities, respectively.

## 12.4  SIGNIFICANCE OF POSITIVE-, NEGATIVE-, AND ZERO-SEQUENCE COMPONENTS

The vectors are equal in magnitude and 120° apart in phase, in which the time order of arrival of the phase vectors at a fixed axis of reference corresponds to the generated voltage. This really means that if a set of positive-sequence voltage is applied to the stator windings of the alternator, the direction of rotation of stator field is same as the rotor. On the other hand, if we fed negative-sequence voltage to the stator windings of the alternator, the direction of rotation of stator field is opposite to that of the rotor. The zero-sequence voltage is single-phase voltage and therefore, they give rise to an alternating field in space. Since the three-phase windings are 120° apart in space, at any particular instant, the three vector fields due to the three phases are 120° apart and therefore, assuming complete symmetry of the windings, the net flux in the air gap will be zero.

From Figure 12.1, the following relations between the original unbalanced vectors and their corresponding symmetrical components can be written as

$$V_a = V_{a1} + V_{a2} + V_{a0} \tag{12.1}$$

$$V_b = V_{b1} + V_{b2} + V_{b0} \tag{12.2}$$

$$V_c = V_{c1} + V_{c2} + V_{c0} \tag{12.3}$$

## 12.5 OPERATOR (α)

It has a magnitude of unity and rotation through 120° magnitude, that is, when any vector is multiplied by α, the vector magnitude remains same but is rotated anticlockwise through 120°.

Assuming phase a as the reference as shown in Figure 12.1, the following relations between the symmetrical components of phases b and c in terms of phase a can be written with the use of operator. Thus

$$\alpha = 1 \angle 120°$$

In the complex form,

$$\alpha = \cos 120° + j \sin 120°$$
$$= -0.5 + \frac{j\sqrt{3}}{2}$$
$$= -0.5 + j0.866$$

$$\alpha^2 = \alpha \cdot \alpha$$
$$= 1\angle 120° \cdot 1\angle 120°$$
$$= 1\angle 240°$$
$$= \cos 240° + j\sin 240°$$
$$= -0.5 - j0.866$$

$$\alpha^3 = \alpha \cdot \alpha \cdot \alpha$$
$$= 1\angle 120° \cdot 1\angle 120° \cdot 1\angle 120°$$
$$= 1\angle 360°$$
$$= \cos 360° + j\sin 360°$$

Therefore,

$$\alpha^3 = 1$$

or

$$(\alpha^3 - 1) = 0$$

or

$$(\alpha - 1)(\alpha^2 + \alpha + 1) = 0$$

Since $\alpha \neq 1$ as $\alpha$ is a complex quantity as defined above,

$$\alpha^2 + \alpha + 1 = 0$$

Now we go back to deriving relations between the symmetrical components of phases b and c in terms of symmetrical components of phase a. From Figure 12.1,

$$V_{b1} = \alpha^2 V_{a1}$$

This means in order to express $V_{b1}$ in terms of $V_{a1}$, $V_{a1}$ should be rotated anticlockwise through 240°.

Similarly,

$$V_{c1} = -V_{a1}$$

For negative-sequence vectors,

$$V_{b2} = \alpha V_{a2} \quad \text{and} \quad V_{c2} = \alpha^2 V_{a2}$$

For zero-sequence vectors,

$$V_{a0} = V_{b0} = V_{c0}$$

and

$$V_a = V_{a1} + V_{a2} + V_{a0} \tag{12.4}$$

$$V_b = V_{b1} + V_{b2} + V_{b0} \tag{12.5}$$

$$V_c = V_{c1} + V_{c2} + V_{c0} \tag{12.6}$$

Substituting these relations in Equations 12.4 through 12.6,

$$V_a = V_{a1} + V_{a2} + V_{a0} \tag{12.7}$$

$$V_b = \alpha^2 V_{a1} + \alpha V_{a2} + V_{a0} \tag{12.8}$$

$$V_c = \alpha V_{a1} + \alpha^2 V_{a2} + V_{a0} \tag{12.9}$$

Equations 12.7 through 12.9 can be put in matrix form

$$\begin{bmatrix} V_a \\ V_b \\ V_c \end{bmatrix} = \begin{bmatrix} 1 & 1 & 1 \\ 1 & \alpha^2 & \alpha \\ 1 & \alpha & \alpha^2 \end{bmatrix} \begin{bmatrix} V_{a1} \\ V_{a2} \\ V_{a0} \end{bmatrix} \tag{12.10}$$

and

$$\begin{bmatrix} V_{a1} \\ V_{a2} \\ V_{a0} \end{bmatrix} = \begin{bmatrix} 1 & 1 & 1 \\ 1 & \alpha^2 & \alpha \\ 1 & \alpha & \alpha^2 \end{bmatrix}^{-1} \begin{bmatrix} V_a \\ V_b \\ V_c \end{bmatrix}$$

$$\therefore \begin{bmatrix} V_{a1} \\ V_{a2} \\ V_{a0} \end{bmatrix} = \frac{1}{3} \begin{bmatrix} 1 & \alpha & \alpha^2 \\ 1 & \alpha^2 & \alpha \\ 1 & 1 & 1 \end{bmatrix} \begin{bmatrix} V_a \\ V_b \\ V_c \end{bmatrix} \tag{12.11}$$

In terms of separate equations, we may write

$$V_{a1} = \frac{1}{3}(V_a + \alpha V_b + \alpha^2 V_c) \tag{12.12}$$

$$V_{a2} = \frac{1}{3}(V_a + \alpha^2 V_b + \alpha V_c) \tag{12.13}$$

$$V_{a0} = \frac{1}{3}(V_a + V_b + V_c) \tag{12.14}$$

Similarly, these relations for current are given as

$$I_{a1} = \frac{1}{3}(I_a + \alpha I_b + \alpha^2 I_c) \tag{12.15}$$

$$I_{a2} = \frac{1}{3}(I_a + \alpha^2 I_b + \alpha I_c) \tag{12.16}$$

$$I_{a0} = \frac{1}{3}(I_a + I_b + I_c) \tag{12.17}$$

In the equations above, $V_a$, $V_b$, and $V_c$ may be line-to-ground voltage, line-to-neutral voltage, line-to-line voltage at a point in the network or they may be the generated or induced voltage, in fact any set of three voltage revolving at the same rate which may exist in the three-phase system. Similarly three current could be phase current, line current, and the current following into a fault from the line conductors.

## 12.6 VOLTAGE OF THE NEUTRAL

The potential of the neutral, when it is grounded through some impedance or is isolated, will not be at ground potential under unbalanced conditions such as unsymmetrical faults. Potential of the neutral is given as $V_n = -I_n Z_n$, where $Z_n$ is the neutral grounding impedance and $I_n$ is the neutral current.

For a three-phase system,

$$
\begin{aligned}
I_n &= I_a + I_b + I_c \\
&= (I_{a1} + I_{a2} + I_{a0}) + (\alpha^2 I_{a1} + \alpha I_{a2} + I_{a0}) + (\alpha I_{a1} + \alpha^2 I_{a2} + I_{a0}) \\
&= I_{a1}(1 + \alpha + \alpha^2) + I_{a2}(1 + \alpha + \alpha^2) + 3I_{a0} \\
&= 3I_{a0} \quad \because (\alpha^2 + \alpha + 1) = 0
\end{aligned}
$$

$$ \therefore V_n = -3I_{a0}Z_n \tag{12.18} $$

Neutral current is three times that of the zero-sequence current. Here negative sign indicates the voltage of the neutral point is lower than the ground potential.

Since the positive- and negative-sequence components of current through the neutral are absent, the drop due to these current are also zero. In addition, for a balanced set of current or voltage, the neutral is at ground potential; therefore, for positive- and negative-sequence networks, neutral of the system will be taken as reference.

## 12.7 SEQUENCE NETWORK EQUATIONS

These equations will be delivered for an unloaded alternator with neutral solidly grounded (Figure 12.2), assuming that the system is balanced, that is, the generated voltage are of equal magnitude and displaced by 120°.

Since the sequence impedances per phase are same for all the three phases and we are considering initially a balanced system, the analysis will be done on single-phase basis.

1. *Positive-sequence network.* The positive-sequence component of voltage at the fault point is the positive-sequence-generated voltage

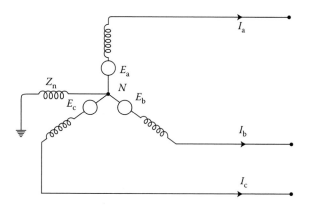

FIGURE 12.2    A balanced three-phase system.

minus the drop due to positive-sequence current in positive-sequence impedance (as positive-sequence current does not produce drop in negative and zero-sequence impedance). The corresponding sequence network for the unloaded alternator is shown in Figure 12.3.

$$V_{a1} = E_a - I_{a1}Z_1 \qquad (12.19)$$

2. *Negative-sequence network.* The negative-sequence component of voltage at the fault point is the generated negative-sequence voltage minus the drop due to negative-sequence current in negative-sequence impedance (as negative-sequence current does not produce

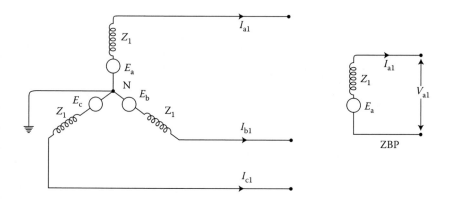

FIGURE 12.3    Positive-sequence network.

drop in positive- and zero-sequence impedance). The corresponding sequence network for the unloaded alternator is shown in Figure 12.4.

$$V_{a2} = E_{a2} - I_{a2}Z_2$$

The negative-sequence voltage generated is zero because the phase sequence of negative-sequence components is opposite to the actual phase sequence. Therefore,

$$V_{a2} = -I_{a2}Z_2 \qquad (12.20)$$

3. *Zero-sequence network.* For zero-sequence voltage,

$$E_{a0} = 0$$

$$V_{a0} = V_n - I_{a0}Z_{g0} = -3I_{a0}Z_n - I_{a0}Z_{g0} = -I_{a0}(Z_{g0} + 3Z_n)$$

where $Z_{g0}$ is the zero-sequence impedance of the generator and $Z_n$ is the neutral impedance.
  And

$$Z_0 = Z_{g0} + 3Z_n$$

$$V_{a0} = -I_{a0}Z_0 \qquad (12.21)$$

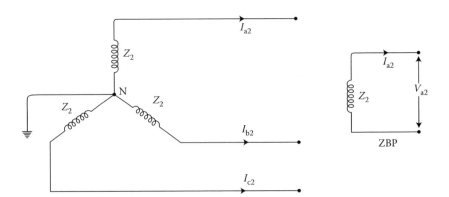

FIGURE 12.4   Negative-sequence network.

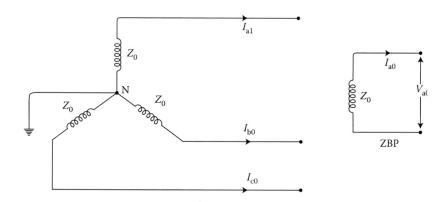

FIGURE 12.5   Zero-sequence network.

The corresponding sequence network for the unloaded alternator is shown in Figure 12.5.

## 12.8 SEQUENCE IMPEDANCES OF POWER SYSTEM ELEMENTS

The concept of impedances of various elements of power system (e.g., generators, transformers, transmission lines, etc.) to positive-, negative-, and zero-sequence current is of considerable importance in determining the fault current in a three-phase unbalanced system. A complete consideration of this topic does not fall within the scope of this book, but a short preliminary explanation may be of interest here. The following three main pieces of equipment will be considered:

1. Synchronous generators

2. Transformers

3. Transmission lines

1. *Synchronous generators.* The positive-, negative-, and zero-sequence impedances of rotating machines are generally different. The positive-sequence impedance of a synchronous generator is equal to the synchronous impedance of the machine. The negative-sequence impedance is much less than the positive-sequence impedance. The zero-sequence impedance is a variable item and if its value is not given, it may be assumed to be equal to the positive-sequence impedance. In short,

Negative-sequence impedance < Positive-sequence impedance

Zero-sequence impedance = Variable item

= May be taken equal to positive-sequence impedance if its value is not given

It may be worthwhile to mention here that any impedance Ze in the earth connection of a star-connected system has the effect to introduce an impedance of three Ze per phase. It is because the three equal zero-sequence current, being in phase, do not sum to zero at the star point, but they flow back along the neutral earth connection.

2. *Transformers.* Since transformers have the same impedance with reversed phase rotation, their positive- and negative-sequence impedances are equal; this value being equal to the impedance of the transformer. However, the zero-sequence impedance depends upon earth connection. If there is a through circuit for earth current, zero-sequence impedance will be equal to positive-sequence impedance otherwise it will be infinite. In short,

Positive-sequence impedance = Negative-sequence impedance

= Impedance of transformer

Zero-sequence impedance = Positive-sequence impedance, if there is circuit for earth current

= Infinite, if there is no through circuit for earth current

3. *Transmission lines.* The positive- and negative-sequence impedances of a line are the same; this value being equal to the normal impedance of the line. This is expected because the phase rotation of the current does not make any difference in the constants of the line. However, the zero-sequence impedance is usually much greater than the positive- or negative-sequence impedance.

In short,

Positive-sequence impedance = Negative-sequence impedance

= Impedance of the line

Zero-sequence impedance = Variable item

                   = May be taken as three times the positive-sequence impedance if its value is not given

## 12.9 ANALYSIS OF UNSYMMETRICAL FAULTS

In the analysis of unsymmetrical faults, the following assumptions will be made:

1. The generated electromotive force system is of positive sequence only.

2. No current flows in the network other than due to fault, that is, load current are neglected.

3. The impedance of the fault is zero.

4. Phase "a" shall be taken as reference phase.

## 12.10 SINGLE LINE-TO-GROUND FAULT (L–G)

Figure 12.6 represents a three-phase unloaded alternator.
    Let the fault take place on phase a. The boundary condition is

$$I_b = 0 \tag{12.22}$$

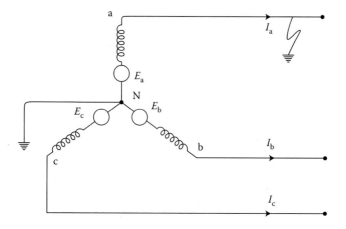

FIGURE 12.6  A solidly grounded, unloaded alternator: L–G fault on phase a.

$$I_c = 0 \tag{12.23}$$

$$V_a = 0 \tag{12.24}$$

The sequence network equations are

$$V_{a1} = E_a - I_{a1}Z_1 \tag{12.25}$$

$$V_{a2} = -I_{a2}Z_2 \tag{12.26}$$

$$V_{a0} = -I_{a0}Z_0 \tag{12.27}$$

The solution of these six equations will give six unknowns $V_{a1}$, $V_{a2}$, $V_{a0}$, and $I_{a1}$, $I_{a2}$, $I_{a0}$.

Also we know

$$I_{a1} = \frac{1}{3}(I_a + \alpha I_b + \alpha^2 I_c) \tag{12.28}$$

$$I_{a2} = \frac{1}{3}(I_a + \alpha^2 I_b + \alpha I_c) \tag{12.29}$$

$$I_{a0} = \frac{1}{3}(I_a + I_b + I_c) \tag{12.30}$$

Putting $I_b = 0$ and $I_c = 0$ in Equations 12.28 through 12.30, we get

$$I_{a1} = \frac{1}{3}I_a$$

$$I_{a2} = \frac{1}{3}I_a$$

$$I_{a0} = \frac{1}{3}I_a$$

Therefore, $I_{a1} = I_{a2} = I_{a0} = (1/3)I_a$.

Also we know

$$V_a = V_{a1} + V_{a2} + V_{a0}$$

Now substituting the values of $V_{a1}$, $V_{a2}$, $V_{a0}$ from the sequence network equation,

$$0 = E_a - I_{a1}Z_1 - I_{a2}Z_2 - I_{a0}Z_0 \quad\quad (12.31)$$

$$\because I_{a1} = I_{a2} = I_{a0}$$

Equation 12.31 becomes

$$E_a - I_{a1}(Z_1 + Z_2 + Z_0) = 0$$

$$I_{a1} = \frac{E_a}{(Z_1 + Z_2 + Z_0)} \qu\quad (12.32)$$

From Equation 12.32, it is clear that to simulate an L–G fault, all the three sequence networks are required, and since the current are all equal in magnitude and phase angle, therefore, the three sequence networks must be connected in series. The voltage across each sequence network corresponds to the same sequence component of $V_a$. The interconnection of sequence network for L–G fault is shown in Figure 12.7 (Table 12.1).

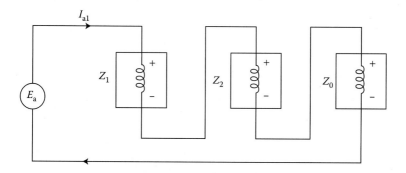

FIGURE 12.7   Interconnection of sequence network for L–G fault.

TABLE 12.1    Probability of Different Types of Fault

| Types of Fault | Probability |
| --- | --- |
| Single line-to-ground fault | 80%–90% |
| Line-to-line fault | 10%–15% |
| Double line-to-ground fault | 5%–10% |
| Three-phase fault | 1%–2% |

The assumption made in arriving at Equation 12.32 is that the fault impedance is zero. However, if the fault impedance is $Z_f$, then expression becomes

$$I_{a1} = \frac{E_a}{(Z_1 + Z_2 + Z_0 + 3Z_f)} \tag{12.33}$$

It may be added here that if the neutral is not grounded, then zero-sequence impedance will be infinite and the fault current is zero. This is expected because now no path exists for the flow of fault current.

## 12.11  LINE-TO-LINE FAULT

As shown in Figure 12.8, the line-to-line fault takes place on phases b and c. The boundary condition is

$$I_b + I_c = 0 \tag{12.34}$$

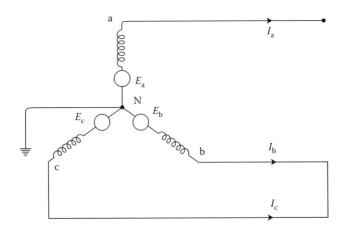

FIGURE 12.8    L–L fault on an unloaded and neutral grounded alternator.

$$I_a = 0 \tag{12.35}$$

$$V_b = V_c \tag{12.36}$$

The sequence network equations are

$$V_{a1} = E_a - I_{a1}Z_1 \tag{12.37}$$

$$V_{a2} = -I_{a2}Z_2 \tag{12.38}$$

$$V_{a0} = -I_{a0}Z_0 \tag{12.39}$$

The solution of these six equations will give six unknowns. Using the relations,

$$I_{a1} = \frac{1}{3}(I_a + \alpha I_b + \alpha^2 I_c)$$

$$I_{a2} = \frac{1}{3}(I_a + \alpha^2 I_b + \alpha I_c)$$

$$I_{a0} = \frac{1}{3}(I_a + I_b + I_c)$$

Substituting for $I_a$, $I_b$, and $I_c$,

$$I_{a1} = \frac{1}{3}(I_a + \alpha I_b + \alpha^2 I_c) = \frac{1}{3}(\alpha I_b - \alpha^2 I_b) = \frac{1}{3}I_b(\alpha - \alpha^2)$$

$$I_{a2} = \frac{1}{3}(0 + \alpha^2 I_b + \alpha I_c) = (\alpha^2 I_b - \alpha I_b) = \frac{1}{3}I_b(\alpha^2 - \alpha)$$

$$I_{a0} = \frac{1}{3}(0 + I_b - I_b) = 0$$

For line-to-line fault, the zero-sequence component of current is absent. Therefore,

$$I_{a1} = -I_{a2} \tag{12.40}$$

And we know

$$V_b = (\alpha^2 V_{a1} + \alpha V_{a2} + V_{a0})$$
$$V_c = \alpha V_{a1} + \alpha^2 V_{a2} + V_{a0}$$

Substituting these relation in Equation 12.36:

$$(\alpha^2 V_{a1} + \alpha V_{a2} + V_{a0}) = \alpha V_{a1} + \alpha^2 V_{a2} + V_{a0}$$

$$V_{a1}(\alpha^2 - \alpha) = V_{a2}(\alpha^2 - \alpha)$$

$$V_{a1} = V_{a2} \tag{12.41}$$

That is, positive-sequence component of voltage equals the negative-sequence component of voltage. This also means that the two sequence networks are connected in opposition.

Now making use of the sequence network equation and the Equation 12.41,

$$E_a - I_{a1}Z_1 = -I_{a2}Z_2$$

or

$$E_a - I_{a1}Z_1 = I_{a1}Z_2$$

$$I_{a1} = \frac{E_a}{Z_1 + Z_2} \tag{12.42}$$

Since $Z_0$ is absent, no zero-sequence component is available in L–L fault. The interconnection of sequence network for simulation of L–L fault is shown in Figure 12.9.

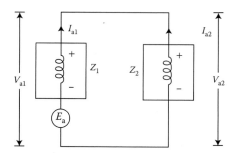

FIGURE 12.9   Interconnection of sequence network for L–L fault.

The assumption made in arriving at Equation 12.42 is that the fault impedance is zero. However, if the fault impedance is $Z_f$, then expression becomes:

$$I_{a1} = \frac{E_a}{Z_1 + Z_2 + Z_f}$$

$(12.43)$

## 12.12  DOUBLE LINE-TO-GROUND FAULT

As shown in Figure 12.10, the double line-to-ground fault takes place on phases b and c.

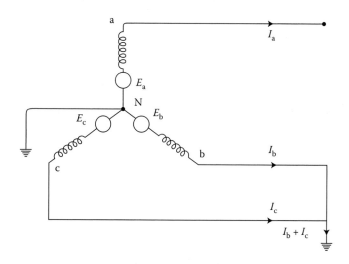

FIGURE 12.10   L–L–G fault on an unloaded and neutral grounded alternator.

The boundary condition is

$$I_a = 0 \qquad (12.44)$$

$$V_b = 0 \qquad (12.45)$$

$$V_c = 0 \qquad (12.46)$$

The sequence network equations are

$$V_{a1} = E_a - I_{a1}Z_1 \qquad (12.47)$$

$$V_{a2} = -I_{a2}Z_2 \qquad (12.48)$$

$$V_{a0} = -I_{a0}Z_0 \qquad (12.49)$$

The solution of these six equations will give six unknowns. We know

$$V_{a1} = \frac{1}{3}(V_a + \alpha V_b + \alpha^2 V_c)$$

$$V_{a2} = \frac{1}{3}(V_a + \alpha^2 V_b + \alpha V_c)$$

$$V_{a0} = \frac{1}{3}(V_a + V_b + V_c)$$

Substituting for $V_a$, $V_b$, and $V_c$ from Equations 12.45 and 12.46,

$$V_{a1} = \frac{1}{3}V_a$$

$$V_{a2} = \frac{1}{3}V_a$$

$$V_{a0} = \frac{1}{3}V_a$$

that is,

$$V_{a1} = V_{a2} = V_{a0} \qquad (12.50)$$

Using this relation of voltage and substituting in the sequence network equations:

$$V_{a1} = V_{a2}$$

$$E_a - I_{a1}Z_1 = -I_{a2}Z_2$$

$$I_{a2} = \frac{I_{a1}Z_1 - E_a}{Z_2} \qquad (12.51)$$

Similarly,

$$V_{a2} = V_{a0}$$

$$-I_{a2}Z_2 = -I_{a0}Z_0$$

$$I_{a0} = I_{a2}\frac{Z_2}{Z_0} \qquad (12.52)$$

Now from Equation 12.46,

$$I_a = (I_{a1} + I_{a2} + I_{a0}) = 0$$

Substituting the values of $I_{a2}$ and $I_{a0}$ from Equations 12.51 and 12.52,

$$I_{a1} + \frac{I_{a1}Z_1 - E_a}{Z_2} + \left(\frac{I_{a1}Z_1 - E_a}{Z_2}\right) \cdot \frac{Z_2}{Z_0} = 0$$

$$I_{a1} + \frac{I_{a1}Z_1}{Z_2} - \frac{E_a}{Z_2} + \frac{I_{a1}Z_1}{Z_0} - \frac{E_a}{Z_0} = 0$$

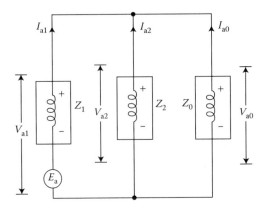

FIGURE 12.11    Interconnection of sequence network for L–L–G fault.

$$I_{a1}\left(1 + \frac{Z_1}{Z_2} + \frac{Z_1}{Z_0}\right) = E_a\left(\frac{1}{Z_2} + \frac{1}{Z_0}\right)$$

$$I_{a1} = \frac{E_a((Z_0 + Z_2)/Z_2 Z_0)}{(1 + (Z_1/Z_2) + (Z_1/Z_0))}$$

$$I_{a1} = \frac{E_a(Z_0 + Z_2)}{Z_2 Z_0 + Z_1 Z_0 + Z_1 Z_2}$$

$$I_{a1} = \frac{E_a(Z_0 + Z_2)}{Z_2 Z_0 + Z_1(Z_2 + Z_0)}$$

$$I_{a1} = \frac{E_a}{Z_1 + (Z_2 Z_0/(Z_0 + Z_2))} \tag{12.53}$$

From Equation 12.53, it is clear that all the three sequence networks are required to simulate L–L–G fault and also that the negative and zero-sequence networks are connected in parallel. Interconnection of sequence network for (L–L–G) as shown in Figure 12.11.

## 12.13  L–L–L FAULT/THREE-PHASE FAULT/SYMMETRICAL FAULT

Three-phase fault (dead short circuit) is the most dangerous fault because the fault impedance is minimum in case of L–L–L fault.

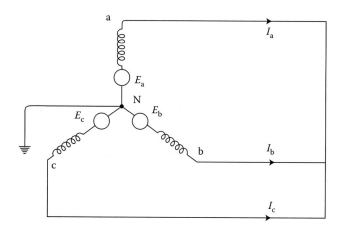

FIGURE 12.12   A three-phase neutral grounded and unloaded alternator three-phase shorted.

As shown in Figure 12.12, boundary conditions are

$$V_a = V_b = V_c \tag{12.54}$$

$$(I_a + I_b + I_c) = 0 \tag{12.55}$$

Since $|I_a| = |I_b| = |I_c|$ and if $I_a$ is taken as reference,

$$I_c = \alpha I_a \quad \text{and} \quad I_b = \alpha^2 I_a$$

Using the relation,

$$I_{a1} = \frac{1}{3}(I_a + \alpha I_b + \alpha^2 I_c)$$

$$I_{a2} = \frac{1}{3}(I_a + \alpha^2 I_b + \alpha I_c)$$

$$I_{a0} = \frac{1}{3}(I_a + I_b + I_c)$$

Substituting the values of $I_b$ and $I_c$,

$$I_{a1} = \frac{1}{3}(I_a + \alpha^3 I_a + \alpha^3 I_a)$$

$$I_{a1} = I_a \tag{12.56}$$

$$I_{a2} = \frac{1}{3}(I_a + \alpha^4 I_a + \alpha^2 I_a)$$

$$I_{a2} = \frac{1}{3}I_a(1 + \alpha^4 + \alpha^2) = \frac{1}{3}I_a(1 + \alpha^3 \cdot \alpha + \alpha^2)$$

$$I_{a2} = \frac{1}{3}I_a(1 + \alpha + \alpha^2)$$

$$I_{a2} = 0 \tag{12.57}$$

$$I_{a0} = \frac{1}{3}(I_a + I_b + I_c)$$

$$I_{a0} = 0 \tag{12.58}$$

Which means that for a three-phase fault, zero as well as negative-sequence components of current are absent and the positive-sequence component of current is equal to the phase current.

Now using the voltage boundary relation

$$V_{a1} = \frac{1}{3}(V_a + \alpha V_b + \alpha^2 V_c)$$

$$V_{a1} = \frac{1}{3}V_a(1 + \alpha + \alpha^2) = 0 \tag{12.59}$$

Since

$$V_{a1} = 0 = E_a - I_{a1}Z_1$$

$$I_{a1} = \frac{E_a}{Z_1} \tag{12.60}$$

Interconnection of sequence network for (L–L–L) is shown in Figure 12.13.

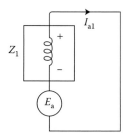

FIGURE 12.13   Interconnection of sequence network for three-phase fault.

## WORKED EXAMPLES

### EXAMPLE 12.1

100 MVA, 20 kV generator has subtransient reactance which is the same as the positive- and negative-sequence reactance, and it is 20%. Its neutral is grounded through a reactance 0.32 $\Omega$ and the zero-sequence reactance of the generator $X_{g0}$ is 5%. The generator is operating at rated voltage at no load and disconnected from system when a line-to-ground fault occurs, find the subtransient current in fault phase. Neglect reactance.

**Solution**

Let the line-to-neutral voltage at the fault point before the fault be 1 pu:

$$\text{Fault current } I_{f(pu)} = \frac{3E_a}{x_{1(pu)} + x_{2(pu)} + x_{o(pu)}}$$

$$\text{Reactance of the neutral } X_{n(pu)} = X_{n.ohm} \frac{(MVA)_{base}}{(kV)^2_{base}} = \frac{0.32 \times 100}{20^2}$$

$$= 0.08 \text{ pu}$$

$$X_{o(pu)} = X_{g0(pu)} + 3X_{n(pu)}$$

$$= 0.05 + 0.24$$

$$= 0.29 \text{ pu}$$

$$X_{1(pu)} = X_{2(pu)} = 0.2 \text{ pu}$$

$$I_{f(pu)} = \frac{3 \times 1.0}{0.2 + 0.2 + 0.29} = 4.34 \text{ pu}$$

$$I_{f(amp)} = I_{f(pu)} \times I_{base} = \frac{4.34 \times (VA)_{base}}{\sqrt{3}V_{base}} = 4.34 \times \frac{100 \times 10^6}{\sqrt{3} \times 20 \times 10^3}$$

$$= 12.52 \text{ kA}$$

## EXAMPLE 12.2

With reference to Example 12.1, determine the subtransient current in fault phase when a line-to-line fault occurs. Neglect resistance.

**Solution**

Let the line-to-neutral voltage at the fault point before the fault be 1 pu:

$$\text{Fault current } I_{f(pu)} = \frac{\sqrt{3}E_a}{x_{1(pu)} + x_{2(pu)}} = \frac{\sqrt{3} \times 1}{0.2 + 0.2} = 4.33 \text{ pu}$$

$$I_{f(amp)} = I_{f(pu)} \times I_{base} = \frac{4.33 \times (VA)_{base}}{\sqrt{3}V_{base}} = 4.33 \times \frac{100 \times 10^6}{\sqrt{2} \times 20 \times 10^3} = 12.5$$

## EXAMPLE 12.3

Positive-sequence current for a line-to-line fault of a 2-kV system is 1400 A, current for double line-to-ground fault is 2220 A. Determine the zero-sequence impedance of the system. Neglect resistance.

**Solution**

$$\text{Fault current for L–L fault,} \quad I_f(amp) = \frac{E_a}{X_1 + X_2}$$

$$= 1400$$

Here the line voltage is 2 kV. But we consider the phase voltage as a reference voltage,

$$\therefore E_a = \frac{2000}{\sqrt{3}} \text{ V}$$

$$\therefore \frac{2000}{\sqrt{3}(X_1 + X_2)} = 1400\,\text{A}$$

$$X_1 + X_2 = 0.82\,\Omega$$

or

$$X_1 + X_2 = \frac{0.82}{2} = 0.41$$

Then, fault current for L–L–G fault

$$I_f(\text{L–L–G}) = \frac{E_a}{X_1 + (X_2 \| X_0)} = 2220\,\text{A}$$

$$\therefore \frac{2000}{\sqrt{3}[0.41 + (0.41 \| X_0)]} = 2220$$

$$\therefore X_0 = 0.15\,\Omega$$

## EXAMPLE 12.4

Zero-sequence current of a line-to-ground fault is $(j2.4)$ pu. Determine the current through neutral during fault and neutral potential if neutral reactance is 5%.

### Solution

Current through the neutral is

$$I_n = 3I_{a_0} = 3 \times j2.4 = j7.2\,\text{pu}$$

Neutral potential $V_n = I_n X_n = (j7.2 \times 0.52) = j0.360\,\text{pu}$

## EXAMPLE 12.5

At a 220-kV substation of a power system, it is given that 3-$\phi$ fault level is 4000 MVA and line-to-ground fault level is 5000 MVA.

Determine (1) the positive-sequence and (2) zero-sequence driving point reactance at the bus. Neglect resistance.

**Solution**

1. Fault current for L–L–L fault is

$$I_f = \frac{E_a}{X_1}$$

MVA during fault, $(MVA)_{sc} = 3E_a \times I_f = 4000$

$$\therefore 3E_a \times \frac{E_a}{X_1} = 4000 \times 10^6$$

$$\frac{3\left((220 \times 10^3)/\sqrt{3}\right)^2}{X_1} = 4000 \times 10^6$$

Positive-sequence driving point reactance, $X_1 = 12.1\ \Omega$.

2. Fault current for L–G fault is

$$I_f = \frac{3E_a}{X_1 + X_2 + X_0}$$

Let $X_1 = X_2$

MVA during fault, $(MVA)_{sc} = 3E_a \times I_f = 5000$

$$\therefore 3E_a \left(\frac{3E_a}{2X_1 + X_0}\right) = 5000 \times 10^6$$

$$\therefore \frac{9 \times \left((220 \times 10^3)/\sqrt{3}\right)^2}{2X_1 + X_0} = 5000 \times 10^6$$

$$\therefore 3E_a \left(\frac{3E_a}{2X_1 + X_0}\right) = 5000 \times 10^6$$

$$\therefore \quad \frac{9 \times \left( (220 \times 10^3)/\sqrt{3} \right)^2}{2X_1 + X_0} = 5000 \times 10^6$$

Zero-sequence driving point reactance, $X_0 = 4.84 \ \Omega$.

## EXERCISES

1. What is a 3-$\phi$ unsymmetrical fault? Discuss the different types of unsymmetrical faults that can occur on a 3-$\phi$ system.

2. Discuss the "symmetrical components method" to analyze an unbalanced 3-$\phi$ system.

3. What is operator "$\alpha$"?

4. Express unbalanced phase current in a 3-$\phi$ system in terms of symmetrical components.

5. What do you understand by positive-, negative-, and zero-sequence impedances? Discuss them with reference to synchronous generators, transformers, and transmission lines.

6. Derive an expression for fault current for single line-to-ground fault by symmetrical components method.

7. Derive an expression for fault current for line-to-line fault by symmetrical components method."

8. Derive an expression for fault current for double line-to-ground fault by symmetrical components method.

9. What do you understand by sequence networks? What is their importance in unsymmetrical fault calculations?

10. Write short notes on the following:

    a. Positive-sequence network

    b. Negative-sequence network

    c. Zero-sequence network

# CHAPTER 13

# Circuit Breakers

## 13.1 INTRODUCTION

A circuit breaker (CB) is a device that can operate under normal or abnormal conditions to make or break the circuit by manually, automatically, or remote control. It can act in different voltage levels from low to high and some big CBs have the auto reclosing facility by which they can close their constants automatically when the fault subsides from the system.

The CB cannot operate individually without the help of relay, CT and PT, and other some auxiliary equipments. The function of a relay is to sense the fault in a system and to give a signal to the CB and the CB opens the circuit by tripping automatically. The necessary power of tripping is supplied from a DC source. The time from the occurrence of the fault to the total clearing of the fault is known as fault-clearing time and it is in the order of fraction of a second (two to three cycles).

## 13.2 DIFFERENCE BETWEEN CB AND FUSE

1. Fuse can operate in only low and medium voltage level and its application is limited, whereas the CB can operate in a wide range of voltage from low to high.

2. No fuse has the auto reclosing facility and it needs to be connected manually when the fault subsides. On the other hand, large CB has the auto reclosing capacity and its contacts close automatically when the system comes healthy from its fault condition.

3. Reliability of CB, that is, on the probability of tripping under faulty condition is very high w.r.t fuse.

4. The fuse (except the high-rupturing capacity fuse) has normally no arc extinguishing medium, whereas the CB has different cooling medium such as transformer oil, forced air, and sulfur hexafluoride ($SF_6$) gas to cool down the arc generated at the time of fault.

5. The structure of CB is much more complex w.r.t fuse.

6. The cost of CB is very high and sometimes it is uneconomical to be used in the low-voltage level.

## 13.3 OPERATING PRINCIPLE OF CB

The function of a CB is to isolate the faulty part of the power system in case of abnormal condition such as faults. A protective relay detects abnormal conditions and sends a tripping signal to CB. After receiving the trip signal from the relay the CB isolates faulty part of the power system. The simplified diagram of the CB control for opening operation is shown in Figure 13.1. When a fault occurs in the protected circuit the relay connected to the CT and VT detects the fault, actuates, and closes its contacts to complete the trip circuit. Current flows from the battery in the trip circuit. As the trip coil of the CB energized, the CB operating mechanism is actuated and it operates for the opening operation to disconnect the faulty element. A CB has two contacts—a fixed and a moving. These are placed in a closed

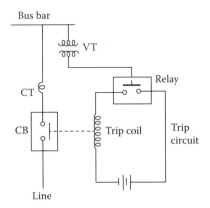

FIGURE 13.1 Simplified diagram of circuit breaker control for opening operation.

chamber containing a fluid insulating medium which quenches the arc formed between the contacts. Under normal condition the contacts remain in closed position. When the CB is required to isolate the fault part, the moving contact moves to interrupt the circuit. On the separation of contacts an arc is formed between them and the current continues to flow from one contact to other through the arc. The circuit is interrupted when the arc is finally extinguished. As a result, the trip circuit of the CB is closed and the moving contact opens from its fixed contacts to open the circuit. The necessary power of tripping is supplied from a DC source whose voltage varies depending on the system voltage and the power ratings of the CB in substations; normally this voltage is taken as 30 V. The DC source is considered as the "heart" of the substation. When the two contacts are disconnected from each other a large amount of voltage which is known as the switching surge voltage is generated across the contacts and due to it a heavy electrostatic stress in created in between the contacts. It leads to ionization of the path and fault current continues to flow. As a result of it large amount of heat loss takes place. Some part of this heat energy is converted to light energy of shorter wave length and this phenomenon is known as the "arc."

## 13.4 ARC PHENOMENON

After the occurrence of the fault, when the contacts of the CB begins to separate an arc is established in the contact gap. The physics behind this arc can be categorically described in two steps:

1. Initiation of arc

2. Maintenance of arc

1. Initiation of arc: The contact gap after separation gets ionized due to following two reasons:

   a. *Thermionic emission.* At the time of separation the contact area sharply decreases and the current density becomes very high. It is worthwhile to remember that the fault current is much higher than the normal. This increased density of current causes the generation of high heat which ultimately leads to thermionic emission at the contact gap.

   b. *Field emission.* During separation of contacts as contact area decreases resistance increases. The high fault current flowing through the resistance causes considerable voltage drop. If the voltage gradient at the contact surface becomes sufficient to dislodge electron. Ionization takes place. The gradient can be more than $10^6$ V/cm. This is known as field emission.

2. Maintenance of arc: The electric arc is a self-sustained discharge of electricity between the electrodes. As the path is ionized the fault current flows even through the air after the contacts are finally separated from each other. This arc increases the temperature of the zone and facilitates further ionization. A part of the power loss taking place during arc is converted to light energy.

## 13.5 PRINCIPLES OF ARC EXTINCTION

The two main causes responsible for generating arc between the contacts of a CB are as follows:

1. Potential difference (PD) between the contacts: When the contacts have a small separation, the pd between them is sufficient to maintain the arc. One way to extinguish the arc is to separate the contacts to such a distance that pd becomes inadequate to maintain the arc. However, this method is impracticable in high-voltage system where a separation of many meters may be required.

2. Ionized particles between contacts: The ionized particles between the contacts tend to maintain the arc. If the arc path is deionized, the arc extinction will be facilitated. This may be achieved by cooling the arc or by bodily removing the ionized particles from the space between the contacts.

## 13.6 METHODS OF ARC EXTINCTION

There are two methods of extinguishing the arc in CBs viz.

### 13.6.1 High Resistance Method

In this method, arc resistance is made to increase with time so that current is reduced to a value insufficient to maintain the arc. Consequently, the current is interrupted or the arc is extinguished. The principal disadvantage of this method is that enormous energy is dissipated in the arc. Therefore, it is employed only in DC CBs and low-capacity AC CBs.

The resistance of the arc may be increased by:

1. *Lengthening the arc.* The resistance of the arc is directly proportional to its length. The length of the arc can be increased by increasing the gap between contacts.

2. *Cooling the arc.* Cooling helps in the deionization of the medium between the contacts. This increases the arc resistance. Efficient cooling may be obtained by a gas blast directed along the arc.

3. *Reducing cross-section of the arc.* If the area of cross-section of the arc is reduced, the voltage necessary to maintain the arc is increased. In other words, the resistance of the arc path is increased. The cross-section of the arc can be reduced by letting the arc pass through a narrow opening or by having smaller area of contacts.

4. *Splitting the arc.* The resistance of the arc can be increased by splitting the arc into a number of smaller arcs in series. Each one of these arcs experiences the effect of lengthening and cooling. The arc may be split by introducing some conducting plates between the contacts.

### 13.6.2 Low Resistance or Current Zero Method

This method is employed for arc extinction in AC circuits only. In this method, arc resistance is kept low until current is zero where the arc extinguishes naturally and is prevented from restriking in spite of the rising voltage across the contacts. All modern high-power AC CBs employ this method for arc extinction.

In an AC system, current drops to zero after every half cycle. At every current zero, the arc extinguishes for a brief moment. Now the medium between the contacts contains ions and electrons so that it has small dielectric strength and can be easily broken down by the rising contact voltage known as restriking voltage. If such a breakdown does occur, the arc will persist for another half cycle. If immediately after current zero, the dielectric strength of the medium between contacts is built up more rapidly than the voltage across the contacts, the arc fails to restrike and the current will be interrupted. The rapid increase of dielectric strength of the medium near current zero can be achieved by

1. Causing the ionized particles in the space between contacts to recombine into neutral molecules.

2. Sweeping the ionized particles away and replacing them by unionized particles.

3. The dielectric strength of the medium is proportional to length of the gap between contacts.

4. Creating high pressure surrounding the arc which causes higher rate of deionization, etc.

## 13.7 SOME IMPORTANT DEFINITIONS

1. *Arc voltage*. It is the voltage that appears across the contacts of the CB during the arcing period. Its value is low except for the current zero because arc can be considered as a short path between the contacts. At current zero, the arc voltage rises rapidly to peak value (2 Em). It tends to maintain the current flow in the form of arc. As the arc path is more or less resistive, the arc voltage is almost in phase with the arc current.

2. *Restriking voltage*. It is the transient voltage that appears across the contacts at or near current zero during arcing period. At current zero, a high-frequency transient voltage appears across the contacts and is caused by the rapid distribution of energy between the magnetic and electric fields associated with the plant and transmission lines of the system. This transient voltage is known as restriking voltage (Figure 13.2). The current interruption in the circuit depends upon this voltage. If the restriking voltage rises more rapidly than the dielectric strength of the medium between the contacts, the arc will persist for another half cycle. On the other hand, if the dielectric strength of the medium builds up more rapidly than the restriking voltage, the arc fails to restrike and the current will be interrupted.

3. *Recovery voltage*. It is the normal frequency (50 Hz) rms voltage that appears across the contacts of the CB after final arc extinction. It is approximately equal to the system voltage. When contacts of CB are opened, current drops to zero after every half cycle. At some current zero, the contacts are separated sufficiently apart and dielectric strength of the medium between the contacts attains a high value due to the removal of ionized particles. At such an instant, the medium between the contacts is strong enough to prevent the breakdown by the restriking voltage. Consequently, the final arc extinction takes

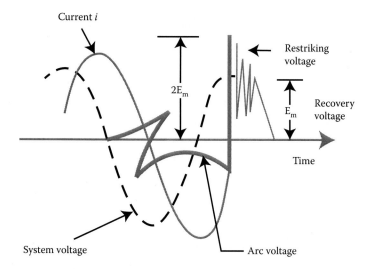

FIGURE 13.2   Arc voltage, restriking voltage, recovery voltage.

place and circuit current is interrupted. Immediately after final current interruption, the voltage that appears across the contacts has a transient part (Figure 13.2). However, these transient oscillations subside rapidly due to the damping effect of system resistance and normal circuit voltage appears across the contacts. The voltage across the contacts is of normal frequency and is known as recovery voltage.

## 13.8  EXPRESSION FOR RESTRIKING VOLTAGE TRANSIENTS

Figure 13.3a shows a short-circuit fault on a feeder beyond the location of the CB. Figure 13.3b shows an equivalent electrical circuit, where $L$ and

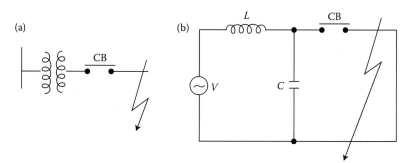

FIGURE 13.3   (a) Fault on a feeder near circuit breaker. (b) Equivalent electrical circuit for analysis of restriking voltage.

$C$ are the inductance and capacitance per phase of the system up to the point of CB location, respectively. The resistance of the circuit has been neglected. During the time of fault, a heavy fault current flows in the circuit. When the CB is closed, the fault current flows through $L$ and the contacts of the CB, the capacitance $C$ being short circuited by the fault. Hence, the circuit of Figure 13.3b becomes completely reactive and the fault current is limited entirely by inductance of the system.

The fault is cleared by opening of the CB contacts. The parting of the CB contacts does not in itself interrupt the current because an arc is established in between the parting contacts and the current continues to flow through the arc. Successful interaction depends upon controlling and finally extinguishing the arc. Extinction of the arc takes place at the instant when current passes through zero.

Since the circuit of Figure 13.3b is completely reactive, the voltage at the instant of current zero will be at its peak. The voltage across the CB contacts, and therefore across the capacitor $C$, is the arc voltage. In high-voltage circuits, it is usually only a small percentage of the system voltage. Hence, the arc voltage may be assumed to be negligible.

For the analysis of this circuit, the time is measured from the instant of interruption, when the fault current comes to zero. Since the voltage is a sinusoidally varying quantity and is at its peak at the moment of current zero, it is expressed as "$V_m \cos \omega t$."

When the CB contacts are opened and the arc is extinguished, the current $i$ is diverted through the capacitance $C$, resulting in a transient condition. The inductance and the capacitance from a series oscillatory circuit. The voltage across the capacitance which is restriking voltage rises and oscillates.

$$V_m \cos \omega t$$
$$V_m \cos \omega t = V_m$$

The natural frequency of oscillation is given by

$$f_n = \frac{1}{2\pi\sqrt{LC}} \tag{13.1}$$

And the natural angular frequency is

$$\omega_n = \frac{1}{\sqrt{LC}} \tag{13.2}$$

The voltage across the capacitance which is the voltage across the contacts of the CB can be calculated in terms of $L$, $C$, $f_n$, and system voltage. The mathematical expression for the transient condition is as follows:

$$L\frac{di}{dt} + \frac{1}{C}\int i\,dt = V_m \cos\omega t \qquad (13.3)$$

Immediately after the instant of arc extinction, the voltage across the capacitance which is the restriking voltage, oscillates at the natural frequency given by Equation 13.1. Since the natural frequency of oscillation is a fast phenomenon, it persists for only a small period of time. During this short period which is of interest, the change in the power frequency term is very little and, hence negligible, because $\cos\omega t = 1$. Hence, the sinusoidally varying voltage $V_m \cos\omega t$ can be assumed to remain constant at $V_m$ during this short interval of time, that is, the transient period.

Substituting $V_m \cos\omega t \approx V_m$, the Equation 13.3 can be written as

$$L\frac{di}{dt} + \frac{1}{C}\int i\,dt = V_m \qquad (13.4)$$

$$i = \frac{dq}{dt} = \frac{d(cv_c)}{dt} \qquad (13.5)$$

where $v_c$ = voltage across the capacitor = restriking voltage.
Therefore,

$$\frac{di}{dt} = \frac{d^2(cv_c)}{dt^2} = c\frac{d^2v_c}{dt^2} \qquad (13.6)$$

$$\frac{1}{C}\int i\,dt = \frac{q}{c} = v_c \qquad (13.7)$$

Substituting these values in Equation 13.4, we get

$$LC\frac{d^2v_c}{dt^2} + v_c = V_m \qquad (13.8)$$

Taking Laplace transform of both sides of Equation 13.8, we get

$$LCS^2 v_c(s) + v_c(s) = \frac{V_m}{s}$$

where $v_c(s)$ is the Laplace transform of $v_c$

Other terms are zero as initially $q = 0$ at $t = 0$ or

$$v_c(s)[LCS^2 + 1] = \frac{V_m}{s}$$

or $\quad v_c(s) = \dfrac{V_m}{s(LCS^2 + 1)} = \dfrac{V_m}{LCS[s^2 + (1/LC)]}$

$$\omega_n = \frac{1}{\sqrt{LC}}, \quad \therefore \omega_n^2 = \frac{1}{\sqrt{LC}}$$

$$\omega_n = \frac{1}{\sqrt{LC}}, \quad \therefore \omega_n^2 = \frac{1}{\sqrt{LC}} \tag{13.9}$$

Taking the inverse Laplace of Equation 13.9, we get

$$v_c(t) = \omega_n V_m \int_0^t \sin \omega_n t$$

$$= \omega_n V_m \left[ \frac{-\cos \omega_n t}{\omega_n} \right]_0^t$$

As $v_c(t) = 0$ at $t = 0$, constant $= 0$.

$$v_c(t) = V_m(1 - \cos \omega_n t) \tag{13.10}$$

This is the expression for restriking voltage.

The maximum value of the restriking voltage occurs at

$$t = \frac{\pi}{\omega_n} = \pi\sqrt{LC}$$

Hence, the maximum value of restriking voltage $= 2V_m$

$$= 2 \times \text{Peak value of}$$
$$\text{the system voltage}$$

The amplitude factor of the restriking voltage is defined as the ratio of the peak of the transient voltage to the peak value of the system frequency voltage. If losses are ignored, this factor becomes 2.

The rate of rise of restriking voltage (RRRV)

$$\text{RRRV} = \frac{d}{dt}[V_m(1 - \cos\omega_n t)]$$

or $\text{RRRV} = V_m\omega_n\sin\omega_n t$. The maximum value of RRRV occurs when $\omega_n t = \pi/2$, that is, when $t = \pi/2\omega_n$.

Hence, the maximum value of $\text{RRRV} = V_m\omega_n$.

## 13.9 CURRENT CHOPPING

It is the phenomenon of current interruption before the natural current zero is reached. Current chopping mainly occurs in air-blast CBs because they retain the same extinguishing power irrespective of the magnitude of the current to be interrupted. When breaking low current (e.g., transformer magnetizing current) with such breakers, the powerful deionizing effect of air blast causes the current to fall abruptly to zero well before the natural current zero is reached. This phenomenon is known as current chopping and results in the production of high-voltage transient across the contacts of the CB as discussed below.

Suppose the arc current is $i$ when it is chopped down to zero value as shown by point a in Figure 13.4. As the chop occurs at current $i$, the energy stored in inductance is $L i2/2$. This energy will be transferred to the capacitance C, charging the latter to a prospective voltage $e$ given by

$$\frac{1}{2}Li^2 = \frac{Ce^2}{2}$$
$$e = \frac{i\sqrt{L}}{C}\text{V} \tag{13.11}$$

The prospective voltage $e$ is very high as compared to the dielectric strength gained by the gap so that the breaker restrikes. As the deionizing force is still

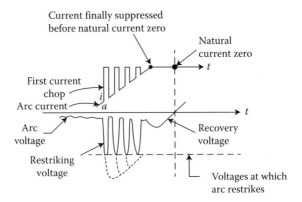

FIGURE 13.4    Current chopping.

in action, chop occurs again but the arc current this time is smaller than the previous case. This induces a lower prospective voltage to reignite the arc. In fact, several chops may occur until a low enough current is interrupted which produces insufficient induced voltage to restrike across the breaker gap. Consequently, the final interruption of current takes place.

## 13.10  RESISTANCE SWITCHING

It has been discussed above that current chopping, capacitive current breaking, etc., give rise to severe voltage oscillations. These excessive voltage surges during circuit interruption can be prevented by the use of shunt resistance $R$ connected across the CB contacts as shown in the equivalent circuit in Figure 13.5. This is known as resistance switching.

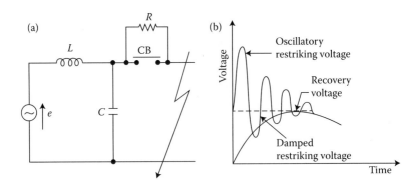

FIGURE 13.5    (a) Equivalent circuit for analysis of resistance switching. (b) Restriking voltage oscillation.

When a fault occurs, the contacts of the CB are opened and an arc is struck between the contacts. Since the contacts are shunted by resistance R, a part of arc current flows through this resistance. This results in the decrease of arc current and an increase in the rate of deionization of the arc path. Consequently, the arc resistance is increased. The increased arc resistance leads to a further increase in current through shunt resistance. This process continues until the arc current becomes so small that it fails to maintain the arc. Now, the arc is extinguished and circuit current is interrupted.

The shunt resistor also helps in limiting the oscillatory growth of restriking voltage. It can be proved mathematically that natural frequency of oscillations of the circuit shown in Figure 13.5a is given by

$$f_n = \frac{1}{2\pi}\sqrt{\frac{1}{LC} - \frac{1}{4R^2C^2}}$$

In order to keep RRRV within the rating of CB, the critical value of resistance R can be determined by making $f_n$, that is,

$$f_n = \frac{1}{2\pi}\sqrt{\frac{1}{LC} - \frac{1}{4R^2C^2}} = 0$$

$$\frac{1}{LC} - \frac{1}{4R^2C^2} = 0 \quad \text{or} \quad \frac{1}{LC} = \frac{1}{4R^2C^2}$$

$$\therefore \qquad R = 0.5\sqrt{L/C}$$

Figure 13.5b shows the oscillatory growth and exponential growth when the circuit is critically damped.

Resistors across breaker contacts may be used to perform one or more of the following functions:

1. To reduce the RRRV and the peak value of restriking voltage.

2. To reduce the voltage surges due to current chopping and capacitive current breaking.

3. To ensure even sharing of restriking voltage transient across the various breaks in multi-break CBs.

## 13.11 CB RATINGS

A CB may be called upon to operate under all conditions. However, major duties are imposed on the CB, when there is a fault on the system in which it is connected. Under fault conditions, a CB is required to perform the following three duties:

1. It must be capable of opening the faulty circuit and breaking the fault current.

2. It must be capable of being closed on to a fault.

3. It must be capable of carrying fault current for a short time, while another CB (in series) is clearing the fault.

Corresponding to the above-mentioned duties, the CBs have three ratings viz.

1. *Breaking capacity.* It is current (rms) that a CB is capable of breaking at given recovery voltage and under specified conditions (e.g., power factor, RRRV). The breaking capacity is always stated at the rms value of fault current at the instant of contact separation. When a fault occurs, there is considerable asymmetry in the fault current due to the presence of a DC component. The DC component dies away rapidly, a typical decrement factor being 0.8 per cycle. Referring to Figure 13.6, the contacts are separated at $AA'$. At this instant, the fault current has

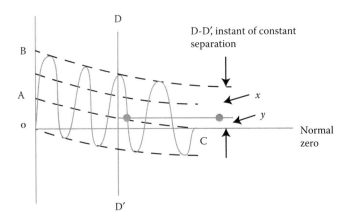

FIGURE 13.6 Short-circuit current waveform.

$x$ = Maximum value of AC component

$y$ = DC component

Symmetrical breaking current = rms value of AC component

$$= \frac{x}{\sqrt{2}}$$

Asymmetrical breaking current = rms value of combined sum of AC and DC components.

Generally the breaking capacity in MVA by taking into account the rated breaking current and rated service voltage. Thus, if $I$ is the rated breaking current in amperes and $V$ is the rated service line voltage in volts, then for a three-phase circuit,

$$\text{Breaking capacity} = 3 \times V \times I \times 10^{-6} \text{ MVA}$$

In India (or Britain), it is a usual practice to take breaking current equal to the symmetrical breaking current. However, American practice is to take breaking current equal to asymmetrical breaking current.

2. *Making capacity.* There is always a possibility of closing or making the circuit under short-circuit conditions. The capacity of a breaker to "make" current depends upon its ability to withstand and close successfully against the effects of electromagnetic forces. These forces are proportional to the square of maximum instantaneous current on closing. Therefore, making capacity is stated in terms of a peak value of current instead of rms value. The peak value of current (including DC component) during the first cycle of current wave after the closure of CB is known as making capacity. It may be noted that the definition is concerned with the first cycle of current wave on closing the CB. This is because the maximum value of fault current possibly occurs in the first cycle only when maximum asymmetry occurs in any phase of the breaker. In other words, the making current is equal to the maximum value of asymmetrical current. To find this value, we must multiply symmetrical breaking current by $\sqrt{2}$ to convert this from rms to peak, and then by 1.8 to include the "doubling effect" of maximum asymmetry. The total multiplication factor becomes $\sqrt{2} \times 1.8 = 2.55$.

Making capacity = 2.55 × Symmetrical breaking capacity

3. *Short-time rating.* The short-time rating of a CB depends upon its ability to withstand the electromagnetic force effects and the temperature rise. The oil CBs have a specified limit of 3 s when the ratio of symmetrical breaking current to the rated normal current does not exceed 40. However, if this ratio is more than 40, then the specified limit is 1 s.

It is the period for which the CB is able to carry fault current while remain closed.

### 13.11.1 Normal Current Rating

It is the rms value of current which the CB is capable of carrying continuously at its rated frequency under specified conditions.

The important condition for normal working of an oil CB is that the temperature of oil should not be more than 40°C and that of contacts should not be more than 35°C.

## 13.12 AUTORECLOSING

The fault which exist on the system. Depending upon the time are classified as follows:

1. Transient fault

2. Semi-permanent fault

3. Permanent fault

It has been found that above 80% fault are transient fault, and 12% are semi-permanent faults.

Transient faults are exist only for a short time. It can be removed faster still if the line is disconnected from the system momentarily so that the arcs blows out. After the arc is deionized, the line can be reclosed to restore normal service. Semi-permanent fault may occur due to twig falling on the power conductor or the bird spanning the power conductor. The reclosing could be restored with some delay so that the cause of the fault could be dispensed with during a time delay trip and the line could be reclosed to restore normal service. In case of permanent fault, reclosing does not help as it is to be attended or removed, and the line is to be taken out till the fault is cleared. Therefore, if the fault is not cleared after first reclosure, a double or triple shot reclosing is required. If the fault is still persist, the line is taken out of service. Autoreclosing may be single- or three-phase type. Mostly one-phase autoreclosing breakers are preferred, as most of the transmission faults are single phase to ground fault. Auto reclosing in one phase also improves stability as the power remains transmitted through the two healthy phases when one phase is interrupted. Mostly, single shot or double shot auto reclousure breakers are available. The breakers may be autoreclosing type or delayed autoreclosing type.

## 13.13  CLASSIFICATION OF CBs

There are several ways of classifying the CBs. However, the most general way of classification is on the basis of medium used for arc extinction. The medium used for arc extinction is usually oil, air, $SF_6$, or vacuum. Accordingly CBs may be classified into the following:

1. Oil CBs

2. Air-blast CBs

3. $SF_6$ CBs

4. Vacuum CBs

## 13.14  OIL CBs

In such CBs, some insulating oil (e.g., transformer oil) is used as an arc quenching medium. The contacts are opened under oil and an arc is struck between them.

The heat of the arc evaporates the surrounding oil and dissociates it into a substantial volume of gaseous hydrogen at high pressure. The hydrogen gas occupies a volume about one thousand times that of the oil decomposed. The oil is, therefore, pushed away from the arc and an expanding hydrogen gas bubble surrounds the arc region and adjacent portions of the contacts (Figure 13.7).

The arc extinction is facilitated mainly by two processes:

1. The hydrogen gas has high heat conductivity and cools the arc, thus aiding the deionization of the medium between the contacts.

2. The gas sets up turbulence in the oil and forces it into the space between contacts, thus eliminating the arcing products from the arc path. The result is that arc is extinguished and circuit current interrupted.

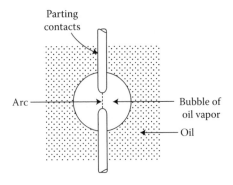

FIGURE 13.7    Gas bubble surrounds the arc region and adjacent portions of the contacts of an oil CB.

*Advantages:* The advantages of oil as an arc quenching medium are as follows:

1. It absorbs the arc energy to decompose the oil into gases which have excellent cooling properties.

2. It acts as an insulator and permits smaller clearance between live conductors and earthed components.

3. The surrounding oil presents cooling surface in close proximity to the arc.

*Disadvantages:* The disadvantages of oil as an arc quenching medium are

1. It is inflammable and there is a risk of a fire.

2. It may form an explosive mixture with air.

3. The arcing products (e.g., carbon) remain in the oil and its quality deteriorates with successive operations. This necessitates periodic checking and replacement of oil.

## 13.15  TYPES OF OIL CBs

The oil CBs find extensive use in the power system. This can be classified into the following types.

### 13.15.1  Bulk-Oil CBs

Bulk-oil CBs use a large quantity of oil. The oil has to serve two purposes.

Firstly, it extinguishes the arc during opening of contacts. Secondly, it insulates the current conducting parts from one another and from the earthed tank.

Such CBs may be classified into

1. Plain break oil CBs

2. Arc control oil CBs

### 13.15.2 Low-Oil CBs

Low-oil CBs use minimum amount of oil. In such CBs, oil is used only for arc extinction; the current conducting parts are insulated by air or porcelain or organic insulating materials.

## 13.16 PLAIN-BREAK OIL CBs

A plain-break oil CB involves the simple process of separating the contacts under the whole of the oil in the tank. There is no special system for arc control other than the increase in length caused by the separation of contacts. The arc extinction occurs when a certain critical gap between the contacts is reached.

It has a very simple construction. It consists of fixed and moving contacts enclosed in a strong weather-tight earthed tank containing oil up to a certain level and an air cushion above the oil level. The air cushion provides sufficient room to allow for the reception of the arc gases without the generation of unsafe pressure in the dome of the CB. It also absorbs the mechanical shock of the upward oil movement. Figure 13.8 shows a double break plain oil CB. It is called a double break because it provides two breaks in series.

The arc extinction is facilitated by the following processes:

1. The hydrogen gas bubble generated around the arc cools the arc column and aids the deionization of the medium between the contacts.

2. The gas sets up turbulence in the oil and helps in eliminating the arcing products from the arc path.

3. As the arc lengthens due to the separating contacts, the dielectric strength of the medium is increased.

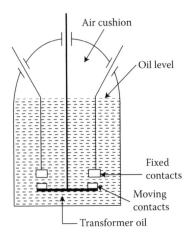

Air cushion

Oil level

Fixed
contacts

Moving
contacts

Transformer oil

FIGURE 13.8   Double break plain oil circuit breaker.

*Disadvantages:*

1. There is no special control over the arc other than the increase in length by separating the moving contacts. Therefore, for successful interruption, long arc length is necessary.

2. These breakers have long and inconsistent arcing times.

3. These breakers do not permit high-speed interruption.

Due to these disadvantages, plain-break oil CBs are used only for low-voltage applications where high breaking capacities are not important. It is a usual practice to use such breakers for low-capacity installations for voltage not exceeding 11 kV.

## 13.17   ARC CONTROL OIL CBs

In case of plain break oil CB discussed above, there is very little artificial control over the arc. Therefore, comparatively long arc length is essential in order that turbulence in the oil caused by the gas may assists in quenching it. However, it is necessary and desirable that final arc extinction should occur while the contact gap is still short. For this purpose, some arc control is incorporated and the breakers are then called arc control CBs.

There are two types of such breakers, namely

1. Self-blast oil CBs

2. Forced-blast oil CBs

### 13.17.1 Self-Blast Oil CBs

In this type of CB, the gases produced during arcing are confined to a small volume by the use of an insulating rigid pressure chamber or pot surrounding the contacts. Since the space available for the arc gases is restricted by the chamber, a very high pressure is developed to force the oil and gas through or around the arc to extinguish it. The magnitude of pressure developed depends upon the value of fault current to be interrupted. As the pressure is generated by the arc itself, such breakers are sometimes called self-generated pressure oil CBs.

The design of the chamber or pot should be such that the pressure developed is sufficient to quench the arc even at low values of current but not so much as to break the pot on heavy current. This has lead to manufacture of a variety of pots, such as

1. Plain explosion pot

2. Cross-jet explosion pot

3. Self-compensated explosion pot

#### 13.17.1.1 Plain Explosion Pot

It is a rigid cylinder of insulating material and encloses the fixed and moving contacts (Figure 13.9). The moving contact is a cylindrical rod passing through a restricted opening (called throat) at the bottom. When a fault occurs, the contacts get separated and an arc is struck between them. The heat of the arc decomposes oil into a gas at very high pressure in the pot. This high pressure forces the oil and gas through and round the arc to

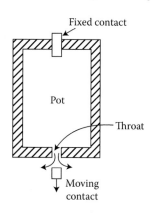

FIGURE 13.9   Plain explosion pot.

extinguish it. If the final arc extinction does not take place while the moving contact is still within the pot, it occurs immediately after the moving contact leaves the pot. It is because emergence of the moving contact from the pot is followed by a violent rush of gas and oil through the throat producing rapid extinction. The principal limitation of this type of pot is that it cannot be used for very low or for very high fault current. With low fault current, the pressure developed is small, thereby increasing the arcing time. On the other hand, with high fault current, the gas is produced so rapidly that explosion pot is liable to burst due to high pressure. For this reason, plain explosion pot operates well on moderate short-circuit current only where the rate of gas evolution is moderate.

### 13.17.1.2 Cross-Jet Explosion Pot

This type of pot is just a modification of plain explosion pot and is illustrated in Figure 13.10. It is made of insulating material and has channels on one side which act as arc splitters. The arc splitters help in increasing the arc length, thus facilitating arc extinction. When a fault occurs, the moving contact of the CB begins to separate. As the moving contact is withdrawn, the arc is initially struck in the top of the pot. The gas generated by the arc exerts pressure on the oil in the back passage. When the moving contact uncovers the arc splitter ducts, fresh oil is forced across the arc path. The arc is, therefore, driven sideways into the "arc splitters" which increase the arc length, causing arc extinction. The cross-jet explosion pot is quite efficient for interrupting heavy fault current. However, for

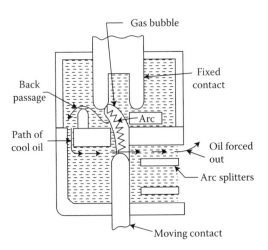

FIGURE 13.10   Cross-jet explosion pot.

low fault current, the gas pressure is small and consequently the pot does not give a satisfactory operation.

### 13.17.1.3 Self-Compensated Explosion Pot

This type of pot is essentially a combination of plain explosion pot and cross-jet explosion pot. Therefore, it can interrupt low as well as heavy short-circuit current with reasonable accuracy. Figure 13.11 shows the schematic diagram of self-compensated explosion pot. It consists of two chambers, the upper chamber is the cross-jet explosion pot with two arc splitter ducts, while the lower one is the plain explosion pot. When the short-circuit current is heavy, the rate of generation of gas is very high and the device behaves as a cross-jet explosion pot. The arc extinction takes place when the moving contact uncovers the first or second arc splitter duct. However, on low short-circuit current, the rate of gas generation is small and the tip of the moving contact has the time to reach the lower chamber. During this time, the gas builds up sufficient pressure as there is very little leakage through arc splitter ducts due to the obstruction offered by the arc path and right angle bends. When the moving contact comes out of the throat, the arc is extinguished by plain pot action. It may be noted that as the severity of the short-circuit current increases, the device operates less and less as a plain explosion pot and more and more as a cross-jet explosion pot. Thus the tendency is to make the control self-compensating over the full range of fault current to be interrupted.

### 13.17.2 Forced-Blast Oil CB

The major limitation of such breakers is that arcing times tend to be long and inconsistent when operating against current considerably less than the rated current. It is because the gas generated is much reduced at low

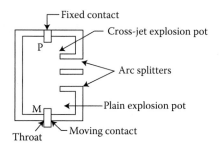

FIGURE 13.11 Self-compensated explosion pot.

values of fault current. This difficulty is overcome in forced blast oil CBs in which the necessary pressure is generated by external mechanical means independent of the fault current to be broken.

In this breaker, oil pressure is created by the piston-cylinder arrangement. The movement of the piston is mechanically coupled to the moving contact. When a fault occurs, the contacts get separated by the protective system and an arc is struck between the contacts. The piston forces a jet of oil toward the contact gap to extinguish the arc.

*Advantages:*

1. Since, oil pressure developed is independent of the fault current to be independent, the performance at low current is more consistent than with self-blast oil CBs.

2. The quantity of oil required is reduced considerably.

## 13.18 LOW-OIL CBs

The oil in the bulk oil CB serves two purposes.

Firstly, it acts as an arc quenching medium and secondly, it insulates the live parts from earth. It has been found that only a small percentage of oil is actually used for arc extinction while the major part is utilized for insulation purposes. For this reason, the quantity of oil in bulk oil CBs reaches a very high figure as the system voltage increases. Consequently besides increase in the expenses, tank size and weight of the breaker the fire risk and maintenance problems are also increase. This factors led to the development of a low-oil CB. A low-oil CB employs solid materials for insulation purposes and uses a small quantity of oil which is just sufficient for arc extinction.

### 13.18.1 Construction

Figure 13.12 shows the cross-section of a single phase low-oil CB. There are two compartments separated from each other but both filled with oil. The upper chamber is the circuit-breaking chamber while the lower one is the supporting chamber. The two chambers are separated by a partition and oil from one chamber is prevented from mixing with the other chamber.

1. *Supporting chamber.* It is a porcelain chamber mounted on a metal chamber. It is filled with oil which is physically separated from the oil in the circuit-breaking compartment. The oil inside the supporting

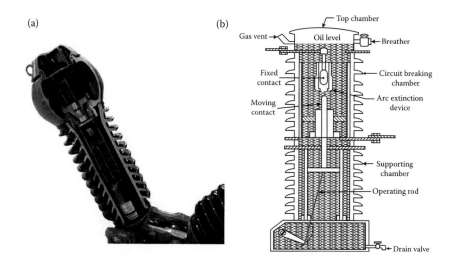

(a)

(b)

Top chamber

Gas vent → Oil level ← Breather

Fixed contact — Circuit breaking chamber

Moving contact — Arc extinction device

Supporting chamber

Operating rod

Drain valve

FIGURE 13.12  (a) Low oil circuit breaker and (b) cross-section of a single phase.

chamber and the annular space formed between the porcelain insulation and bakelized paper is employed for insulation purposes only.

2. *Circuit-breaking chamber.* It is a porcelain enclosure mounted on the top of the supporting compartment. It is filled with oil and has the following parts:

   a. Upper and lower fixed contacts

   b. Moving contact

   c. Turbulator

   The moving contact is hollow and includes a cylinder which moves down over a fixed piston. The turbulator is an arc control device and has both axial and radial vents. The axial venting ensures the interruption of low current, whereas radial venting helps in the interruption of heavy current.

3. *Top chamber.* It is a metal chamber and is mounted on the circuit-breaking chamber. It provides expansion space for the oil in the circuit-breaking compartment.

## 13.18.2  Operation

Under normal operating conditions, the moving contact remains engaged with the upper fixed contact. When a fault occurs, the moving contact is pulled down by the tripping springs and an arc is struck. The arc energy vaporizes the oil and produces gases under high pressure. This action constrains the oil to pass through a central hole in the moving contact and results in forcing series of oil through the respective passages of the turbulator. The process of turbulation is orderly one, in which the sections of the arc are successively quenched by the effect of separate streams of oil moving across each section in turn and bearing away its gases.

*Advantages:* A low-oil CB has the following advantages over a bulk oil CB:

1. It requires lesser quantity of oil

2. It requires smaller space

3. There is reduced risk of fire

4. Maintenance problems are reduced

*Disadvantages:* A low-oil CB has the following disadvantages over a bulk oil CB:

1. Due to smaller quantity of oil, the degree of carbonization is increased

2. There is a difficulty of removing the gases from the contact space in time

3. The dielectric strength of the oil deteriorates rapidly due to high degree of carbonization

## 13.19 MAINTENANCE OF OIL CBs

The maintenance of oil CB is generally concerned with a checking of contacts and dielectric strength of oil. After a CB has interrupted fault current a few times or load current several times, its contacts may get burned by arcing and the oil may loss some of its dielectric strength due to carbonization. This results in the reduced rupturing capacity of the breaker. Therefore, it is a good practice to inspect the CB at regular intervals of 3 or 6 months. During inspection of the breaker, the following points should be kept in view:

1. Check the current carrying parts and arcing contacts. If the burning is severe, the contacts should be replaced.

2. Check the dielectric strength of the oil. If the oil is badly discolored, it should be changed or reconditioned.

3. Check the insulation for possible damage. Clean the surface and remove carbon deposits with a strong and dry fabric.

4. Check the oil level.

5. Check closing and tripping mechanism.

## 13.20 AIR-BLAST CBs

These breakers employ a high pressure air blast as an arc quenching medium. The contacts are opened in a flow of air blast established by the opening of blast valve. The air blast cools the arc and sweeps away the arcing products to the atmosphere. This rapidly increases the dielectric strength of the medium between contacts and prevents from re-establishing the arc. Consequently, the arc is extinguished and flow of current is interrupted.

*Advantages:* An air-blast CB has the following advantages over a low-oil CB:

1. Cheapness and free availability of the interrupting medium, chemical stability and inertness of the air.

2. High-speed operation.

3. Elimination of fire hazard.

4. Short and consistent time and therefore, less burning of contacts.

5. Less maintenance.

6. Stability for frequent operation.

7. Facility for high-speed reclosures.

*Disadvantages:*

1. The air has relatively inferior arc extinguishing properties.

2. The air-blast CBs are very sensitive to the variations in the RRRV.

3. Considerable maintenance is required for the compressor plant which supplies the air blast.

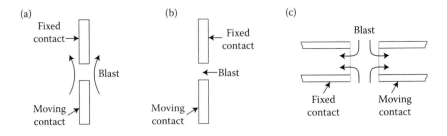

FIGURE 13.13 (a) Axial-blast type. (b) Cross-blast type. (c) Radial-blast type.

The air-blast CBs are find in wide applications in high-voltage installations. Majority of the CBs for voltage beyond 110 kV are of this type.

## 13.21 TYPES OF AIR-BLAST CB

Depending upon the direction of air blast in relation to the arc, air-blast CB may be classified into the following types:

1. *Axial-blast type* in which the air blast is directed longitudinally, that is, in the line with the arc as shown in Figure 13.13a.

2. *Cross-blast type* in which the air blast is directed at right angles to the arc path as shown in Figure 13.13b.

3. *Radial-blast type* in which the air blast is directed radially as shown in Figure 13.13c.

## 13.22 VACUUM CBs

In such breakers, vacuum (VCB, degree of vacuum being in the range from $10^{-7}$ to $10^{-5}$ torr) is used as the arc quenching medium. Since, vacuum offers the highest insulating strength, it has far superior arc quenching properties than any other medium.

### 13.22.1 Construction

Figure 13.14 shows the parts of a typical vacuum CB. It consists of fixed contact, moving contact, and arc shield mounted inside a vacuum chamber. The movable member is connected to the control mechanism by stainless steel bellows. This enables the permanent sealing of the vacuum chamber so as to eliminate the possibility of leak. A glass vessel or ceramic vessel is used as the outer insulating body. The arc shield prevents the deterioration of the internal dielectric strength by preventing metallic vapors

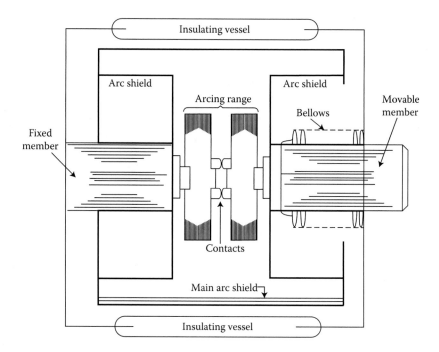

FIGURE 13.14   Vacuum circuit breaker.

falling on the inside surface of the outer insulating cover. The vapor condensing shield is made of synthetic resin.

### 13.22.2  Working Principle

When the breaker operates, the moving contact separates from the fixed contact and an arc is struck between the contacts by the ionization of metal vapors of contacts. However, the arc is quickly extinguished because the metallic vapors, electrons, and ions produced during arc rapidly condense on the surfaces of the CB contacts, resulting in quick recovery of dielectric strength.
*Advantages:*

1. They are compact, reliable, and have longer life.

2. There are no fire hazards.

3. There is no generation of gas during and after operation.

4. They can interrupt any fault current. The outstanding feature of a VCB is that it can break any heavy fault current perfectly just before the contacts reach the definite open position.

5. They require little maintenance and are quiet in operation.

6. They can successfully withstand lightning surges.

7. They have low arc energy.

8. They have low inertia and hence require smaller power for control mechanism.

### 13.22.3 Application

In India, where distances are quite large and accessibility to remote areas is difficult, the installation of such outdoor, maintenance free CBs should proof a definite advantage. Vacuum CBs are being employed for outdoor applications ranging from 22 to 66 kV. Even with limited rating of say 66–100 MVA, they are suitable for a majority of applications in rural areas.

## 13.23  $SF_6$ CBs

$SF_6$ is an electro-negative gas and has a strong tendency to absorb free electrons. The contacts of the breaker are opened in a high pressure flow of $SF_6$ gas and an arc is struck between them. The conducting free electrons in the arc are rapidly captured by the gas to form relatively immobile negative ions. This loss of conducting electrons in the arc quickly builds up enough insulation strength to blow out the arc.

Owing to the following properties, $SF_6$ proves superior to other medium such as oil or air for use in CB:

1. Very high dielectric strength.

2. No-reactive to the other component of the CB.

3. About 100 times more effective than air is extinction of arc.

4. Owing to its high density, its heat transfer property is about 1.6 times that of the air.

5. Can be stored at a relatively smaller pressure than that of air (since thermal time constant of $SF_6$ is low).

$SF_6$ breakers can withstand severe RRRV and this are most suitable for short line faults without switching resistors and can interrupt capacitive current without restriking.

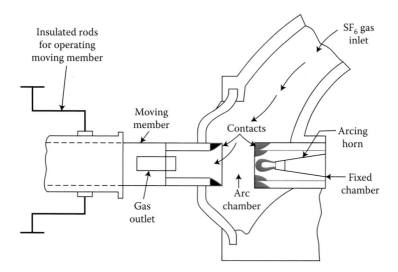

FIGURE 13.15  SF$_6$ circuit breakers.

## 13.23.1 Construction

It consists of an interruption chamber which houses fixed and moving contacts and is connected to SF$_6$ gas reservoir (Figure 13.15). When the breakers contacts open, the valve mechanism permits a high pressure SF$_6$ gas from the reservoir to flow toward the interruption chamber.

The fixed contact is a hollow cylindrical current carrying contact fitted with an arc horn.

The moving contact is also a hollow cylinder with rectangular holes in the sides to permit the SF$_6$ gas to let out through these holes after flowing along and across the arc. The tips of fixed contact, moving contact and arcing horn are coated with copper–tungsten arc-resistant material. SF$_6$ gas being costly, it is reconditioned and reclaimed by suitable auxiliary system after each operation of the breaker.

## 13.23.2 Working

When the breaker in closed position of the breaker, the contacts remain surrounded by SF$_6$ gas at a pressure of about 2.8 bar. As the breaker operates, the moving contact is pulled apart and an arc is struck between the contacts. The movement of the moving contact is synchronized with the opening of a valve which permits SF$_6$ gas at 14 kg/cm$^2$ pressure from the reservoir to the arc interruption chamber. The high pressure flow of SF$_6$ rapidly absorbs the free electrons in the arc path to form immobile negative ions which are ineffective as charge carriers. The result is that the medium between the contacts quickly builds up high dielectric strength and causes the extinction of arc. After the arc extinction the valve is closed by the action of a set of springs.

*Advantages:*

1. Very short arching time, this reduces contacts erosion.

2. No risk of fire (since $SF_6$ gas is noninflammable).

3. No reduction of dielectric strength (since no $CO_2$ is formed).

4. Since the dielectric strength of $SF_6$ gas is two to three times that of air, such breakers can interrupt much longer current.

5. Silent operation.

6. The closed gas enclosure keeps the interior dry so that moisture problem is almost eliminated.

7. Being totally enclosed, these breakers are particularly suitable where explosion hazard exist (e.g., coal mines).

*Disadvantages:*

1. $SF_6$ breakers are costly due to the high cost of $SF_6$.

2. Since $SF_6$ gas has to be reconditioned after every operation of the breaker, additional equipment is required for this purpose.

## 13.24 HIGH-VOLTAGE DC CB

In AC circuit, current passes through natural current zeros, and hence it is possible to design AC CB to interrupt large current. This feature is not available in DC. If a high current is suppressed abruptly in DC, a very high-transient voltage appears across the contacts of the CBs. Therefore, in DC CB, some external circuit have to be provided to bring down the current from full valve to zero, smoothing without suppressing it abruptly. The additional circuit creates artificial current zeros which are utilized for arc interruption as shown in Figure 13.16.

A schematic diagram of high-voltage DC (HVDC) CB is shown in Figure 13.17. HVDC CB consists of a main CB (MCB) and a circuit to produce artificial current zero and to suppress transient voltage. MCB may either be an $SF_6$ or vacuum CB. $R$ and $C$ are connected in parallel with the MCB to reduce $dv/dt$ after the final current zero. $L$ is a saturable reactor in series with the MCB. It is used to reduce $di/dt$ before current zero. $C_p$ and $L_p$ are

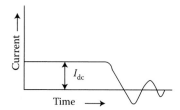

FIGURE 13.16   Artificial current zero in DC.

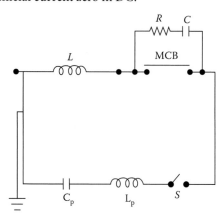

FIGURE 13.17   HVDC circuit breaker.

connected in parallel to produce artificial current zero after the separator of the contacts in the MCB. A nonlinear resistor is used to suppress the transient overvoltage which may be produce across the contacts of the MCB.

Switch $S$, which is a triggered vacuum gap, is switched immediately after the opening of the contacts of the main CB. The capacitor $C_p$ is precharged in the direction as shown in figure when $S$ is closed, the precharged capacitor $C_p$ is discharges through the main CB and sends a current in opposition to the MCB. This will force the main circuit current to become zero with a few oscillations. The arc is interrupted at a current zero.

## WORKED EXAMPLES

### EXAMPLE 13.1

A CB interrupts the magnetizing current of a 50-MVA transformer at 110 kV. The magnetizing current of the transformer is 6% of the full load current. Determine the maximum voltage which may appear across the gap of the breaker when the magnetizing current is interrupted at 50% of its peak value. The stray capacitance is 2000 μF. The inductance is 25 H.

**Solution**

The full load current of the transformer $= \dfrac{50 \times 10^6}{\sqrt{3} \times 110 \times 10^3} = 262.43\,A$

Magnetizing current $= \dfrac{6}{100} \times 262.43 = 15.75\,A$

Current chopping occurs at $0.50 \times 15.75 \times \sqrt{2} = 11.13\,A$

$$\frac{1}{2}Li^2 = \frac{1}{2}CV^2$$

$$\frac{1}{2} \times 25 \times (11.13)^2 = \frac{1}{2} \times 2000 \times 10^{-6} V^2$$

$$V = 1.244\,kV$$

**EXAMPLE 13.2**

For a 110 kV system, the reactance and capacitance up to the location of the CB is 2 Ω and 0.012 μF, respectively. Calculate the following:

1. The frequency of transient oscillation
2. The maximum value of restricting voltage across the contacts of the CB
3. The maximum value of RRRV

**Solution**

1. $L = \dfrac{2}{2\pi \times 50} = 6.36\,mH = 0.00636\,H$

$$f_n = \frac{1}{2\pi\sqrt{LC}} = \frac{1}{2\pi\sqrt{0.00636 \times 0.012 \times 10^{-6}}} = 18.218\,kHz$$

2. The restricting voltage $V_c = V_m[1 - \cos w_n t]$

The maximum value of the restricting voltage $= 2V_m$

$$= 2 \times \frac{110\sqrt{2}}{\sqrt{3}} = 179.63$$

3. The maximum value of RRRV $= w_n V_m$

$$= 2\pi f_n \times \frac{110}{\sqrt{3}} \times \sqrt{2} \times 1000$$

$$= 2\pi \times 18.218 \times 1000 \times \frac{110}{\sqrt{3}}$$

$$\times \sqrt{2} \times 1000 \text{ V/s}$$

$$= 10.28 \text{ kV/}\mu\text{s}$$

## EXAMPLE 13.3

In a short-circuit test on a CB, the following readings were obtained on single frequency transient:

1. Time to reach the peak restriking voltage, 60 μs.
2. The peak restriking voltage, 110 kV

Determine the average RRRV and frequency of oscillation.

### Solution

$$\text{Average RRRV} = \frac{\text{Peak restriking voltage}}{\text{Time to reach peak value}}$$

$$= \frac{110 \text{ kV}}{60 \text{ μs}} = 1.83 \text{ kV/μs} = 1.83 \times 10^6 \text{ kV/s}$$

Natural frequency of oscillations,

$$f_n = \frac{1}{2 \times \text{Time to reach peak value}}$$

$$= \frac{1}{2 \times 60 \times 10^{-6}} = 8.33 \text{ kHz}$$

## EXAMPLE 13.4

In a 132 kV system, the reactance and capacitance up to the location of CB is 6 Ω and 0.020 μF, respectively. A resistance of 500 Ω is connected across the contacts of the CB determine the following:

1. Natural frequency of oscillation
2. Damped frequency of oscillation

3. Critical value of resistance which will give no transient oscillation
4. The value of resistance which will give damped frequency of oscillation, one-fourth of the natural frequency of oscillation

**Solution**

$$L = \frac{X_L}{2\pi f} = \frac{6}{2\pi 50} = 0.019 \text{ H}$$

1. Natural frequency of oscillation $= \dfrac{1}{2\pi} \dfrac{1}{\sqrt{LC}}$

$$= \frac{1}{2\pi\sqrt{0.019 \times 0.020 \times 10^{-6}}}$$

$$= 8.143 \text{ kHz}$$

2. Frequency of damped oscillation,

$$f = \frac{1}{2\pi}\sqrt{\frac{1}{LC} - \frac{1}{4c^2R^2}}$$

$$= 1.825 \text{ kHz}$$

3. The value of critical resistance,

$$R = \frac{1}{2}\sqrt{\frac{L}{C}} = 487.34 \ \Omega$$

4. The damped frequency of oscillation is

$$\frac{1}{4} \times 8.14 \text{ kHz} = 2035 \text{ Hz}$$

$$2035 = \frac{1}{2\pi}\sqrt{\frac{1}{LC} - \frac{1}{4c^2R^2}}$$

$$R = 503.22 \ \Omega$$

## EXERCISES

1. What is a circuit breaker? Describe its operating principle.

2. Discuss the arc phenomenon in a circuit breaker.

3. Explain the various methods of arc extinction in a circuit breaker.

4. Define and explain the following terms as applied to circuit breakers:

   a. Arc voltage

   b. Restriking voltage

   c. Recovery voltage

5. Describe briefly the action of an oil circuit breaker. How does oil help in arc extinction?

6. Discuss the advantages and disadvantages of oil circuit breakers.

7. Explain with neat sketches the construction and working of the following circuit breakers:

   a. Plain explosion pot

   b. Cross-jet explosion pot

   c. Self-compensated explosion pot

8. Explain the difference between bulk oil circuit breakers and low-oil circuit breakers.

9. Discuss the constructional details and operation of a typical low-oil circuit breaker? What are its relative merits and demerits?

10. Discuss the principle of operation of an air-blast circuit breaker. What are the advantages and disadvantages? of using air as the arc quenching medium?

11. Explain briefly the following types of air-blast circuit breakers:

    a. Axial-blast type

    b. Cross-blast type

12. What are the important components common to most of circuit breakers? Discuss each component briefly.

13. Write a short note on

    a. The rate of restriking voltage indicating its importance in the arc extinction

    b. Auto-reclosing

14. Discuss the phenomenon of

    a. Current chopping

    b. Capacitive current breaking

15. Write short notes on the following:

    a. Resistance switching

    b. Circuit breaker ratings

    c. Circuit interruption problems

16. What is the difficulty in the development of high-voltage DC circuit breaker? How does this difficulty is overcome?

# CHAPTER 14

# Different Types of Relays

## 14.1 INTRODUCTION

Relay is a device which is used to sense fault. It sends information to the circuit breaker (CB), which interrupts the faulty circuit as early as possible (generally within two or three cycles). If the fault exists for a long time, it results in

1. Discontinuity in the service

2. Massive damage to the system

3. Spreading of the effects of fault into the healthy part of the system

It is a fault-sensing device which serves the information of the fault to the CB for disconnecting the faulty part from the healthy part of the system.

## 14.2 ESSENTIAL QUALITIES OF PROTECTION

Basic requirements of a good protection system are those qualities which are essential for satisfactory operation of a protective scheme comprising of relay, CB, and other protective devices. The fundamental qualities are as follows.

1. *Speed.* The faulty part of the system should be isolated within a minimum possible time after the occurrence of a fault. If a particular type of fault is allowed to exist in the system for a abnormally long time, it may result more faults in addition to the existing one. So the speed of the protective system should be reasonably high. The operating

time of a protective relay is usually one or half cycle. For distribution systems, the operating time is allowed to be more than one cycle.

2. *Selectivity.* It is the quality by which a relay can locate the faulty part correctly and disconnect it by tripping the nearest CB not interfering into the healthy part of the system. The relay should also be able to discriminate between a fault and transient condition like power surges or inrush of transformer's magnetizing current.

3. *Sensitivity.* The less value of the actuating parameters of a relay and VA burden of that relay the more is its sensitivity. The parameters may be current, voltage, impedance, frequency, phase angle, etc. Highly sensitive relay is a very expensive one.

4. *Reliability.* The reliability means that relay system must function in case of any type of fault without fail. A highly reliable system is so designed that it can act also on assumption. A simple system is more reliable. Reliability of a protection scheme should be at least 95%. To achieve high degree of reliability greater attention should be given to the design, installation, maintenance, and testing of various elements of the protective scheme.

5. *Simplicity.* From the maintenance point of view, a relay system should be simple. It cuts down cost and increases reliability. But a simple system sacrifices more amount of sensitivity; so a compromise is made judiciously between simplicity, reliability, and sensitivity.

6. *Economy.* Lastly on electrical system must be economical. For an ideal protection, system cost is very high. On the other hand, we have to place reliability before economy in case of protection of generator, power transformer, transmission line, etc. Actually here also a compromise is adopted.

7. *Stability.* A protective system should be remaining stable even when a large current is flowing through its protective zone due to an external fault, which does not lie in its zone.

## 14.3 CLASSIFICATION OF RELAY

Depending on the construction and operation, protective relays can be broadly classified into three categories:

1. Electromechanical

2. Static relay

3. Numerical relay

Electromechanical relays are further classified into two categories:

1. Electromagnetic relays

2. Thermal relays

Electromagnetic relays are also classified into two categories:

1. Electromagnetic attraction type

2. Electromagnetic induction type

According to the speed of operation relay also can be classified into four categories:

1. Instantaneous relay

2. Definite time lag relay

3. Inverse-time lag relay

4. Inverse definite minimum time lag relay (IDMT)

Further relay can be classified into six categories depending on their function:

1. Under voltage

2. Overcurrent

3. Directional reverse current relay

4. Directional reverse power relay

5. Differential relay

6. Distant relay

## 14.4 BASIC RELAY TERMINOLOGY

1. *Operating torque.* It is the torque for which the relay contacts close.

2. *Restraining torque.* It is the torque which opposes the closure of the relay contacts.

3. *Pickup level.* It is the threshold or boundary value of the actuating parameter (current, voltage, frequency, phase angle, etc.) above which the relay operates.

4. *Dropout or reset level.* It is the value of the actuating parameter below which the relay goes back to its normal position. The ratio of the dropout or reset value is called dropout or reset ratio. Its value is less than or equal to 1.

5. *Burden.* It is the power (VA) consumed by relay circuit.

6. *Operating time.* It is the time calculated from the moment when the actuating parameter attains its pickup value until the relay operates.

7. *Reach of distance protection of line.* The limiting distance covered by the protection, the faults beyond which are not within the reach of the protection and should be covered by other relay.

8. *Over reach.* The operations of distant relay for a fault beyond its set protected distance is called it's over reaches.

9. *Under reach.* Failure of distant relay to operate within set protected distance is known as under reach.

10. *Actuating parameter.* The electrical quantity, that is, current, voltage, frequency, impedance, etc. Either alone or in conjunction with other electrical quantities required for the functioning of the relay is known as the actuating parameter of the relay.

11. *Setting.* The actual value of actuating parameter at which relay is designed to operate under given conditions is known as setting of relay.

12. *Reset time.* It is the time that elapses from the moment the actuating quantity falls below its reset value to the instant when the relay come back to its initial position.

13. *Auxiliary relay.* Auxiliary relays assist the protective relays. They repeat the operations of protective relays and control switches. The auxiliary

relay may be instantaneous or may have a time lag and may operate within large limits of the actuating quantity.

14. *Seal-in-relay.* This is a kind of an auxiliary relay. It is energized by the contacts of the main relay. Its contacts are placed in parallel with that of the main relay and designed to relieve the contacts of the main relay from their current carrying duty. The seal-in contacts are usually heavier in comparison with the main relay.

15. *Backup relay.* A backup relay operates after a slight delay, if the main relay fails to operate.

16. *Flag or target.* Flag is a device that gives visual indication whether a relay has operated or not.

17. *Unit system of protection.* A unit system of protection is one which is able to detect and respond to faults occurring only within its own zone of protection. It does not respond to the faults occurring beyond its own zone of protection. Examples are differential protection of alternators.

18. *Residual current.* It is the algebraic sum of all current in a multiphase system. It is denoted by $I_{res}$. In a three-phase system,

$$I_{res} = I_A + I_B + I_C$$

## 14.5 ZONES OF PROTECTION

The total electrical system is divided into several protective zones such as generator, low-voltage switchgear, transformer, high-voltage switchgear, and transmission line. Every zone is overlapped with its adjacent zone. The various protective zones of a typical power system are shown in Figure 14.1.

When a fault arises in a particular zone, the CB nearest to the fault will only be disconnected leaving the rest of the system healthy. If the switchgear in a particular zone fails in for any reason, the overlapping zone will act to subside the fault if necessary provision of backup protection is made. The protected zone is that part of a power system protective scheme for each individual equipment such as generator protection, busbar protection, transformer protection, etc. It usually contains on or at the most two elements of the power system. The zones are arranged to overlap so that no part of the system remains unprotected. Overlapping must be done in both sides of a particular CB, otherwise blind spots will be generated as shown in Figure 14.1.

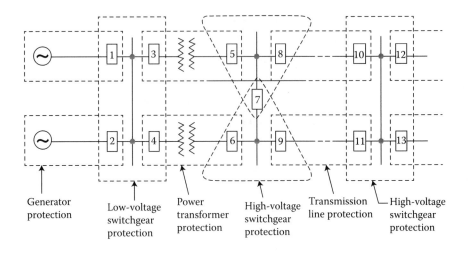

FIGURE 14.1   Zones of protection.

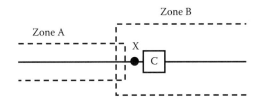

FIGURE 14.2   Dead spot (X) in protective zone.

It can be seen that for a fault at X, the CB of zone B, including breaker C will be tripped, however, this does not interrupt the flow of fault current from zone A. So here X is a dead spot. If the overlapping is done keeping C in the overlapping area, this type of spot will not be generated (Figure 14.2).

If a fault occurs in the overlapping zone, more CBs (of both adjacent zones) than the maximum necessary to isolate the faulty element of the system would trip. A relativity low extent of overlap reduces the probability of faults in this region, and consequently, tripping of too many breakers does not occur frequently.

## 14.6  PRIMARY AND BACKUP PROTECTION

The system is divided in a number of protection zones as explained earlier, each having its own protection scheme. If a fault occurs in a particular zone, it is the duty of the primary relays of that zone to isolate the faulty element. The primary relay fails to operate; there is a backup protective scheme to clear the fault as second line of defense.

The backup relays are made independent of those factors which might cause primary relays to fail. A backup relay operates after a time delay to give the primary relays sufficient time to operate. When a backup relay operates, a larger part of the power system is disconnected from the source, but this is unavoidable.

There are three types of backup relays.

## 14.6.1 Remote Backup

In this type of backing up, the backup relays are placed in the neighboring station. It is the cheapest and simplest and so widely used for protection of transmission lines. It does not fail due to factors causing failure of the primary protection scheme.

## 14.6.2 Relay Backup

Here an additional relay is required for the backup protection. It is also known as local backup. It trips the same CB if the primary relay fails. This scheme is costly. The principle of operation of backup relays should be different from those of the primary protection. They should be supplied from separate current transformer (CT) and potential transformer (PT).

## 14.6.3 Breaker Backup

This is also a kind of local backup. This type of backup is necessary for a bus-bar system where a number of CBs are connected to it. When a protective relay operates in response to a fault but the CB fails to trip, the fault treated as a bus-bar fault. In such a situation, it becomes necessary that all other CB on that bus bar should trip. After a time delay, the main relay closes the contact of a backup relay which trips all other CBs on the bus bar if the proper breaker does not trip coil is energized.

## 14.7 CLASSIFICATION OF PROTECTIVE SCHEMES

A protective scheme is used to protect equipment or a selection of the line. It includes one or more relays of the same or different types. Protective schemes can be classified into four major categories.

### 14.7.1 Overcurrent Protection

This scheme of protection is used in case of the protection of distribution lines, large motors, equipments, etc. It includes one more overcurrent relays. An overcurrent relay operates when the current exceeds its pickup value.

### 14.7.2 Distance Protection

This protection is used for transmission line mainly, usually up to 132 kV. The distance relays measure the distance in terms of impedance between the relay location and fault point. The relay operates if the fault takes place within the protected zone. There are various kinds of distance relays such as impedance relay, mho relay, and reactance relay. An impedance relay measures the impedance; a reactance relay measures reactance and mho relay measures admittance.

### 14.7.3 Differential Protection

This scheme of protection is used for the protection of generators, transformers, motors of very large size, bus zone, etc. CTs are placed on both sides of the each winding of a machine. The output of their secondary's are applied to the relays coil. The relay compares the current entering a machine winding and leaving the same. Under normal condition, these two current are same and their difference is zero. But under abnormal condition, there will be difference in current which actuates the relay. In bus zone protection, relays are placed on both sides of the bus bar.

### 14.7.4 Carrier-Current Protection

This scheme of protection is mainly used in case of lines above 132 kV. A carrier signal in the range of 50–500 kc/s is generated for the purpose. A transmitter and a receiver are installed at each end of a transmission line to be protected. Information regarding the direction of the fault current is transmitted from one end of the line section to the other. Depending on the information, relays placed at each end trip if the fault lies within their protected section. Relays do not trip in case of external faults. The relays are of distance type.

## 14.8 CONSTRUCTION AND OPERATING PRINCIPLES OF RELAY

### 14.8.1 Electromagnetic Induction Relay

Induction relays use electromagnetic induction principle for their operation. Their principle of operation is same as that of a single-phase induction motor. Hence, they can be used for AC current only. Two types of construction of this relays are fairly standard.

### 14.8.1.1 Induction Disk Relay

There are two types of construction of induction disk relays namely the shaded pole type and Watt-hour meter type.

1. *Shaded pole construction.* In shaded pole type (Figure 14.3) of relay, there is a C-shaded electromagnet. It is energized by the coil fed from the secondary winding of the protection CT. One half of the each pole of the electromagnet is surrounded by a copper band known as shading ring. The shaded portion of the pole produces a flux which is displaced in space and times cut the disk and produce eddy current in it. Torque is produced by the interaction of each flux with the eddy current produced by the other flux. The resultant torque causes the disk to rotate. There is a break magnet whose field can apply breaking action on the movement of the disk. An arm is fitted with the movement of the disk to close the trip circuit. The length of travel of this arm should not be large. Otherwise operating time will be large.

2. *Watt-hour type construction.* This structure gets name from the fact that is looks like Watt-hour meter, that is, the energy meter. Here, there are two magnets as shown in Figure 14.4—one is the upper magnet and other is lower magnet. The fault current coming from CT secondary flows in the primary winding and it induces electromotive force as well as current $I_2$ in the secondary just like transformer. Each magnet produces an alternating flux that cuts the disk. To obtain a phase displacement between two fluxes produce by the upper and lower magnets, their coils may be energized by two different sources. If they are energized by the same source, the resistances and reactances of the two circuits are made different, so that there will be sufficient phase difference between two fluxes. The flux $\phi_2$ in

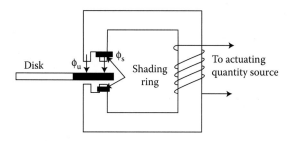

FIGURE 14.3    Shaded pole-type induction disk relay.

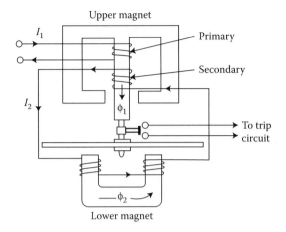

FIGURE 14.4   Watt-hour meter type construction.

the lower magnet will interact with flux $\phi_1$ (phase displaced by an angle $\theta$) to produce the torque acting on the disk and the trip circuit is closed in same manner like shaded pole relays. The net torque is proportional to $|\phi_1||\phi_2|\sin\theta$. VA burden in these relays is somewhat higher in comparison with that of the attraction relays. It is of the order of 2.5 VA. These are robust and reliable. The only disadvantage of these relays is that the opening of the secondary winding makes the relays inoperative.

14.8.1.1.1 Torque Production in an Induction Relay   In both shaded pole type and Watt-hour meter type, there are two fluxes. These are sinusoidally varying fluxes with phase angle $\theta$ between them.
     Let

$$\phi_1 = |\phi_1|\sin\omega t \quad (|\phi_1| \text{ being the maximum value of } \phi_1)$$

$$\phi_2 = |\phi_2|\sin(\omega t + \theta) \quad (|\phi_2| \text{ being the maximum value of } \phi_2)$$

The fluxes induce current $i\phi_1$ and current $i\phi_2$ in the aluminum disk. The direction of this current will be according to left hand rule.
     Now assuming the self-inductance of the disk being negligible, the disk current will be in phase with their voltage:

$$i_{\phi_1} \propto \frac{d\phi_1}{dt}$$

$$i_{\phi_1} \propto \frac{d\left(|\phi_1| \sin \omega t\right)}{dt}$$

$$i_{\phi_1} \propto |\phi_1| \cos \omega t \qquad (14.1)$$

$$i_{\phi_2} \propto |\phi_2| \cos(\omega t + \theta) \qquad (14.2)$$

The fluxes interacting with current produce forces with direction indicated in Figure 14.5 above.

The net force,

$$F = |F_2 - F_1| \propto |\phi_2 i_{\phi_1} - \phi_1 i_{\phi_2}|$$

$$\begin{aligned}
F = |F_2 - F_1| &\propto (|\phi_2| \sin (\omega t + \theta) \cdot |\phi_1| \cos \omega t - |\phi_1| \sin \omega t \cdot |\phi_2| \cos(\omega t + \theta) \\
&\propto |\phi_1||\phi_2|[\sin (\omega t + \theta) \cos \omega t - \sin \omega t \cos(\omega t + \theta)] \\
&\propto |\phi_1||\phi_2| \sin \theta
\end{aligned}$$

$$\text{Average force} \propto |\phi_1||\phi_2| \sin \theta$$

The greater the phase angle $\theta$ between the fluxes, the greater is the net force applied to the disk. Obviously, the maximum force will be produced, when the two fluxes are 90° out of phase.

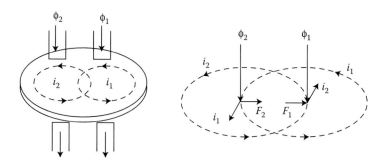

FIGURE 14.5   Troque produced in an induction relay.

14.8.1.1.2 Induction Cup Relay    The induction cup relay (Figure 14.6) may have two, four, or more number of poles. These poles are energized by the current from CT. There is a stationary iron core and the center over which a hollow metallic cup (rotor conductor) is placed. This cup is free to rotate. It just resembles an induction motor. By the method of induction eddy current are produced in metallic cup. These current interact with the flux produced to avoid continuous rotation. A control spring is attached to the spindle of rotating cup. The rotating cup carries an arm that closes the contact of trip circuit.

The basic theory is same for the disk relay and cup relay. In both cases, the net torque varies as $\phi_1 \phi_2 \sin \theta$. But the cup relays are more efficient torque producer and hence they are faster. A modern induction cup relay may have an operating time in the order of 0.01 s.

### 14.8.1.2 Electromagnetic Attraction Relay

Electromagnetic attraction relays are also classified into the following categories.

14.8.1.2.1 Attracted Armature Type    These are simplest type of relays shown in Figure 14.7. These relays have a coil or an electromagnet energized by coil. The coil is energized by operating quantity which may be proportional to circuit current or voltage. A plunger or armature is subjected to the action of magnetic field produced by operating quantity. It is basically a single actuating quantity relay. Attracted armature relay responds to both AC and DC because torque is proportional to $I^2$. These relays are fast relays. They have fast operation and fast reset because of small length

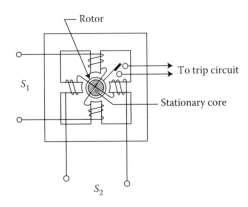

FIGURE 14.6    Induction cup relay.

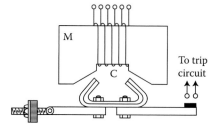

FIGURE 14.7 Simple construction of an attracted armature type relay.

of travel and light moving parts. They are described as instantaneous but their operating time does vary with current. Slow operating and resetting time can be obtained by decaying of flux in the magnetic circuit by fitting a Cu ring around the magnet. The eddy current produced in the ring opposes the very cause as per Lenz's law so effective flux is reduced. Operating time may be as slow as 0.1 s and resetting time may be as slow as 0.5 s. On the other hand, very high operating speed is possible. A modern relay has an operating time of 0.5 ms. As fault current increases operating time decreases, so they have the inverse relationship which leads to a hyperbolic curve (Figure 14.8).

The only disadvantage of this type of relays is that they are so fast that they can operate even in transient state. VA burden depends upon their construction. For a typical relay, it is in order of 0.2–0.6 VA. These relay are compact, robust, and reliable.

*Operating Principle:* The electromagnetic force exerted on the moving elements is proportional to the square of the operating current.

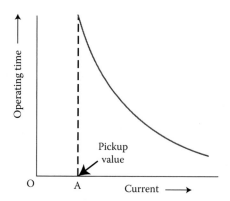

FIGURE 14.8 Time current characteristics.

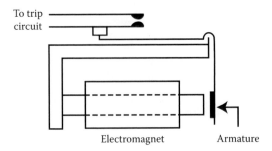

Electromagnet                    Armature

FIGURE 14.9    Hinged armature type construction.

Let $F$ = net force, $K_1$ = constant, $I$ = current in operating coil, and $K_2$ = restraining force including friction.

$$F = K_1 I^2 - K_2$$

On the verge of the operation $F = 0$, $I = \sqrt{K_2/K_1}$ = constant.

By adjusting the restraining force, one can adjust the relay for new current setting.

The following are the different types of construction of attracted armature type relays.

1. *Hinged armature type relay.* Figure 14.9 shows a hinged armature type construction. In the hinged armature type relay, there is an electromagnet which is excited by the current of secondary winding of the protection CT. When energized, it drags the hinged armature and thus the trip circuit becomes closed.

2. *Plunger type relay.* Figure 14.10 shows a plunger type relay. In plunger type of relay, the energized electromagnet creates the necessary force to drag the plunger down to close the trip circuit.

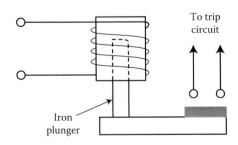

FIGURE 14.10    Plunger type relay.

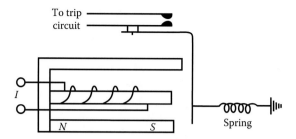

FIGURE 14.11   Polarized moving iron type relay.

3. *Polarized moving iron type relay.* Figure 14.11 shows a polarized moving iron type relay. In this relay, the armature made of iron is attracted by the electromagnet energized by the abnormal current supplied by CT and thus the trip circuit is closed. In normal condition, the force is not sufficient to move the iron piece.

14.8.1.2.2   Balanced Beam Relay   Figure 14.12 shows a balance beam relay. This type of balance beam relay consists of a horizontal beam pivoted centrally with one electromagnet attached to either side. The beam remains in horizontal position till the operating force become more than restraining force. The beam is just given a slight mechanical bias by means of spring such that in normal condition the contacts are open. When operating, torque increases the beam tilts and the contacts closes.

*Operating Principle:* Neglecting spring effect, the net torque

$$T = K_1 I^2 - K_2 I_2^2$$

on the verge of operation, $T > 0$,

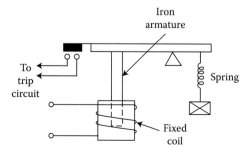

FIGURE 14.12   Balance beam relay.

$$\frac{I_1}{I_2} = \sqrt{\frac{K_2}{K_1}} = \text{Constant}$$

These relays are fast and instantaneous. It has the time in the order of one cycle. Its VA burden varies from 0.2 to 0.6 VA.

### 14.8.1.3 Thermal Relays

The operation of these relays is based on the principle of electrothermal effect of the actuating quantity. Such type of the relays is mainly used for the protection of small motors against overloading and unbalanced current.

There are mainly three types of thermal relays namely: (a) bimetallic thermal, (b) unimetallic thermal, and (c) bimetallic spiral type.

Figure 14.13a shows the bimetallic thermal relay in which two metal strips of different coefficient of thermal expansion are joined together. In abnormal condition of the line, the current of the CT secondary flows through heating coil and thereby heating the bimetallic strip. One strip expands more than the other, resulting in bending of the strip. After bending, it deflects and closes the relay contacts.

Bimetallic strip is in spiral (Figure 14.13b) form. The unequal expansions of the two metals cause the unwinding of the spiral, which results in closure of the contacts. In the third type, the unimetallic strips (Figure 14.13c) are also used as thermal elements in a hair-pin-like shape. When the strip gets heated, it expands and closes the contacts.

For the protection of the three-phase motors, three bimetallic strips are used. They are energized by the current from the three phases. Their contacts are arranged in such a way that if any one of the spirals moves differently from the others, due to an unbalance exceeding 12%, their contacts meet and cause the CB to trip.

### 14.8.1.4 Static Relay

When the relay circuit has no moving parts within it, it is designated as a static relay. In this relays, there is one master relay followed by a slave relay. The master relay actually performs the function of measurement and comparison of the actuating quantity. The slave relay only closes the contacts depending on the information obtained from the master relay. These relays are economical to use because of their low cost. Normally thyristor are used as slave relay. A static relays also known as solid-state

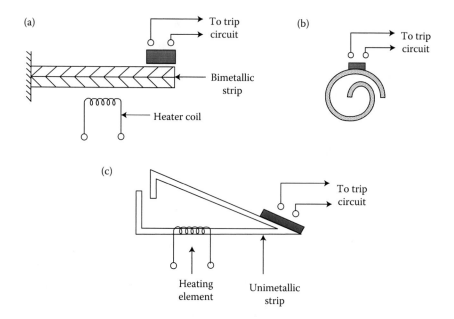

FIGURE 14.13 (a) Bimetallic thermal relay. (b) Bimetallic spiral type thermal relay. (c) Unimetallic thermal relay.

relay employs semiconductor diodes, transistors, zener diodes, thyristors, logic gates, etc., as its components. Sometimes integrated circuits are also used in place of transistors. They are reliable and compact. Now a days, the use of static relay is slowly taking the place of popular induction relays.

## 14.9 OVERCURRENT PROTECTION

The overcurrent relays operate when the load current either in normal or abnormal condition exceeds a preset value, known as pickup value. These relays are used extensively for the protection of distribution lines, large motors, power equipment, etc., a scheme that incorporates such relays for the protection of an element of a power system is called overcurrent protection scheme. An overcurrent scheme of protection may use a number of overcurrent relays. Electromagnetic or induction relays are commonly used in this scheme.

## 14.10 TIME–CURRENT CHARACTERISTICS

The different types of time current characteristics available for the overcurrent relays are as follows.

1. *Definite time overcurrent relay.* This relay operates after a predetermined value of time, when the current exceeds the pickup value of the relay. The operating time of such relays does not depend upon the magnitude of current. The desired definite time can be set by suitable mechanism of the relay. Figure 14.14 shows the time current characteristic of this relay.

2. *Instantaneous relay.* These relays operate almost instantaneously after the current attaining the pickup value. The operating time is less than 0.1 s. They have fast operation and the operating time does not change with the magnitude of current. This characteristic can be achieved with the help of the hinged armature relays.

3. *Inverse-time overcurrent relay.* As shown in the above Figure 14.14, in the inverse time, overcurrent relays the operating time inversely varies with the magnitude of current. It means the operating time decreases as the current increases. The characteristics can be achieved with induction type relays by using a suitable core which does not saturate for a large value of fault current.

4. *IDMT relay.* This type of relay gives an inverse time–current characteristic at lower values of the fault current and definite time characteristic at higher values of the fault current. IDMT relays are widely for the protection of distribution lines. The particular time–current relationship for IDMT relays is given by

$$t = \frac{0.14}{(I^{0.02} - 1)}$$

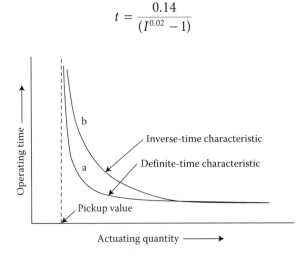

FIGURE 14.14   Definite time- and inverse-time characteristics of overcurrent relays.

The general expression for time–current characteristic of overcurrent relay is given by

$$t = \frac{k}{(I^n - 1)}$$

This can be achieved by using a core of the electromagnet which gets saturated for current slightly greater than the pickup current.

5. *Very inverse-time overcurrent relay.* This relay gives more inverse characteristic than that of an IDMT relay. Its time current characteristic lies between IDMT and extremely inverse-time relays. It has better selectivity than IDMT relays. Its standard time current relationship is given by

$$t = \frac{13.5}{(I - 1)}$$

6. *Extremely inverse-time overcurrent relay.* An extremely inverse-time current relay gives a time–current characteristic more inverse than that of the very inverse and IDMT relays. It has the time current relationship given by

$$t = \frac{80}{(I^2 - 1)}$$

These types of relays are very suitable for the protection of machines against overheating, protection of alternators, power transformers, earthing transformers, expensive cables, and railway trolley wires. They have highest selectivity.

## 14.11  CURRENT SETTING

The pickup level of current of a relay can be set at any desired value. This is known as current setting of the relay. It is usually achieved by the use of tapings on the relay operating coil. The taps are brought out to a plug bridge as shown in Figure 14.15. It permits to alter the number of turns on the relay coil. This changes the torque on the disk and hence the time of operation of the relay. The values assigned to each tap are expressed in terms of percentage of full load rating of CT with which the relay is

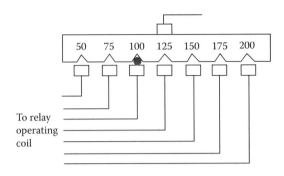

FIGURE 14.15   Plug bridge.

associated and represented the value above which the disk commences to rotate and finally closes the trip circuit.

Pickup current = Rated secondary current of CT × Current setting

For example, let current setting is at 125%. If the relay fed from the CT of ratio 200/5, then pickup current will be 1.25 × 5 = 6.25 A.

## 14.12  PLUG SETTING MULTIPLIER

Plug setting multiplier (PSM) defined as the ratio of current in the CT secondary and the relay current setting.

$$PSM = \frac{\text{Secondary current of CT}}{\text{Relay current setting}}$$
$$= \frac{\text{Actual fault current of CT primary}}{\text{Relay current setting} \times \text{CT ratio}}$$

Figure 14.16 shows the curve between time of operation and PSM of a typical relay.

## 14.13  TIME MULTIPLIER SETTING

The operating time of the relay can be set at a desired value. In induction disk type relay, the angular distance by which the moving parts of the relay travels for closing the contacts can be adjusted to get different operating time. There are 10 steps in which time can be set. Time multiplier setting (TMS or TSM) is used for these steps of time setting. The values of TSM (see Figure 14.17) are 0.1, 0.2, 0.3,..., 0.9, 1.

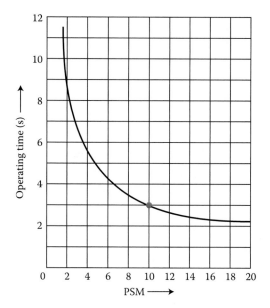

FIGURE 14.16    Standard IDMT characteristic.

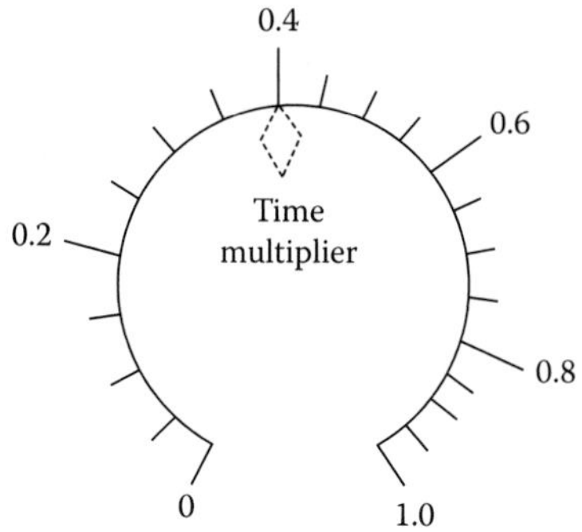

FIGURE 14.17    Time multiplier setting.

Suppose that at a particular value of the current or PSM, the operating time is 4 s with TMS = 1. The operating time for the same current with TMS = 0.5 will be $4 \times 0.5 = 2$ s. The operating time with TMS = 0.2 will be $(4 \times 0.2) = 0.8$ s.

## 14.14  OVERCURRENT PROTECTION SCHEME

Overcurrent protection scheme are widely used for the protection of distribution lines. For proper selectivity of the relays, the following systems are required.

1. *Time-graded system.* In time graded system, definite time overcurrent relays are used.

   In this operation, the relays are graded using a definite time interval of approximately 0.5 s. The operating time of the relay is adjusted in increasing order from the far end of the feeder, which is shown in Figure 14.18. In Figure 14.18, the relay $R_3$ at the extremity of the network is set to operate in the fastest possible time, while its upstream relay $R_2$ is set 0.5 s higher. Relay operating time sequentially at 0.5 s, intervals on each section moving back toward the source. With fast CBs and modern relays, it is now possible to reduce the time gap to 0.4 or 0.3 s.

   When a fault occurs beyond $c$, all relays come into action as the fault current flows through all of them. The least time setting is for the relay placed at $c$. So it operates after 0.5 s, and the fault is cleared. Now the relays at A and B are reset. If the relay or CB at C fails to operate after 1 and 1.5 s, CB at B and CB at A will trip, respectively. The drawback of this scheme is that for fault near the power source, the operating time is more. If a fault near the power source, it involves a large current and hence it should be cleared quickly. But this scheme takes the longest time in clearing the heaviest fault, which is understandable because the heaviest fault is the most destructive.

2. *Current-graded system.* In this scheme, the pickup values of current gradually increase toward the source. The relays used here are fast instantaneous relays. The operating time for all the relays is kept same, to protect different sections of the feeder, as shown in Figure 14.19. Ideally the relay at B should operate for faults within B and C. But it should not operate for faults beyond C. Similarly, the

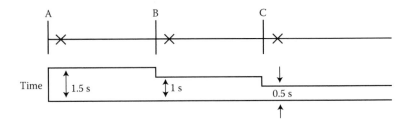

FIGURE 14.18  Time graded overcurrent protection of a feeder.

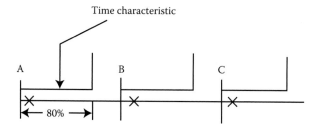

FIGURE 14.19   Instantaneous overcurrent protection of a feeder.

relay at A should trip for fault A and B. The relay at C should trip for fault beyond C.

The ideal operation is not achieved due to the following reasons:

a.  The relay at A is not able to differentiate between faults very close to B which may be on either side of B. If a fault in the section BC is very close to the section B, the relay at A understands that it is in section AB. This happens due to the fact that there is very little difference in fault current if a fault occurs at the end of the section AB or in the beginning of the section BC.

b.  The magnitude of the current cannot be accurately determined, as all the circuit parameters may not be known.

Consequently to obtain proper discrimination, relays are set to protect only a part of the feeder; usually about 80%. Since this scheme cannot protect the entire feeder, this system is not used along. It may be used in conjunction with IDMT (Figure 14.20).

The advantage of this system as compared to the time graded scheme is that the operating time is less near the power source.

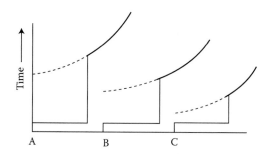

FIGURE 14.20   Combined instantaneous and IDMT protection.

3. *Combination of current and time grading.* This scheme is widely used for the protection of the distribution line. IDMT relays are employed in this scheme; they have the combined features of the current and time grading. IDMT relays have current as well as time setting arrangements. The current setting of the relay is made according to the fault current level of the particular section to be protected. The relays are set to pick up progressively at higher current levels, toward the source. Time setting is also done in a progressively increasing order toward the source. The difference in operating time of two adjacent relays is kept 0.5 s. If fault current is higher (in case of fault in that particular section), then operating time will be less, but if the fault takes place in other section then fault current is less operating time will be more.

## 14.15 DIRECTIONAL POWER OR REVERSE POWER RELAY

An electromechanical directional relay is shown in Figure 14.21a. This relay is energizes by two quantities—voltage and current. The interaction torque produced by $\phi_1$ and $\phi_2$ is given by

$$T = \phi_1 \phi_2 \sin \theta$$

So $T$ varies as $I_1 I_2 \sin \theta$, because the fluxes are produced by $I_1$ and $I_2$. Here $I_2$ is the current produced in the voltage coil and its lags $V$ by 90°.

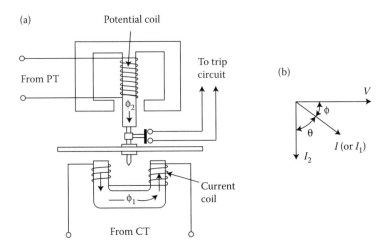

FIGURE 14.21   Induction disk-type directional relay.

On the other hand, $I_1$ is the load current coming from CT secondary lags $V$ by an angle $\phi$ (Figure 14.21b). Therefore,

$$\theta = (90° - \phi)$$
$$T = I_1 I_2 \sin(90° - \phi) \propto I_1 I_2 \cos\phi \propto VI \cos\phi$$

In induction cup, construction can also be used to produce a torque proportional to $VI \cos\phi$.

Torque produced is positively when $\cos\phi$ is positive, that is, $\phi$ is less than 90° (Figure 14.22a) when $\phi$ is more than 90° (between 90° and 180°), the torque is negative. At a particular relay location, when power flows in the normal direction, the relay is connection to produce negative torque. The angle between the actuating quantities supplied to the relay is kept $(180° - \phi)$, to produce negative torque. If due to any reason, the power flows in the reverse direction, the relay product a positive torque and it operates.

In this condition, the angle between the actuating quantities $\phi$ is kept less than 90° to produce a positive torque.

For normal flow of power, the relay is supplied with $V$ and $-I$. For reverse flow, the actuating quantities becomes $V$ and $I$. Torque becomes $VI \cos\phi$, that is, positive. This can be achieved easily by reversing the current coil (Figure 14.22b).

## 14.16 DIRECTIONAL OVERCURRENT RELAY

Figure 14.23 shows a directional overcurrent relay. A directional overcurrent relay operated when the current exceeds a specified value in a

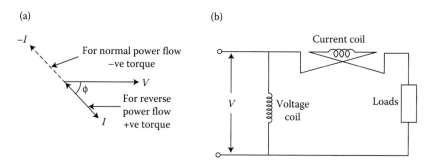

FIGURE 14.22 (a) Phasor diagram for directional relay. (b) Connection of current coil for reverse power relay.

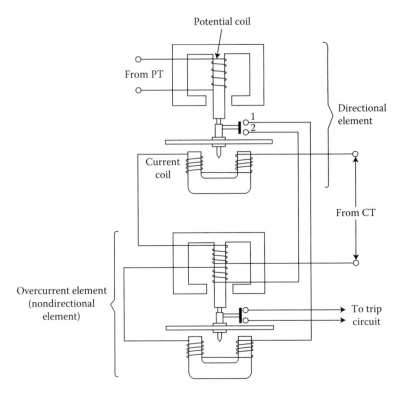

FIGURE 14.23   Directional overcurrent relay.

specified direction. It contains two relaying units. One overcurrent unit and the other a directional unit. For directional control, the secondary winding of the overcurrent unit is kept open. When the directional unit operated, it closes the open contacts of the secondary winding of the overcurrent unit. Thus, a directional feature is attributed to the overcurrent relay. The overcurrent unit may be either Watt-hour meter type or shaded pole type of construction.

## 14.17 PROTECTION OF PARALLEL FEEDER

The word feeder here means the connecting link between two circuits. The feeder could be in the form of transmission line, or this could be a distribution circuit.

Figure 14.24 shows an overcurrent protection scheme for parallel feeder. At the sending end of the feeders (at A and B), nondirectional relays are required. The double-headed arrow symbol indicates a nondirectional

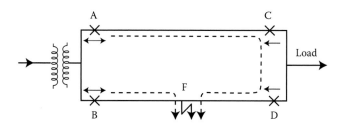

FIGURE 14.24   Protective scheme for parallel feeder.

relay. At other end of the feeders (at C and D), directional overcurrent relays are required. The arrow marks for the directional relay placed at C and D indicate that the relay will operate if the current flows in the direction shown by the arrow. If a fault current occurs at F, the directional relay at D trip, as the direction of the current is reversed. The relay at C, does not trip, as current flows in the normal direction. The relay at B trips for a fault at F. Thus the faulty feeder is isolated and the supply of the healthy feeder is maintained. For faults at feeders, the direction of current at A and B does not change and hence relays used at A and B are nondirectional and also current graded. In other arrangement, if nondirectional relays are used at C and D, then healthy feeder will also be tripped which is highly undesirable.

## 14.18  PROTECTION OF RING MAINS

Compared with radial feeders, the protection of ring feeders is costly and complex. Each feeder requires two relays. Figure 14.25 shows a protection scheme of parallel feeder. A nondirectional relay is required at one end and a directional relay at the other end. If a fault occurs at $F_1$, then only relay at C′ and at D′ will trip. Hence the faulty feeders will be isolated. In all other directional relays, direction of current is such that they will not trip. The nondirectional relays are current and time graded. So for a fault, only the nearest relay will trip, other will not.

## 14.19  EARTH FAULT PROTECTION SCHEME

A fault that involves ground is called earth fault, for example, L–G fault, L–L–G fault, etc. The protection scheme of an element of a power system against earth fault is known as earth fault protection.

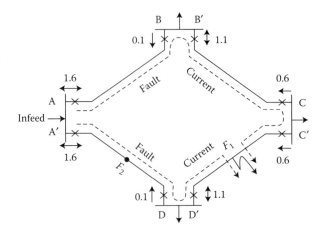

FIGURE 14.25  Protection of ring feeders.

Relays used in earth fault protection are known as earth fault relays. The PSM of earth fault relays varies from 20% to 80% of the CT. These relays are highly sensitive. Figure 14.26a shows current in the secondary of CTs of different phases. Earth fault relays are mainly energized by the residual current in a three-phase system, that is, $(I_A + I_B + I_C)$, which is zero under balanced condition. But in earth fault condition, this current will have a value. If this value is higher than the pickup value of the relay then it operates. Theoretically its current setting may be at any value above zero. But in practical cases, it is normally 20%–30% PSM. The manufacturer provides a range of plug settings for earth fault relay from 20% to 80% of the CT secondary rating in steps of 10%.

The magnitude of the earth fault current depends on the fault impedance. In case of earth fault, the fault impedance depends on the system parameter and on the type of the neutral earthing. The neutral may be solidly grounded or grounded through resistance and reactance. The fault impedance for earth faults is much higher than that for phase faults. Hence, the earth fault current is low compared to phase fault current. An earth fault relay is independent of load current. Its setting is below normal load current.

Figure 14.26b and c show an earth fault relay used for the transformer and alternator, respectively. When an earth fault occurs, zero sequence current flows through the neutral. This current is actually responsible for actuating the relay. Figure 14.26d shows the connection of an earth fault relay using a special type of CT, known as core balance CT or ring CT, which encircles the three-phase conductor.

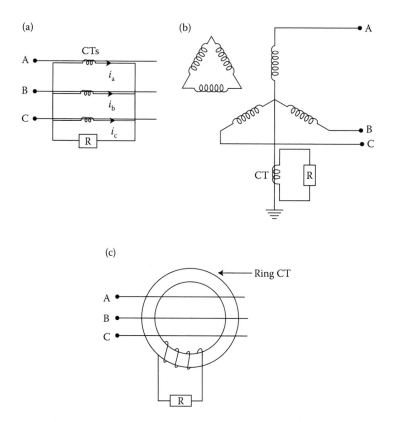

FIGURE 14.26   Various earth fault protective schemes.

## 14.20 DISTANCE PROTECTION SCHEME

Distance protection is mainly used in the extra high-voltage and high-voltage transmission and subtransmission lines. The relays used in this scheme basically measure the impedance or some components of the line impedance at the relay location. The measured quantity is proportional to the line length between the location of the relay and the point where the fault has occurred. The measured quantity proportional to the distance along the measuring relay is known as distance relay. Modern distance relays provide high-speed fault clearance. They are used for the protection of transmission and subtransmission lines at 220, 132, 66, and 33 kV. Sometimes, they are also used at 11 kV. In recent times, carrier current protection is largely used in lines above 132 kV. The relaying used in carrier current protection are distance relays. A distance protection scheme is a non-unit system of protection. A single scheme provides both primary and backup protection.

The distance relays have the following types:

1. Impedance relays

2. Reactance relays

3. Mho relays

4. Angle impedance relays

5. Quadrilateral relays

## 14.21 IMPEDANCE RELAY

It is already stated that the impedance relay measures the impedance of the line between relay location and the fault point. The impedance relay is an indirect measurement of the line length and the impedance includes both resistance and the reactance of the said line segment.

### 14.21.1 Operating Principle

In impedance relays, current (from CT) is compared with voltage (from PT) at the relay location. The current produces a positive torque (operating torque) and voltage produces a negative torque (restraining torque). The equation can be written as

$$T = K_1 I^2 - K_2 V^2 - K_3$$

where $K_1$, $K_2$, and $K_3$ are constants, $K_3$ being the torque due to the control spring effect. Neglecting the spring effect, we have

$$T = K_1 I^2 - K_2 V^2$$

For the operation of the relay, the following operation should be satisfied.

$$K_1 I^2 > K_2 V^2$$

$$\text{or} \quad \frac{V^2}{I^2} < \frac{K_1}{K_2}$$

$$\text{or} \quad \left(\frac{V}{I}\right) < K \quad \text{where } K \text{ is a constant}$$

$$\text{or} \quad Z < K$$

The relay operates if the measured impedance $Z$ is less than a given present value.

### 14.21.2 Characteristic

The characteristic of impedance relay can be shown in Figure 14.27a, in terms of voltage and current at the relay location. The curve is slightly bent due to the control spring effect. The R-X diagram (Figure 14.27b) of the impedance relay is also shown below. Here $Z = K$ represents a circle and $Z < K$ indicates the area within the circle. It is the operating zone of the relay. The radius of the circle is setting of the relay. Here $\phi$ is the phase angle between $V$ and $I$. The operating time is constant, irrespective of the fault location within the protected section.

### 14.21.3 Directional Units Used with Impedance Relays

Impedance relays are basically nondirectional relays. As its characteristic is a circle, the relay will trip for a fault point lying within the circle, irrespective of the fact that the fault point lies either in forward direction or in the reverse direction. In the R-X diagram, the relay will trip for a fault point at F which is behind the relay location. It is always desired that the relay should trip in the forward direction only. So a directional relay is used in series. At any location, three impedance relays along with a directional unit is employed as shown in Figure 14.28. Here zone 2 and zone 3 will backup zone 1 if the relay fails at zone 1. The operating time of these relays is different from each other. Here $t_3 > t_2 > t_1$.

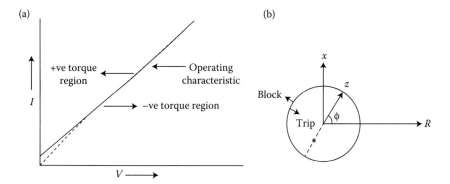

FIGURE 14.27 (a) Operating characteristic of an impedance relay. (b) Operating characteristic of an impedance relay on the $R$-$X$ diagram.

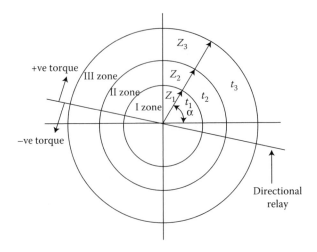

FIGURE 14.28 Characteristics of three-zone impedance relays with directional unit.

## 14.22 REACTANCE RELAY

The reactance type distance relay is an overcurrent relay with directional element that either aids or opposes the overcurrent element. In other words, a reactance relay is an overcurrent relay with directional restraint. The directional element is so designed that its maximum torque angle is 90°, that is, $\alpha = 90°$.

### 14.22.1 Operating Characteristic

The operating characteristic of reactance type distance relay can be derived as follows:

$$T = K_1 I^2 - K_2 VI \cos(\theta - \alpha) - K_3$$

$$T = K_1 I^2 - K_2 VI \sin\theta - K_3 \; (\theta \text{ is positive when } I \text{ lags})$$

where $T$ is the torque, $V$ is the voltage, $I$ is the current, $K_1$, $K_2$ are the constants, and $K_3$ is the spring constant.

$K_1 I^2$ represents the characteristic of overcurrent relay, and $K_2 VI \sin\theta$ represents the characteristic of directional element.

At balance point

$$K_1 I^2 = K_2 VI \sin\theta + K_3$$

Dividing both sides by $I^2$, we get

$$K_1 = K_2 \frac{V}{I} \sin\theta + \frac{K_3}{I^2}$$

or

$$\frac{V}{I} \sin\theta = Z\sin\theta = \frac{K_1}{K_2} - \frac{K_3}{K_2 I^2}$$

Neglecting control spring effect,

$$X = \frac{K_1}{K_2} = \text{Constant}$$

This is an equation of straight line and operating characteristic will be as shown in Figure 14.29.

In this type of relay, the resistance component of the impedance has no effect on its operation and the relay responds solely to the reactance component. The structure that is generally used for reactance type distance relay is induction cup and double induction loop structure.

## 14.23  MHO TYPE DISTANCE RELAY

The significance of the name of mho relay is that the mho characteristic when plotted in the admittance instead of impedance axis indicates a straight line. The mho relay supersedes impedance and reactance relays,

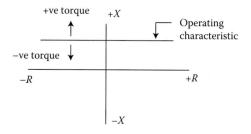

FIGURE 14.29   Operating characteristic of a reactance relay.

due to the fact that it has more stable swings than reactance or impedance relay.

This relay has a voltage restraining element that opposes the directional element. The operation of this relay is same as that of impedance type distance relay but with a difference that no separate directional unit is required because mho units are inherently directional.

The torque equation is given as

$$T = K_1 VI \cos(\theta - \alpha) - K_2 V^2 - K_3$$

At balance point:

$$K_2 V^2 = K_1 VI \cos(\theta - \alpha) - K_3$$

Dividing both side by $K_2 VI$, we get

$$\frac{V}{I} = Z = \frac{K_1}{K_2} \cos(\theta - \alpha) - \frac{K_3}{K_2 VI}$$

If the control spring effect is neglected,

$$Z = \frac{K_1}{K_2} \cos(\theta - \alpha)$$

The operating characteristic of such a relay, which is circular one, is shown in Figure 14.30.

## 14.24 UNIVERSAL TORQUE EQUATION

Most of the protection relay consists of some arrangement of electromagnet which have current winding, voltage winding or both. Current winding produce magnetic fluxes and torque is developed by interaction between the fluxes of same windings or between fluxes of both the windings.

Torque developed by current winding $= K_1 I^2$

Torque developed by voltage winding $= K_2 V^2$

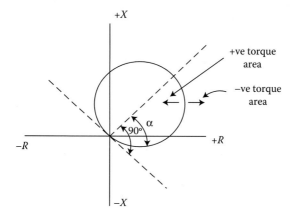

FIGURE 14.30   Operating characteristic of mho relay.

If both the current and voltage winding are employed, torque developed by the interaction between fluxes is

$$= K_3 VI \cos(\theta - \alpha)$$

where $\theta$ is the angle between $V$ and $I$, and $\alpha$ is the value of torque angle (known as relay maximum torque angle). It is the design control of the relay.

Where all elements are present, the torque will be produce by all the three cases, so that the total torque produce in general case is

$$T = K_1 I^2 + K_2 V^2 + K_3 VI \cos(\theta - \alpha) + K_4$$

where $K_1$, $K_2$, $K_3$ are the tap settings or constants of $I$ and $V$, and $K_4$ is the mechanical constraint due to a spring or gravity.

By assigning plus or minus sign to some of the terms and letting others be zero and sometime adding some terms having a combination of voltage and current, the operating characteristics of all types of relays can be obtained.

For example, for overcurrent relay $K_2 = 0$, $K_3 = 0$ and the spring torque will be $K_4$. Similarly for directional relay $K_1 = 0$, $K_2 = 0$.

## 14.25  DIFFERENTIAL RELAYS

Differential relay is a suitably connected overcurrent relay which operates when the phasor difference of current at the two ends of a protected element exceeds a pre determined value. Most of the differential relays are of current differential types.

The following are the various type of differential relays:

1. Simple (basic) differential relay

2. Percentage (biased) differential relay

3. Balanced (opposed) voltage differential relay

## 14.26 SIMPLE DIFFERENTIAL RELAY

Figure 14.31 shows the arrangement of an overcurrent relay connected to work on the simple differential principle. In this arrangement, a pair of CT are fitted on either ends of the element to be protected and secondary winding of CTs are connected in series so that they carry induced current in the same direction.

Under normal conditions, where there is no fault or there is external fault, the current in two CTs secondary are equal and relay operating coil, therefore does not carry any current.

Whenever there is an interal fault, current in the two secondaries of CTs (fitted on either end) are different, the relay operating coil gets energized by the current equal to their difference, and the trip circuit is completed to operate the CBs.

For the simple differential relay, the stability ratio is very poor, so a percentage differential relay has been developed.

## 14.27 PERCENTAGE DIFFERENTIAL RELAY

The schematic diagram of the percentage-biased differential relay is shown in Figure 14.32. This relay is designed to respond to the differential current

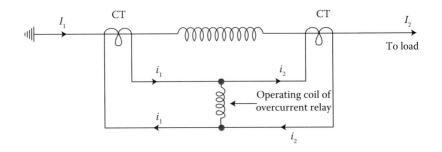

FIGURE 14.31   Principle of simple differential relay.

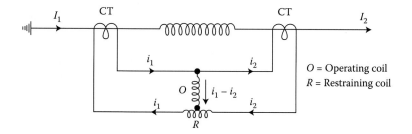

FIGURE 14.32   Principle of percentage-biased differential relay.

in the form of its fractional relation to the current following through the protected section. It is essentially an overcurrent balanced beam type relay with an additional restraining coil. The restraining coil produce a bias force in the opposite direction to the operating force.

Under normal and through conditions, the bias force due to restraining coil is greater than the operating force therefore the relay remains inoperative.

When an internal fault occurs, the operating force exceeds the bias force and consequently the trip contacts are closed to open the CB.

The differential current in the operating coil O is $(i_1 - i_2)$, while the current in the restraining coil R is $((i_1 + i_2)/2)$, since the operating coil is connected to the midpoint of the restraining coil. The number of turns in the restraining coil is $N_r$, so the total ampere turns are $((i_1 N_r/2) + (i_2 N_r/2))$.

Now the ampere turns of the operating coil, $(AT)_o = N_o(i_1 - i_2)$.

Neglecting spring restraint, the relay will operate when

$$(AT)_o > (AT)_r$$

$$N_o(i_1 - i_2) > N_r \left( \frac{i_1 + i_2}{2} \right)$$

or

$$(i_1 - i_2) > \frac{N_r}{N_o} \left( \frac{i_1 + i_2}{2} \right)$$

$$\therefore i_o > K i_r$$

Thus, at the threshold of operation of the relay, the ratio of the differential operating current $(i_o)$ to the restraining current $(i_r)$ is a fixed percentage. And for operation of the relay, the differential operating current must be greater than this fixed percentage of the restraining (through fault) current. Hence this relay is called percentage differential relay. The percentage differential relay is also known as "biased differential relay."

## 14.28 BALANCED VOLTAGE DIFFERENTIAL RELAY

Figure 14.33 shows a balanced voltage differential relay. In this scheme of protection, two similar CTs are connected at either end of the element to be protected by means of pilot curves. The secondaries of CT are connected in series with a relay in such a manner that under normal condition their induced electromotive forces are in opposition.

Under normal condition, equal current ($i_1 = i_2$) flow in both the primary windings. Therefore, the secondary voltage of the two transformers are balanced against each other and no current will flow through relay operating coil.

Whenever fault occurs in the protected zone, the current in the two primaries will differ from one another (i.e., $i_1 \neq i_2$) and their secondary voltage will no longer be in balance. Consequently, a circulating current will flow through the operating coil, causing the trip circuit to close.

*Disadvantages:*

1. In order to achieve the accurate balance between CT pairs, a multi-gap transformer construction is required.

2. Owing to the capacitance of pilot wires, this system is suitable for protection of cables of relatively short lengths. On long cables, the charging current may be sufficient to operate the relay even if a perfect balance of CT is attained.

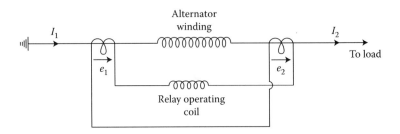

FIGURE 14.33  Principle of balanced voltage differential relay.

The above disadvantages have been overcome in Translay (modified) balanced voltage system.

## 14.29 TRANSLAY RELAY

Translay relay is similar to overcurrent relay except that (1) secondary is not closed on itself and also (2) the central limb is provided with copper loop or ring. Figure 14.34 shows a simplified diagram of a Translay scheme.

Under healthy condition, when the current at the two ends of the feeder is the same, the primaries of the two relays carry the same current including equal voltage in the secondaries, which are so connected that their voltage are in opposition. Hence no current flows in the two secondary circuits under the condition and so no torque results.

Whenever the current leaving the feeder differs the current entering the feeders, unequal voltage are induced in the secondaries. Consequently, a circulating current flows in the secondary circuit causing torque to be exerted on the disk of each relay. Since the direction of secondary current will be opposite in the two relays, therefore torque in one relay will tend to close the trip circuit while, in other relay, the torque will hold the movement in the operated condition. It may be noted that the resulting operating torque depends upon the position and nature of the fault in the protected zone and at least one element of either relay will operate under any fault condition. A Translay relay is extensively used in Translay protection applied to feeders. This relays used embrace the function of transformer as well as relay.

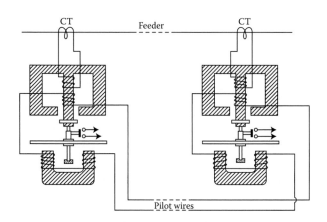

FIGURE 14.34   Simplified diagram of a Translay scheme.

## WORKED EXAMPLES

### EXAMPLE 14.1

The current setting of an IDMT relay is 5 A. The relay has a plug setting of 125% and the TMS of 0.6. The CT ratio is 400/5. Determine the operating time of the relay for a fault current 4000 A. At TMS = 1, operating time at various PSH are given in Table 14.1.

**Solution**

$$CT \text{ ratio} = \frac{\text{Secondary current}}{\text{Relay current setting}}$$

$$= \frac{(\text{Fault current(primary current)}) / CT \text{ ratio}}{\text{Relay current settings}}$$

$$= \frac{4000/80}{6.25} = 8$$

The operating time from the given table at PSM 8 is 3 s with TMS = 1.

Therefore, with PSM = 8, TMS = 0.6, operating time $t_{op}$ = 0.6 × 3 = 1.8 s.

### EXAMPLE 14.2

A 20-MVA transformer, which may be called upon to operate at 25% overload, feeds 11-kV bus bars through a CB; other CB supply outgoing feeders. The transformer CB is equipped with 1000/5 A CTs and the feeder CBs with 500/5 A CTs and all sets of CTs feed induction-type overcurrent relays. The relays on the feeder CBs have a 125% plug seeting and a 0.4 time setting. If a three-phase fault current of 7500 A flows from the transformer to one of the feeders, find the operating time of the feeder relay, the minimum plug setting of the transformer relay, and its time setting assuming a discriminative time margin of 0.5 s. The time–current characteristic of the relays is same as shown in Figure 14.16.

TABLE 14.1   Operating Times at Various PSM

| PSM | 2 | 4 | 5 | 8 | 10 | 20 |
|---|---|---|---|---|---|---|
| Operating time in seconds | 10 | 5 | 4 | 3 | 2.8 | 2.4 |

**Solution**

1. Feeder:

$$\text{Secondary current} = 7500 \times \frac{5}{500} = 75 \, \text{A}$$

Relay current setting $= 125\%$ of $5 \, \text{A} = 1.25 \times 5 = 6.25 \, \text{A}$

$$\text{PSM} = \frac{\text{Secondary current}}{\text{Relay current setting}} = \frac{75}{6.25} = 12$$

From the curve in Figure 14.16, the operating time at PSM of 12 for a TMS of $1 = 2.8 \, \text{s}$
Since TMS of the relay $= 0.4$, operating time of the relay $= 0.4 \times 2.8 = 1.12 \, \text{s}$

2. Transformer:

$$\text{Overload current} = \frac{(1.25 \times 20) \times 10^3}{\sqrt{3} \times 11} = 1312 \, \text{A}$$

$$\text{Secondary current} = 1312 \times \frac{5}{1000} = 6.56 \, \text{A}$$

$$\text{PSM} = \frac{6.56}{\text{PS} \times 5}$$

where PS means plug setting of the relay.

Since the transformer relay must not operate to overload current, its PSM must be less than 1, that is, $\text{PS} \times 5 > 6.56$. Thus, $\text{PS} > 6.56/5 > 1.31$ or $131\%$.

The plug setting are restricted to standard values in intervals of $25\%$, so the nearest value is $150\%$.

$$\text{Secondary fault current} = 7500 \times \frac{5}{1000} = 37.5 \, \text{A}$$

Relay current setting $= 150\%$ of $5 \, \text{A} = 1.5 \times 5 \, \text{A} = 7.5 \, \text{A}$

$$\text{PSM} = \frac{\text{Secondary current}}{\text{Relay current setting}} = \frac{37.5}{7.5} = 5$$

The operating time from the curve in figure at PSM of 5 and TMS of 1 = 4.7 s. But,

Actual operating time required

= Operating time of feeder relay + Discriminative time margin

= 1.12 s + 0.5 s

= 1.62 s

Hence required TMS = 1.62/4.7 = 0.345.

### EXERCISES

1. What is protective relay? Explain its function in an electrical system.

2. Discuss the fundamental requirements of protective relaying.

3. Describe briefly some important types of electromagnetic attraction relays.

4. Derive the equation for torque developed in an induction relay.

5. Write a brief note on relay timing.

6. Define and explain the following terms as applied to protective relaying:

    a. Pickup value

    b. Current setting

    c. Plug-setting multiplier

    d. Time-setting multiplier

7. Sketch a typical time/PSM curve.

8. Explain with the help of neat diagram the construction and working of:

    a. Nondirectional induction type overcurrent relay

    b. Induction type directional power relay

9. Describe the construction and principle of operation of an induction type directional overcurrent relay.

10. Explain the working principle of distance relays.

11. Write a detailed note on differential relays.

12. Describe the Translay scheme of protection.

# CHAPTER 15

# Protection of Alternators and Transformers

## 15.1 INTRODUCTION

The modern power system comprises of several elements that include alternators, transformers, induction motors, bus bar, transmission lines, and other equipment. It is suitable and necessary to protect each element from a mixture of fault conditions. The protective relays can be productively applied to detect the unlawful conduct of any circuit element and initiate disciplinary measures. As a matter of convenience, this chapter deals with the protection of alternator and transformers only.

## 15.2 PROTECTION OF ALTERNATORS

The generator or alternator is the most important and costly equipment of the power system. It is subjected to most number of faults and the nature of the faults is most severe. So the protection of alternator is complex and elaborate. A modern generator is provided with the following protections:

1. Stator protection:

    a. Percentage differential protection

    b. Protection against stator inter-turn faults

    c. Stator overheating protection

2. Rotor protection:

    a. Field ground fault protection

    b. Loss of excitation protection

    c. Protection against rotor overheating because of unbalanced $3\phi$ stator current

3. Miscellaneous:

    a. Overvoltage protection

    b. Overspeed protection

    c. Protection against motoring

    d. Protection against vibration

    e. Bearing overheating protection

    f. Protection against auxiliary failure

    g. Protection against voltage regulator failure

## 15.3 STATOR PROTECTION

### 15.3.1 Percentage Differential Protection

This protection is used for the generators above 1 MW. It protects generators against internal winding faults, that is, phase-to-phase and phase-to-ground faults. This is also called bias differential protection. The polarities of the secondary voltages of current transformers (CTs) at a particular moment for an external fault are shown in Figure 15.1a. This polarity is also true for normal direction of current flow in healthy condition.

In the operating coil, the current sent by the upper CT is cancelled by the current sent by the lower CT and the relay does not operate. For an internal fault, the polarity of the secondary voltage of the upper CT is reversed, as shown in Figure 15.1b, in that particular phase affected by the fault. Now, the operating coil carries the sum of the current sent by the upper CT and the lower CT, it operates and trips the circuit breaker (CB). This protection does not respond to external faults and overloads.

This protection is also known as circulating current protection or Merz–Price protection.

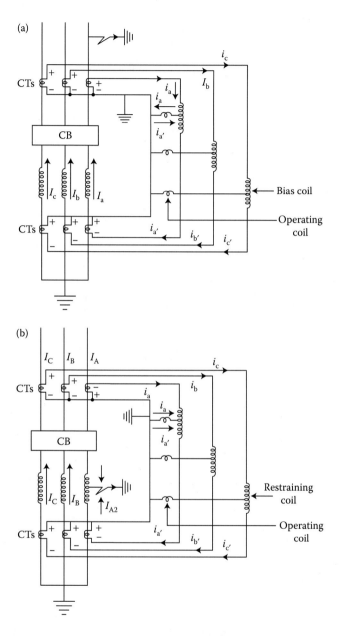

FIGURE 15.1 (a) Percentage of differential protection for external fault condition (instantaneous current directions shown for external fault condition). (b) Percentage of differential protection for generator (instantaneous current directions shown for internal fault condition).

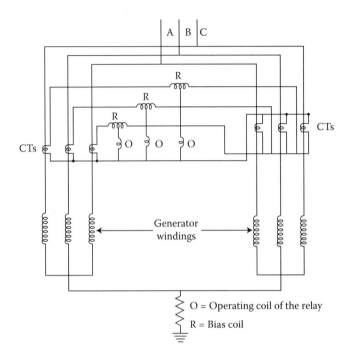

FIGURE 15.2  Transverse percentage of differential protection for multi-winding generator.

### 15.3.2 Protection against Stator Inter-turn Faults

Longitudinal percentage differential protection does not detect stator inter-turn faults. A transverse percentage differential protection shown in Figure 15.2 is employed for the protection of the generator against stator inter-turn faults. This type of protection is used for generators having parallel windings separately brought out to the terminals. The coil of the modern large steam turbine-driven generators usually have only one turn per phase per slot and hence they do not need inter-turn fault protection. But hydro generators having parallel windings in each phase employ such protection which thus provides backup protection and detects inter-turn faults. This scheme is also known as split-phase protection.

### 15.3.3 Stator Overheating Protection

Overheating of the stator may be caused by the failure of the cooling system, overloading, or core faults like short-circuited laminations and failure of core bolt insulation. Modern generators employ two methods to detect overheating both being used in large generators (above 2 MW).

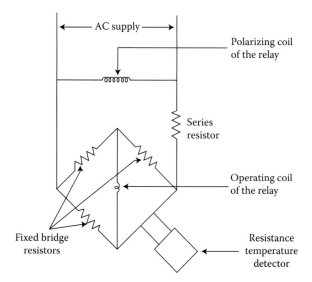

FIGURE 15.3   Stator-overheating protection.

In one method, the inlet and outlet temperature of the cooling medium which may be hydrogen/water are compared for detecting overheating. In the other method, the temperature sensing elements are embedded in the stator slots to sense the temperature. Figure 15.3 shows a stator over-heating relaying scheme. When the temperature exceeds a certain present maximum temperature limit, the relay sounds an alarm. The scheme employs a temperature detector limit, relay, and Wheatstone bridge for the purpose. The temperature sensing elements may either be thermistors, thermocouples, or resistance temperature indicators. They are embedded in the stator slots at different locations. These elements are connected to a multiway selector switch which checks each one in turn for a period long enough to operate an alarm relay.

For small generators, a bimetallic strip heated by the secondary current of the CT is placed in the stator circuit. This relay will not operate for the failure of cooling system.

## 15.4  ROTOR PROTECTION

### 15.4.1  Field Ground-Fault Protection

As the field circuit is operated undergrounded, a single ground fault does not affect the operation of the generator or cause any damage. However, a single rotor fault to earth increases the stress to the ground in the field when stator transient induces an extra voltage in the field winding. Thus, the probability of occurrence of the second ground fault is increased. In case a second ground fault occurs, a part of the field winding is bypassed, thereby increasing the current through the remaining portion of the field winding. This causes an unbalance in the air-gap fluxes, thereby creating an unbalance in the magnetic forces on opposite sides of the rotor. The unbalancing in the magnetic forces makes the rotor shaft eccentric. This also causes vibrations. Even though the second ground fault may not bypass enough portion of the field winding to cause magnetic unbalance, the arcing at the fault causes local heating which slowly distorts the rotor producing eccentricity and vibration. Figure 15.4 shows the schematic diagram of rotor earth protection.

For this protection, a DC voltage is impressed between the field circuit and earth through polarized moving iron relay. It is not necessary to trip the machine when a single field earth fault occurs. Usually an alarm is sounded. Then immediate steps are taken to transfer the load from the faulty generator and to shut it down as quickly as possible to avoid further problem.

### 15.4.2 Loss of Excitation Protection

When the excitation of a generator is lost, it speeds up slightly and operates as an induction generator. Round rotor generators do not have damper windings and hence they are not suitable for such an operation. The rotor is overheated quickly due to heavy induced current in the rotor

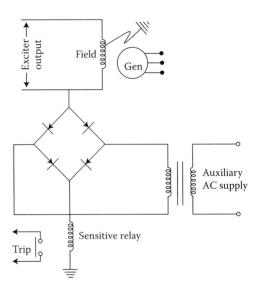

FIGURE 15.4   Earth fault protection.

iron. The rotors of salient pole generators are not overheated because they have damper windings which carry induced current. The stators of both salient and non-salient pole generators are overheated due to wattles current drawn by the machines as magnetizing current from the system. The stator overheating does not occur as quickly as rotor overheating. A large machine may upset the system stability because it draws reactive power from the system when it runs as an induction generator, whereas it supplies reactive power when it runs as a generator. A machine provided with a quick-acting automatic voltage regulator and connected to a very large system may run for several minutes as an induction generator without harm.

Field failure may be caused by the failure of excitation. A protective scheme employing offset mho or directional impedance relay.

### 15.4.3 Protection against Rotor Overheating Because of Unbalanced 3φ Current

The unbalanced condition of an alternator may arise due to the following reasons:

1. When a fault occurs in the stator winding

2. An unbalanced external fault which is not cleared quickly

3. Open circuiting of a phase

4. Failure of one contact of the C.B

Due to the unbalancing, the negative sequence component of current is produced. It causes double frequency current to be induced in the rotor iron. This results in severe overheating of the rotor.

The time for which the rotor can be allowed to withstand such a condition is related by the expression:

$$I_2^2 t = K$$

where $I_2$ is the negative sequence component of the current, $t$ is the time, and $K$ is the constant that depends on the type of generating set and cooling system.

$$K = 7 \text{ for turbo generator with direct cooling}$$
$$= 60 \text{ for salient pole hydro generator}$$

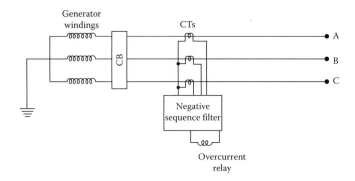

FIGURE 15.5   Protection against unbalanced stator current.

Figure 15.5 shows a protective scheme using negative sequence filter and relay. The overcurrent relay used in the negative-phase sequence protection has a long over rating time with a facility of range setting to permit its characteristic to be matched to $I_2^2 t$ characteristic of a machine. A typical time range of the relay is 0.2–2000 s. It has a typical construction with a special electromagnet. It has shaded pole construction with a Mu-metal shunt. The negative sequence filter gives an output proportional to $I_2$. It actuates an alarm as well as the time current relay which has a very inverse characteristic. The alarm unit also starts a timer which is adjustable from 8% to 40% of negative sequence component. The timer makes a delay in the alarm to prevent the alarm from sounding unnecessarily on unbalanced loads of short duration.

## 15.5 MISCELLANEOUS

### 15.5.1 Overvoltage Protection

Overvoltage protection may be caused by a defective voltage regulator or it may occur due to sudden loss of electrical load on generator. When a load is lost, there is an increase in speed and hence the voltage also increases. Overvoltage relays are provided.

### 15.5.2 Overspeed

A turbo generator is provided with a mechanical overspeed device. The speed governor normally controls its speed. It is designed to prevent any speed rise even with 100% load rejection. An emergency centrifugal overspeed device is also incorporated to trip emergency steam valves when the speed exceeds 110%.

### 15.5.3 Protection against Motoring

When the steam supply is cut off the generator runs as a motor. The steam turbine gets overheated because insufficient steam passes through the turbine to carry away the heat generated by windage loss. Therefore, a protective relay is required for the protection of the steam turbine. Generally, a sensitive reverse power relay is available to give protection of the alternator against motoring.

### 15.5.4 Field Suppression

When a fault occurs in the generator winding the CB trips and the generators is isolated from the system. However, the generator still continues to feed the fault as long as the excitation is maintained, and the damage increases.

Therefore, it is desirable to suppress the field as quickly as possible. The field cannot be destroyed immediately. The energy associated with the flux must be dissipated into an external device. To achieve this, the field winding is connected to a discharging resistor to absorb the stored energy. The discharged resistor is connected in parallel with the field winding before opening the field CB.

## 15.6 TRANSFORMER PROTECTION

Transformer faults may be classified as follows:

1. External faults

2. Internal faults

### 15.6.1 External Faults

In case of external faults, the transformer must be disconnected if other protective devices meant to operate for such faults, fails to operate within a predetermined time. For external faults, time-graded overcurrent relays are employed as backup protection. Also in case of sustained overload conditions, the transformer should not be allowed to operate for long duration. Thermal relays are used to detect overload condition and give an alarm.

### 15.6.2 Internal Faults

Internal faults are classified into two groups.

1. *Short circuit in the transformer windings and connections.* These are electrical faults of serious nature and are likely to cause immediate damage. Such faults are detectable at the winding terminals by unbalances in voltage or current. These types of fault include line to ground or line to line and inter-turn faults on high-voltage (HV) and low-voltage (LV) windings.

2. *Incipient fault.* Initially, such faults are of minor nature but slowly might develop into major faults. Such faults include poor electrical connections, core faults, failure of coolant, regulator faults, bad load sharing between transformers, etc.

## 15.7 PERCENTAGE DIFFERENTIAL PROTECTION

This protection is also known as circulating current protection or Merz–Price protection. It is used specially for short circuits within the transformer of rating 5 MVA or above. This scheme is employed for the protection of transformers against internal short circuits. It is not capable of detecting incipient faults. Figure 15.6 shows the schematic diagram of percentage differential protection for a $Y - \Delta$ transformer. The current entering end has been marked as positive. The end at which current is leaving has been marked negative. $O$ and $R$ are the operating and restraining coils of the relay, respectively. The connection are made in such a way that under normal conditions or in case of external faults, the current flowing in the relay operating coil due to CTs of the primary side. Consequently relay does not operate under such conditions.

If a fault (earth fault or short circuit) occurs on the winding the polarity of the induced voltage of the CT of the secondary side is reversed. Now

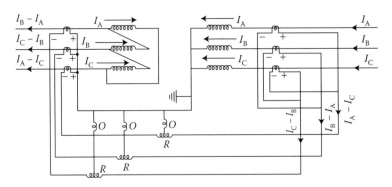

FIGURE 15.6　Percentage of differential protection for a $Y - \Delta$ transformer.

the current in the operating coil from CTs of both sides is in the same direction and causes the operation of the relay. In the star side of the transformer, CTs are connected in delta or vice versa.

## 15.8 OVERHEATING PROTECTION

Overheating of transformer is usually caused by sustained overloads or short circuits and very occasionally by the failure of the cooling system. The relay protection is also not provided against this contingency, and thermal accessories are generally used to sound an alarm.

## 15.9 RATE OF RISE OF PRESSURE RELAY

This device is capable of detecting rapid rise of pressure, rather than absolute pressure. Its operation is quicker than the pressure relief valve.

It is employed in transformers that are provided with gas cushions instead of conservators as shown in Figure 15.7. A modern pressure relay contains a metallic bellows full of silicone oil. The bellows is placed in the transformer oil. The relay is placed at the bottom of the tank where maintenance jobs can be performed conveniently. It operates on the principle of rate or increase of pressure. It is usually design to trip the transformer.

## 15.10 OVERCURRENT PROTECTION

Overcurrent relays are used for the protection of transformers of rating 100 kVA and below 5 MVA. An earth fault tripping element is also provided with overcurrent protection. For small transformers, OC relays are used for both overload and fault protection. An extremely inverse relay is desirable for overload for light faults, with instantaneous OC relay for heavy faults.

## 15.11 OVERFLUXING PROTECTION

The magnetic flux increases when voltage increases. This results in increased iron loss and magnetizing current. The core and core bolts get heated and the lamination insulation is affected. Protection against overfluxing is required where overfluxing due to sustained overvoltage can occur. The reduction in frequency also increases the flux density and consequently, it has similar effects as those due to overvoltage.

The expression of flux in a transformer is given by

$$\phi = K\frac{E}{f}$$

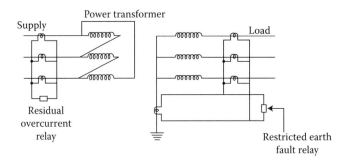

FIGURE 15.7    Earth fault protection for power transformer.

where $\phi$ is the flux, $f$ is the frequency, $E$ is the applied voltage, and $K$ is the constant. Therefore, to control the flux, the ratio $E/f$ is controlled.

Electronic circuits with suitable relays are available to measure $E/f$ ratio. Usually 10% overfluxing can be allowed without damage. If $E/f$ exceeds 1:1, overfluxing protection operates. Overfluxing does not require high-speed tripping and hence instantaneous operation is undesirable when momentary disturbances occur.

## 15.12  EARTH FAULT PROTECTION

A simple overcurrent and earth fault relay does not provide good protection for a star-connected winding, particularly when the neutral point is earthed through impedance. Restricted earth fault protection, as shown in Figure 15.7 provides better protection. This scheme is used for the winding of the transformer connected in star, where the neutral point is either solidly earthed or earthed through impedance. The relay used is of high impedance type to make the scheme stable for external faults.

For delta connection or ungrounded star winding of the transformer, residual overcurrent relay is employed. The relay operates only for a ground fault in the transformer.

The differential protection of the transformer is supplemented by restricted earth fault protection in case of a transformer with its neutral grounded through resistance. For such a case, only about 40% of the winding is protected with a differential relay pickup setting as low as 20% of the CT rating.

## 15.13  BUCHHOLZ RELAY

Buchholz relay is a gas-actuated relay installed in oil-immersed transformers for protection against all kinds of faults. Named after its inventor,

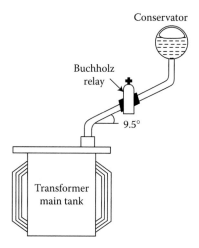

FIGURE 15.8   Transformer tank, Buchholz relay, and conservator.

Buchholz, it is used to give an alarm in case of incipient (i.e., slow develop-
ing) faults in the transformer and to disconnect the transformer from the
supply in the event of severe internal faults. It is usually installed in the
pipe connecting the conservator to the main tank as shown in Figure 15.8.
It is a universal practice to use Buchholz relays on all such oil-immersed
transformers having ratings in excess of 750 kVA.

Figure 15.9 shows the constructional details of a Buchholz relay. It takes
the form of a domed vessel placed in the connecting pipe between the main
tank and the conservator. The device has two elements. The upper element
consists of a mercury type switch attached to a float. The lower element contains a mercury switch mounted on a hinged type flap located in the direct
path of the flow of oil from the transformer to the conservator. The upper
element closes an alarm circuit during incipient faults, whereas the lower
element is arranged to trip the CB in case of severe internal faults.

In case of incipient faults within the transformer, the heat due to fault
causes the decomposition of some transformer oil in the main tank. The
products of decomposition contain more than 70% of hydrogen gas. The
hydrogen gas being light tries to go into the conservator and in the process gets entrapped in the upper part of relay chamber. When a predetermined amount of gas gets accumulated, it exerts sufficient pressure on the
float to cause it to tilt and close the contacts of mercury switch attached
to it. This completes the alarm circuit to sound an alarm. If a serious fault
occurs in the transformer, an enormous amount of gas is generated in the
main tank. The oil in the main tank rushes toward the conservator via

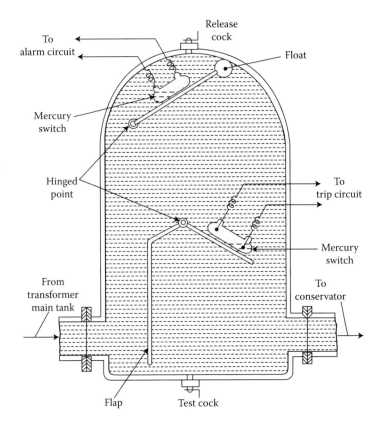

FIGURE 15.9 Buchholz relay.

the Buchholz relay and in doing so tilts the flap to close the contacts of mercury switch. This completes the trip circuit to open the CB controlling the transformer.

*Advantages:*

1. It is the simplest form of transformer protection.

2. It detects the incipient faults at a stage much earlier than is possible with other forms of protection.

*Disadvantages:*

1. It can only be used with oil-immersed transformers equipped with conservator tanks.

2. The device can detect only faults below oil level in the transformer. Therefore, separate protection is needed for connecting cables.

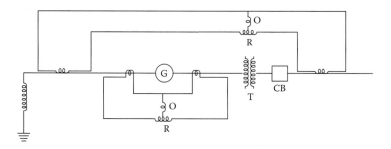

FIGURE 15.10   Differential protection of generator transformer unit.

## 15.14 GENERATOR TRANSFORMER UNIT PROTECTION

In a modern system, each generator is directly connected to delta connected primary winding of the power transformer. The star-connected secondary winding is HV windings and it is connected to the HV bus through a CB. In addition to normal protection of the generator and transformer, an overall biased differential protection is provided to protect both the generator and transformer as one unit. Figure 15.10 shows an overall differential protection. Usually harmonic restrained is not provided because the transformer is only connected to the bus bar at full voltage. However, there is a possibility of a small inrush current when a fault near the bus bar is cleared, suddenly restoring the voltage.

## WORKED EXAMPLES

### EXAMPLE 15.1

The neutral point of a three-phase, 20-MVA, 11-kV alternator is earthed through a resistance of 5 Ω. The relay is set to operate when there is an out of balance current of 1.5 A. The CTs have a ratio of 1000/5. Determine (1) the percentage of protected winding, and (2) value of earthing resistance required to protect 90% of the winding (Figure 15.11).

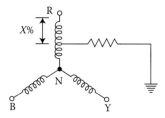

FIGURE 15.11   Neutral point of a three-phase alternator is earthed through a resistance.

**Solution**

1. The minimum current for which relay operators during fault condition $= 1.5 \times (1000/5) = 300$ A
Let $x\%$ be unprotected winding.

$$\text{EMF induced} = x\% \times \frac{11}{\sqrt{3}} \text{kV}$$

$$= \frac{x}{100} \times \frac{11 \times 10^3}{\sqrt{3}}$$

$$= \frac{110x}{\sqrt{3}}$$

where EMF is the electromotive force.

$$\text{Fault current through it} = \frac{110x/\sqrt{3}}{R}$$

$$= \frac{110x/\sqrt{3}}{5} = \frac{22x}{\sqrt{3}} \text{A}$$

Now fault current $= 300$ A. Therefore,

$$\frac{22x}{\sqrt{3}} = 300$$

$$x = 23.6\%$$

Percentage of winding protected $= (100 - 23.6)\% = 76.4\%$.

2. Now, 90% of the winding is protected.
10% of the winding is unprotected.
Unprotected winding $x = 10\% = (1/10)$

$$\therefore \text{EMF induced in 10\% winding} = \frac{11,000}{\sqrt{3}} \times \frac{1}{10}$$

$$= \frac{1100}{\sqrt{3}} \text{V}$$

When the fault current becomes 300 A, the relay will trip.

$$\therefore \quad \frac{1100/\sqrt{3}}{R} = 300$$

$$R = 2.12\,\Omega$$

## EXAMPLE 15.2

A 3-$\phi$ transformer of 33,000/6600 line volts is connected in star/delta. The protective transformers on LV side have ratio 300/5. What should be the CT ratio on HV side?

**Solution**

Phase voltage on the 33,000 V side is

$$V_{ph} = \frac{V_L}{\sqrt{3}}$$

$$= \frac{33,000}{\sqrt{3}}\,V$$

$$= 19,052\,V$$

Phase voltage on the 6600 V side is

$$V_{ph} = V_L = 6600\,V$$

Now turns ratio is

$$\frac{n_1}{n_2} = \frac{V_1}{V_2} = \frac{19,052}{6600} = 2.887$$

Finding the ratio of CT on HV side we have,

$$\frac{I_1}{I_2} = \frac{I_1}{300/5} = \frac{I_1}{60} = \frac{n_2}{n_1}$$

$$I_1 = \frac{60}{2.887} = 20.8\,A$$

Therefore, CT ratio on HV side is 20.8:1.

**EXERCISES**

1. Discuss the important faults on an alternator.

2. Explain with a neat diagram the application of Merz–Price circulating current principle for the protection of alternator.

3. Describe with a neat diagram the balanced earth protection for small-size generators.

4. What factors cause difficulty in applying circulating current principle to a power transformer?

5. Describe the construction and working of a Buchholz relay.

6. Describe the Merz–Price circulating current system for the protection of transformers.

7. Write short notes on the following:

   a. Earth-fault protection for alternator

   b. Earth-fault protection for transformers

# CHAPTER 16

# Traveling Wave

## 16.1 INTRODUCTION

A transmission line is a distributed parameter circuit with the unique ability to support traveling waves of voltage and current. Finite velocity of electromagnetic field propagation is present in a circuit which have distributed parameters. In such a circuit, the changes in voltage and current do not occur simultaneously in all parts of the circuit but spread out in the form of traveling waves and surges. When a transmission line is suddenly connected to a voltage source by closing a switch, the whole of the line is not energized all at once. This is due to the presence of distributed constants (inductance and capacitance in a loss free line). When a switch $S$ is closed, the inductance $L_1$ acts as an open circuit and capacitance $C_1$ as short circuit instantaneously. The same instant next section cannot be charged because the voltage across capacitor $C_1$ is zero. So unless the capacitor $C_1$ is charged to the some value whatsoever, charging of the capacitor $C_2$ through $L_2$ is not possible which, of course, will take some finite time. The same line of reasoning applies to the third section, fault section, and hence along. Thus, it has been ascertained that the potential at the successive section builds up step by step. This gradual buildup of potential over the transmission line conductor can be regarded, as though a voltage wave is traveling from one terminal to another final stage, and the gradual loading of the capacitances is due to associated current wave. The current wave, which is accompanied by a voltage wave, sets up a magnetic field in the surrounding space. At junction and terminations, these surges undergo reflections and refractions. In an extensive network with many lines and junctions, the number of traveling waves initiated by a single incident wave will mushroom at a considerable rate as the wave split and multiple reflections occur. It is rightful that the entire energy of the resulting waves cannot exceed the energy of the incident wave. However, it is possible for the voltage buildup at certain junctions due to reinforcing action of several waves. For a perfect study of the phenomenon, the use of Bewely lattice diagram or digital data processor is necessary. The study of traveling wave, therefore, plays an important role in knowing the voltage and current at all points in a power system. It assists in the design of insulators, protective equipment, the insularity of the terminal equipment, and overall insulation coordination.

## 16.2 SURGE IMPEDANCE AND VELOCITY OF PROPAGATION

Thus far, we have analyzed the transient behavior of several circuits with lumped parameters. However, there are some parts of a power system where this approach is inadequate. The most obvious instance is the transmission line. Here the $L$, $C$, and $R$ are uniformly distributed over the length of the line. For steady-state operation of the line, the transmission line could be exemplified by their actual circuit, that is, distributed parameters. We say that for a 50-Hz supply and short transmission line, the sending end current equals receiving end current and the change in voltage from sending end to receiving end is smooth. This is not so when transmission line is subjected to a transient. To understand the traveling wave phenomenon, overtransmission line has been represent in Figure 16.1 by a large number of $L$ and $C$ π sections. The line is assumed to be lossless. Let us consider $C$ as the capacitance per-unit length and $L$, the inductance per-unit length. As already noted in the presiding para, the gradual establishment of a line voltage can be considered as due to voltage wave traveling from the supply source and toward the far end, and the progressive charging of the line capacitances will account for the associated current wave.

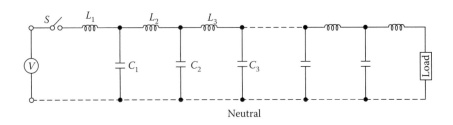

FIGURE 16.1    Equivalent π-section of a long transmission line.

Suppose that the wave after time $t$ has travelled through a distance $x$. Since, we have assumed lossless lines, whatever is the value of voltage and current waves at the kickoff, they remain same throughout the change of location. Take a distance $dx$ which is moved by the waves in time $dt$. The electrostatic flux is related with the voltage wave and the electromagnetic flux with the current wave. The electrostatic flux which is equal to the charge between the conductors of the line up to a distance $x$ given by

$$q = CVx \qquad (16.1)$$

The current in the conductor is determined by the rate at which the charge flows into and out of the line.

$$I = \frac{dq}{dt} = CV\frac{dx}{dt} \qquad (16.2)$$

Here, $dx/dt$ is the velocity of the traveling wave over the line conductor and is represented by $v$, then

$$I = VCv \qquad (16.3)$$

Likewise, the electromagnetic flux linkages created around the conductors due to the current following in them up to a distance of $x$ is given by

$$\psi = ILx \qquad (16.4)$$

The voltage is the rate at which the flux linkages link with the conductor:

$$V = \frac{d\psi}{dt} = IL\frac{dx}{dt} = ILv \qquad (16.5)$$

Dividing Equation 16.5 by 16.3,

$$\frac{V}{I} = \frac{LIv}{VCv} = \frac{LI}{VC}$$

or

$$\frac{V^2}{I^2} = \frac{L}{C}$$

or

$$\frac{V}{I} = \sqrt{\frac{L}{C}}$$

$$Z_n = \sqrt{\frac{L}{C}} \tag{16.6}$$

The expression is the ratio of voltage and current which has the dimensions of impedance and is therefore here designated as surge impedance of the line. It is also called the natural impedance because this impedance has nothing to do with the load impedance, but depends only on the line constant. The value of this impedance is 400–600 $\Omega$ for an overhead line and 40–60 $\Omega$ for a cable.

Multiplying Equations 16.3 and 16.5:

$$VI = CVv \times LIv$$

$$I = LCv^2$$

$$v = \frac{1}{\sqrt{LC}}$$

Now expressions for $L$ and $C$ for overhead lines are

$$L = 2 \times 10^{-7} \ln\frac{d}{r} \, \text{H/m}$$

$$C = \frac{2\pi\epsilon}{\ln(d/r)} \, \text{F/m}$$

$$v = \cfrac{1}{\sqrt{2 \times 10^{-7} \ln(d/r) \cdot [(2\pi \times 8.854 \times 10^{-12})/\ln(d/r)]}}$$
$$= 3 \times 10^{-8} \text{ m/s}$$

This is the velocity of light. This means the velocity of propagation of the traveling waves over a transmission line equals the velocity of light. In actual practice because of the resistance of the line, the velocity of approximately 250 m/μs is assumed. The velocity of propagation over the cables will be smaller than that of the overhead lines because in case of overhead lines $\epsilon_r = 1$ while for cables $\epsilon_r > 1$.

## 16.3 REFLECTION AND REFRACTION OF WAVES

If a traveling wave arrives at a point where the impedance suddenly changes, the wave is partly transmitted and partly reflected. Loading points, line–cable junctions, and even faults constitute such discontinuities. Independent waves meeting along a line will combine in accordance with polarity to provide different voltage and current level at the meeting point. It is convenient to adopt a standard sign convention and in what follows, forward waves of current and voltage are given the same polarity. If the wave is being reflected, the corresponding current and voltage waves are given opposite polarity. This may be illustrated by considering waves of current and voltage being transmitted along a line of characteristic impedance $Z_C$ terminated by an impedance $Z$ (Figure 16.2).

Let $V$ and $I$ represent inclined waves, $V_T$ and $I_T$ represent the transmitted or refracted waves and $V_R$ and $I_R$ the reflected waves. The following relations hold good for incident, transmitted and reflected voltage, and current waves.

$$V = IZ_C \tag{16.7}$$

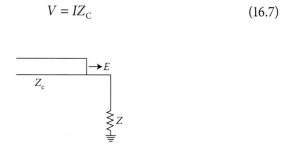

FIGURE 16.2   Line terminated through impedance $Z$.

$$V_T = I_T Z \tag{16.8}$$

$$V_R = -I_R Z_C \tag{16.9}$$

The negative sign Equation 16.9 is because of the fact that $V_R$ and $I_R$ are traveling in the negative direction of $x$ or backwards on the same line. The transmitted voltage and a current will be, respectively, the algebraic sum of incident and reflected voltage and current waves.

$$V_T = V + V_R \tag{16.10}$$

$$I_T = I + I_R \tag{16.11}$$

Substituting the values of $I$, $I_R$, and $I_T$ from Equations 16.7 through 16.9 in Equation 16.11, we have

$$\frac{V_T}{Z} = \frac{V}{Z_C} - \frac{V_R}{Z_C} \tag{16.12}$$

From Equations 16.10 and 16.12, we have

$$\frac{V_T}{Z} = \frac{V}{Z_C} - \frac{V_T - V}{Z_C}$$

or

$$V_T \times \frac{Z_C}{Z} + V_T = 2V$$

or

$$V_T \left( 1 + \frac{Z_C}{Z} \right) = 2V$$

or

$$V_T = V \cdot \frac{2Z}{Z + Z_C} \tag{16.13}$$

The coefficient $2Z/(Z + Z_C)$ is called transmitting coefficient or refraction coefficient. It is denoted by $\beta$.

$$V_T = \beta \cdot V$$

Now,

$$I_T = \frac{V_T}{Z} = \frac{2V}{Z + Z_C} = \frac{2IZ_C}{Z + Z_C} \tag{16.14}$$

$$V_R = V_T - V = \frac{2ZV}{Z + Z_C} - V = V \cdot \frac{Z - Z_C}{Z + Z_C} \tag{16.15}$$

$$V_R = V \cdot \frac{Z - Z_C}{Z + Z_C} \tag{16.16}$$

The coefficient $(Z - Z_C)/(Z + Z_C)$ is called reflection coefficient. It is denoted by $\alpha$.

$$V_R = \alpha \cdot V$$

Now,

$$1 + \alpha = 1 + \frac{Z - Z_C}{Z + Z_C} = \frac{Z + Z_C + Z - Z_C}{Z + Z_C} = \frac{2Z}{Z + Z_C} = \beta \tag{16.17}$$

## 16.4 RECEIVING END TRANSMISSION OPERATING ON A NO-LOAD CONDITION

If the line is open circuited (Figure 16.3) at the receiving end, $Z$ is infinite. An open circuit at the end of a line demands that the current at that point must be zero.

As $I_R = 0$, electromagnetic energy stored by inductor $= (1/2)LI^2 = 0$ and electrostatic energy stored by $C = (1/2)CV^2$.

According to law of conservation of energy, energy can never be destroyed but only can be converted from one form to another, that is, electromagnetic energy is converted into electrostatic energy, so electrostatic energy

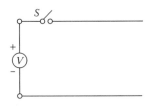

FIGURE 16.3    Case of an open-ended line.

increases. Electrostatic energy is increased due to increase in voltage. Let $e$ be increase in voltage.

$$\frac{1}{2}LI^2 = \frac{1}{2}Ce^2$$

or

$$LI^2 = Ce^2$$

$$e = I\sqrt{\frac{L}{C}} = IZ_C = V$$

that is, voltage at the open-circuited receiving end increases by incident voltage $V$. Therefore, the total potential of the open end when the wave reaches this end is

$$V + V = 2V$$

As soon as the incident current wave $I$ reaches the open end, the current at the open end is zero. That means a current wave of $I$ magnitude travels back over the transmission line. This means for an open-end line, a current wave is reflected with negative sign and coefficient of reflection unity. The variation of current and voltage waves over the line is explained in Figure 16.4.

After the voltage and current waves are reflected back from the open end, they reach the source end, the voltage over the line becomes $2V$ and the current is zero. The voltage at source end cannot be more than the source voltage $V$; therefore, a voltage wave $-V$ and current wave of $-I$ are

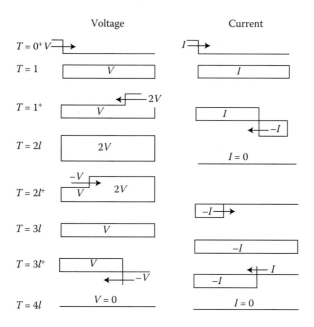

FIGURE 16.4   Variation of voltage and current in an open-ended line.

reflected back into the line. It can be seen that after the waves have travelled through a distance of $4l$, where $l$ is the length of the line, they would have wiped out both the current and voltage wave, leaving the line momentarily in its original state. The above cycle repeats itself.

## 16.5  RECEIVING END OPERATING ON A SHORT CIRCUIT CONDITION

If the line is short circuited (Figure 16.5) at the receiving end, $Z = 0$. The unique characteristic of the short circuit is that voltage across it is zero.

Electromagnetic energy stored by $L$ in magnetic field = $(1/2)LI^2$

Electrostatic energy stored by $C$ in electric field = 0 (as $V = 0$)

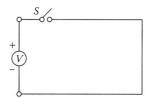

FIGURE 16.5   Case of a short-circuited line.

Energy can never be destroyed but only can be converted from one to another, that is, electrostatic energy is converted to electromagnetic energy. So electromagnetic energy increases.

Let $i$ be the increase in current,

$$\frac{1}{2}CV^2 = \frac{1}{2}Li^2$$

$$CV^2 = Li^2$$

$$i = \frac{V}{\sqrt{L/C}} = \frac{V}{ZC} = I$$

that is, current at short circuit receiving end increases by incident current $I$. As a result the total current at the shorted end, when the current waves reaches the end is $I + I = 2I$ A.

Therefore, for a short-circuit end, a current wave is reflected back with positive sign and coefficient of reflection as unity. Since the voltage at the shorted end is zero, a voltage wave of $-V$ could be considered to have been reflected back into the line with coefficient of reflection as unity. The variation of voltage and current waves over the line is explained in Figure 16.6.

It is seen from above that the voltage wave periodically reduces to zero after it has travelled through a distance of twice the length of the line, whereas after each reflection at either end, the current is built up by an amount $V/Z_n = I$. Theoretically, the reflection will be infinite and therefore the current will reach infinite value. But practically in an actual system, the current will be limited by the resistance of the line.

## 16.6 REFLECTION AND REFRACTION AT A T-JUNCTION

A voltage is traveling over the line with surge impedance $Z_1$, as shown in Figure 16.7.

When it reaches the junction, it looks a change in impedance and, therefore, suffers reflection and refraction. Let $V_{T2}$, $I_{T2}$, and $V_{T3}$, $I_{T3}$ be the voltage and current in the lines having surge impedance $Z_2$ and $Z_3$, respectively. Since $Z_2$ and $Z_3$ form a parallel path as far as the surge wave is concerned, $V_{T2} = V_{T3} = V_T$. Therefore, following relations hold good,

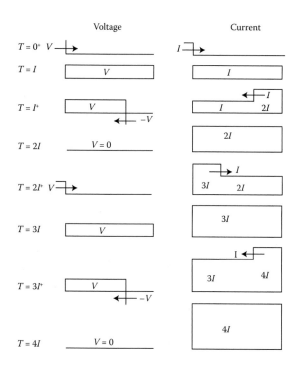

FIGURE 16.6  Variation of voltage and current in a short-circuited line.

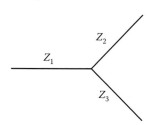

FIGURE 16.7  A bifurcated line.

$$V_T = V + V_R$$

$$V + V_R = V_T$$

$$I = \frac{V}{Z_1}, \quad I_R = -\frac{V_R}{Z_1}$$

$$I_{T2} = \frac{V_T}{Z_2}, \quad I_{T3} = \frac{V_T}{Z_3}$$

$$I + I_R = I_{T2} + I_{T3}$$

$$\frac{V}{Z_1} - \frac{V_R}{Z_1} = \frac{V_T}{Z_2} + \frac{V_T}{Z_3}$$

Substituting for $V_R = V_T - V$,

$$\frac{V}{Z_1} - \frac{V_T - V}{Z_1} = \frac{V_T}{Z_2} + \frac{V_T}{Z_3}$$

$$\frac{2V}{Z_1} = V_T \left[ \frac{1}{Z_1} + \frac{1}{Z_2} + \frac{1}{Z_3} \right]$$

$$V_T = \frac{2V/Z_1}{(1/Z_1) + (1/Z_2) + (1/Z_3)}$$

Similarly, other quantities can be derived.

## WORKED EXAMPLES

### EXAMPLE 16.1

An overhead line with surge impedance of 400 Ω is connected in series with an underground cable having a surge impedance of 100 Ω. If a surge of 50 kV travels from line toward cable, determine the transmitted voltage wave at junction.

**Solution**

$$\text{Transmitting coefficient, } \beta = \frac{2Z}{Z + Z_C}$$

$$\text{Transmitted voltage, } V_T = V \frac{2Z}{Z + Z_C} = 50 \times \frac{2 \times 100}{100 + 400} = 20 \text{ kV}$$

### EXAMPLE 16.2

A surge of 100 kV is traveling in a line of impedance 600 Ω arrives at a junction with two lines of impedance 800 and 200 Ω. Determine the transmitted voltage and currents.

**Solution**

In the problem, $Z_1 = 600\ \Omega$, $Z_2 = 800\ \Omega$, $Z_3 = 200\ \Omega$, and $V = 100$ kV.

Transmitted voltage, $V_T = \dfrac{2V/Z_1}{(1/Z_1) + (1/Z_2) + (1/Z_3)}$

$$= \dfrac{(2 \times 100)/600}{(1/600) + (1/800) + (1/200)} = 42.10\ \text{kV}$$

Transmitted currents of two lines are $I_{T2}$ and $I_{T3}$, respectively.

$$I_{T2} = \dfrac{V_T}{Z_2} = \dfrac{42.10 \times 1000}{800} = 52.62\ \text{A}$$

$$I_{T3} = \dfrac{V_T}{Z_3} = \dfrac{42.10 \times 1000}{200} = 210.5\ \text{A}$$

## EXAMPLE 16.3

A surge of 20 kV magnitude travels along a lossless cable toward its junction with two identical lossless overhead transmission lines. The inductance and capacitance of the cable are 0.4 mH and 0.5 μF/km, and the same of the overhead transmission lines are 1.5 mH and 0.015 μF/km. Find out the magnitude of voltage at the junction due to surge.

**Solution**

Transmitted voltage, $V_T = V\left(\dfrac{2/Z_{\text{Cable}}}{(1/Z_{L1}) + (1/Z_{L2}) + (1/Z_{\text{Cable}})}\right)$

Now,

$$Z_{\text{Cable}} = \sqrt{\dfrac{L}{c}} = \sqrt{\dfrac{0.4 \times 10^{-3}}{0.5 \times 10^{-6}}} = 28.28\ \Omega$$

Now, impedances of lines are

$$Z_{L1} = Z_{L2} = \sqrt{\dfrac{L}{c}} = \sqrt{\dfrac{1.5 \times 10^{-3}}{0.015 \times 10^{-6}}} = 316.22\ \Omega$$

$$\therefore V_T = 20\left(\frac{2/28.28}{(2/316.22) + (1/28.28)}\right) = 33.93\,\text{kV}$$

**EXERCISES**

1. Explain what is meant by the surge impedance of a line and show upon what factors it depends.

2. Obtain an expression for the surge impedance of a transmission line and for the velocity of propagation of electric waves in terms of the line inductance and capacitance.

3. Derive an expression for the surge impedance of a transmission line. Explain what is meant by surge impedance of a transmission line and derive its value in terms of the line constants. Derive expressions for the values of transmitted and reflected waves of current and voltage relative to those of incident waves at a point where the surge impedance changes from $Z_1$ to $Z_2$.

# CHAPTER 17

# Earthing

## 17.1 INTRODUCTION

The term "earthing" or "grounding" means connecting the noncurrent-carrying parts of the electrical equipment or the neutral point of the supply system to the general mass of earth in such a manner that all times an immediate discharge of electrical energy takes place without danger. The neutral grounding is an important aspect of power system design because the performance of the system in terms of short circuits, stability, protection, etc., is greatly affected by the condition of the neutral.

## 17.2 OBJECTS OF EARTHING

1. To save human life from danger or shock or by death by blowing fuse of any apparatus which becomes leaky.

2. To protect all machines fed from overhead lines from lightning.

3. To protect large buildings from atmospheric lightning.

4. To maintain the line voltage constant (since neutral of every alternator, transformer is earthed).

## 17.3 CLASSIFICATION OF EARTHING

Earthing may be classified as

1. Equipment grounding

2. System grounding

### 17.3.1 Equipment Grounding

The process of connecting noncurrent-carrying metal parts (i.e., metallic enclosure) of the electrical equipment to earth (i.e., soil) in such a way that in case of insulation failure, the enclosure effectively remains at earth potential is called equipment grounding.

We are frequently in touch with electrical equipment of all kinds, ranging from domestic appliances and hand-held tools to industrial motors. We shall illustrate the need of effective equipment grounding by considering a single-phase circuit composed of a 230-V source connected to a motor M (see Figure 17.1). Note that neutral is solidly grounded at the service entrance. In the interest of easy understanding, we shall divide the discussion into three heads:

1. *Ungrounded enclosure.* Figure 17.1 shows the case of ungrounded neutral enclosure. If a person touches the metal enclosure, nothing will happen if the equipment is functioning correctly. But if the winding insulation becomes faulty, the resistance $R_e$ between the motor and the enclosure drops to a low voltage (a few hundred ohms or less). A person having a body resistance $R_b$ would complete the current path as shown in Figure 17.1. If $R_e$ is small, the leakage current $I_L$ through the person's body could be dangerously high. As a result the person would get electric shock which may be fatal. Therefore, this system is unsafe.

2. *Enclosure connected to neutral wire.* It may appear that the above problem can be solved by connecting the enclosure to the grounded neutral wire as shown in Figure 17.2. Now the leakage current flows from the motor, through the enclosure, and straight back to the neutral wire. Therefore, the enclosure remains at earth potential. Consequently, the operator would not experience any electric shock.

   The trouble with this method is that the neutral wire may become open either accidentally or due to a faulty installation. For example, if

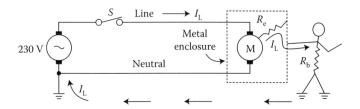

FIGURE 17.1  A person having a body resistance $R_b$ would complete the current path.

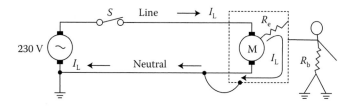

FIGURE 17.2    Connecting the enclosure to the grounded neutral wire.

the switch is inadvertently in series with the neutral rather than the live wire shown in Figure 17.3, the motor can still be turned on and off. However, if someone touched the enclosure when the motor is off, he would receive a severe electric shock. It is because when the motor is off, the potential of the enclosure rises to that of the live conductor.

3. *Ground wire connected to enclosure.* To get rid of this problem, we install a third wire, called ground wire, between the enclosure and the system ground as shown in Figure 17.4. The ground wire may be bare or insulated. If it is insulated, it is colored green.

## 17.3.2  System Grounding

The process of collecting some electric parts of the power system (e.g., neutral point of a star connected system, one conductor of the secondary of a transformer, etc.) to earth is called system grounding.

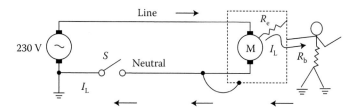

FIGURE 17.3    Switch is inadvertently in series with the neutral.

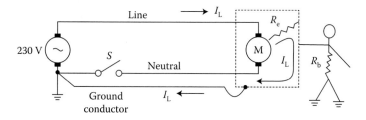

FIGURE 17.4    Ground wire, between the enclosure and the system ground.

## 17.4 ISOLATED NEUTRAL OR UNDERGROUNDED NEUTRAL

A simple three-phase system with isolated neutral is shown in Figure 17.5. The line conductors have capacitances between one another and the earth, the former being delta connected, while the latter star connected. The effect of line capacitances on the grounding characteristic of the system is little and therefore can be neglected. The circuit then reduces to the one shown in Figure 17.6a. First of all consider a three-phase line (perfectly transposed) having some capacitances to ground. In such a line, the charging currents for each line to earth capacitor lead the phase voltage by 90° and are equal.

$$I_{CR} = I_{CY} = I_{CB} = \frac{V_{ph}}{X_C}$$

where $V_{ph}$ is the phase voltage and $X_C$ is the reactance due to the capacitance of the line to ground.

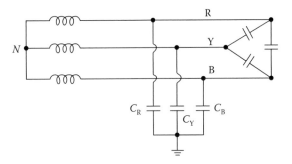

FIGURE 17.5   A simple three-phase system with isolated neutral.

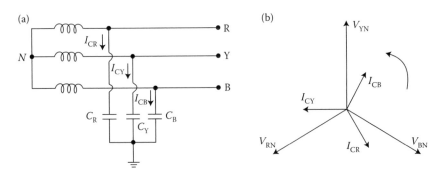

FIGURE 17.6   (a) Three phase system with isolated neutral. (b) Phasor diagram for isolated neutral system.

The capacitive currents $I_{CR}$, $I_{CY}$, and $I_{CB}$ are balanced and their resultant is zero and no current flows to the earth and the potential of neutral is the same as the ground potential. Phasor diagram as shown in Figure 17.6b.

## 17.4.1 Circuit Behavior under Single Line-to-Ground Fault

Now consider a phase to earth fault in line Y say at point F. The circuit then becomes as shown in Figure 17.7a. Under these circumstances, the faulty line takes up the earth potential, while the potentials of remaining two healthy lines R and B rise from phase value to line value. The capacitance current becomes unbalanced and fault current $I_F$ flows through the faulty line into the fault and returns to the system via earth and the earth capacitances $C_R$ and $C_B$. Thus fault current $I_F$ has two components $I_{CR}$ and $I_{CB}$ which flows through capacitances $C_R$ and $C_B$, respectively, under the potential differences of $V_{RY}$ and $V_{BY}$, respectively. The currents lead their respective voltages by 90° and their phasor sum is equal to fault current $I_F$. Phasor diagram is shown in Figure 17.7b.

$$I_{CR} = \frac{V_{RY}}{X_{CR}} = \frac{\sqrt{3}V_{ph}}{X_C}$$

Similarly,

$$I_{CB} = \frac{V_{BY}}{X_{CB}} = \frac{\sqrt{3}V_{ph}}{X_C}$$

Now $I_F$ is equal to phasor sum of $I_{CR}$ and $I_{CB}$. Magnitude of $I_{CR}$ and $I_{CB}$ are equal to angle between them 60°.

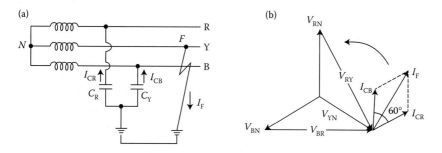

FIGURE 17.7 (a) Isolated neutral system with fault on one phase. (b) Phasor diagram for fault on phase Y.

Therefore, resultant capacitive fault current is given by

$$I_F = 2I_R \cos \frac{60°}{2}$$
$$= 2I_R \cos 30°$$
$$= 2I_R \frac{\sqrt{3}}{2} = \sqrt{3}I_R$$

Therefore,

$$I_F = \sqrt{3}I_{CR} = \sqrt{3} \times \frac{\sqrt{3}V_{ph}}{X_C} = \frac{3V_{ph}}{X_C}$$
$$= 3 \times \text{Per-phase capacitive current under normal condition}$$

When a single line-to-ground fault occurs on an underground neutral system, following effects are produced in the system:

1. The potential of the faulty phase becomes equal to ground potential. However, the voltages of the two remaining healthy phases rise from their normal phase voltage to full line value. This may result in insulation breakdown.

2. The capacitive current in the two healthy phases increase to $\sqrt{3}$ times the normal value.

3. The capacitive fault current $I_F$ becomes three times the normal per-phase capacitive current.

4. The system cannot provide adequate protection against earth faults. It is because the capacitive fault current is small in magnitude and cannot operate protective device.

5. The capacitive fault current flows $I_F$ into earth. Experience shows that $I_F$ in excess of 4 or 5 A is sufficient to maintain an arc in the ionized path of the fault. If this current is once maintained, it may exist even after the earth fault is cleared. This phenomenon of persistent arc is called arcing ground. Due to arcing ground, the system capacity is charged and discharged in a cyclic order. This sets up high frequency oscillation on the whole system, and the phase voltage of

healthy conductors may rise to five to six times its normal value. The over voltages in healthy conductors may damage the insulation in the line.

Due to above disadvantage undergrounded system is not used these days. The modern high-voltage three-phase system employs grounded neutral owing to a number of advantages.

## 17.5 ADVANTAGES OF NEUTRAL GROUNDING

The following are the advantages of neutral grounding:

1. Voltages of the healthy phases with respect to ground remain at normal value.

2. The high voltages due to arcing grounds are eliminated.

3. The protective relays can be used to provide protection against earth faults.

4. The over voltages due to lightning arc discharged to earth.

5. It provides greater safety to personnel and equipment.

## 17.6 METHODS OF NEUTRAL GROUNDING

The methods commonly used for grounding the neutral point of a three-phase system are

1. Solid or effective grounding

2. Resistance grounding

3. Reactance grounding

4. Peterson-coil grounding

### 17.6.1 Solid Grounding

When the neutral point of a three-phase system (e.g., three-phase generator, three-phase transformer, etc.) is directly connected to earth (i.e., soil) through a wire of negligible resistance and reactance, it is called solid grounding or effective grounding. Figure 17.8 shows the solid grounding of the neutral point.

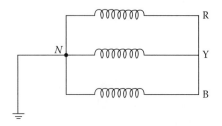

FIGURE 17.8    Solid grounding of neutral.

When there is a ground fault over any phase, the phase to earth voltage of a grounded phase will become zero, but the voltage to earth of the remaining two healthy phases will be the normal phase voltage as in this case neutral point will not shift. Under a line-to-ground fault on phase B, as shown in Figure 17.9a, the neutral and the terminal B are at earth potential. The phasor diagram for such a condition is shown in Figure 17.9b. The reversed phasor is shown at $V_B$. Capacitive current $I_{CR}$ leads $V_{NR}$ by 90° and $I_{CY}$ leads $V_{NY}$ by 90°. The resultant capacitive current $I_C$ will be phasor sum of $I_{CR}$ and $I_{CY}$. It should be noted that in this system, in addition to capacitive current, the supply source also supplies the fault current $I_F$. This current will go to the fault point F through the faulty phase and then return back to supply source through the earth and neutral connection. The fault current $I_F$ lags behind the faulty phase voltage by approximately 90° since the circuit is predominately inductive (due to transformers, machines, and line inductance). The fault current $I_F$ will be in phase opposition to capacitive current $I_C$. Due to this effect the capacitive current $I_C$ will be faulty neutralized by the large fault current. Therefore, no arcing ground phenomenon or over voltage condition can occur.

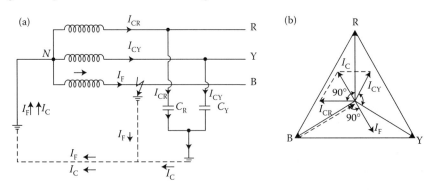

FIGURE 17.9    (a) Solidly grounded system with ground fault on one phase. (b) Phasor diagram for fault on phase B.

In case of solid grounding when there is an earth fault on any phase of the system, the phase to earth voltage of the faulty phase becomes zero. However, the phase to earth voltage of the remaining two healthy phases remains at normal-phase voltage because the potential of the neutral is fixed at earth potential. This permits to insulate the equipment for phase voltage. Therefore, there is a saving in the cost of equipment.

When there is an earth fault on any phase of the system, large fault current flows between the fault point and the grounded neutral. This permits the easy operation of earth relay.

This method also has some limitations:

1. The solid grounding results in heavy earth fault currents. Since the fault has to be cleared by the circuit breaker, the heavy earth fault currents may cause the burning of circuit breaker contacts.

2. The increased earth fault current results in greater interference in neighbouring communication line.

*Application.* This system of grounding is used for voltages up to 33 kV with total power capacity not exceeding 5000 kVA.

### 17.6.2 Resistance Grounding

When it becomes necessary to limit earth fault current, a current limiting device is introduced in the neutral and earth. One method of introducing a current limiting device is resistance earthing or grounding.

The value of $R$ should neither be very low or nor very high. If the value of $R$ is very low, the earth fault current will be large and the system becomes similar to solid grounding system. On the other hand, if the earthing resistance $R$ is very high, the system condition becomes similar to undergrounded system.

When there is a ground fault over any phase, neutral is displaced and the maximum voltage across the healthy phases becomes equal to line-to-line voltage. Figure 17.10a shows a ground fault on phase B of a resistance grounded system. The phasor diagram for such a condition is illustrated in Figure 17.10b. Capacitive current $I_{CR}$ and $I_{CY}$ leads $V_{BR}$ and $V_{BY}$, respectively, by 90°. Fault current $I_F$ lags the phase voltage of the faulted phase by an angle $\phi$, which depends on the grounding resistance $R$ and independent of the system up to the fault point. The fault current $I_F$ can be resolved into two components, one in phase with the faulty phase voltage and the other lagging the faulty phase voltage by 90° ($I_F \sin \phi$). The lagging component of

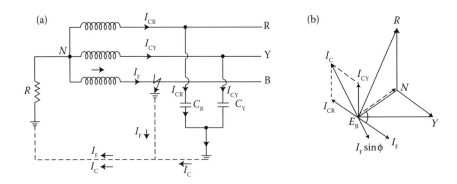

FIGURE 17.10 (a) Resistance grounded system with ground fault on one phase. (b) Phasor diagram for fault on phase B.

fault current is in phase opposition to the capacitive current $I_C$. By adjusting the value of grounding resistance $R$ to a sufficient low value, it is possible to neutralize the effect of $I_C$ so as to avoid the occurrence of transient oscillations due to the arcing ground.

In case the grounding resistance $R$ is made sufficiently large so that the lagging component of fault current becomes less than capacitive current $I_C$, then the system conditions approach that of the isolated neutral system the risk of high transient voltage occurrence.

*Application.* Resistance grounding is usually employed for the systems operating on voltage exceeding 3.3 kV but not exceeding 33 kV. For circuit below 3.3 kV (i.e., say 400 V distribution networks), the external resistance in the neutral circuit is unnecessary because the voltage available between phase and ground is only 230 V.

### 17.6.3 Reactance Grounding

In this system, a reactance is inserted between the neutral and ground as shown in Figure 17.11. The purpose of reactance is to limit the earth fault

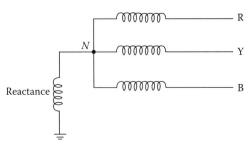

FIGURE 17.11 Reactance grounding.

current. By changing the earthing resistance, the earth fault current can be changed to obtain the condition similar to that of solid grounding.

This method is not used these days because of the following disadvantages:

1. In this system the fault current required to operate the protective device is higher than that of resistance grounding for the same fault conditions.

2. High transient voltages appear under fault conditions.

### 17.6.4 Arc Suppression Coil Grounding (or Resonant Grounding)

We have seen that capacitive currents are responsible for producing arcing grounds. These capacitive currents flow because capacitance exists between each line and earth. If inductance $L$ of appropriate value is connected in parallel with the capacitance of the system, the fault current $I_F$ flowing through $L$ will be in phase opposition to capacitive current $I_C$ of the system. If $L$ is so adjusted that $I_L = I_C$, then resultant current in the fault will be zero. This condition is known as resonant grounding.

When the value of $L$ of arc suppression coil is such that the fault current $I_F$ exactly balances the capacitive current $I_C$, it is called Resonant grounding.

An arc suppression coil (also called Peterson coil) is an iron-cored coil connected between the neutral and earth as shown in Figure 17.12. The reactor is provided with tapings to change the inductance of the coil. By adjusting the tapings on the coil, the coil can be tuned with the capacitance of the system, that is, resonant grounding can be achieved.

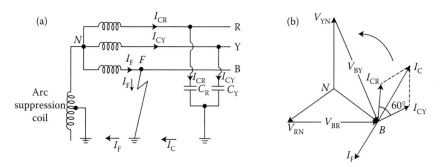

FIGURE 17.12 (a) Resonant grounded system with ground fault on one phase. (b) Phasor diagram for fault on phase B.

On occurrence of a ground fault (say on phase B), a lagging reactive current flows from the faulted phase to the ground and returns to the system through the inductive coil. Simultaneously capacitive current flows from healthy phase to ground. The lagging fault current $I_F$ and leading capacitive current $I_C$ are almost in phase opposition. By a proper selection of the value of inductance $L$ of the arc suppression coil, the two current can be made almost equal so that there is no current through the ground fault and so there will be no arc. The combination of neutral reactance $L$ and line capacitance $C$ acts as a parallel resonant circuit.

$$I_{CR} = I_{CY} = \sqrt{3}V_{ph}\omega C$$

Capacitive current

$$I_C = I_{CR} + I_{CY} = \sqrt{3} \times \sqrt{3}V_{ph}\omega C = 3V_{ph}\omega C$$

For balance condition,

$$I_L = I_C$$
$$\frac{V_P}{W_L} = 3V_{ph}\omega C \quad \text{or} \quad L = \frac{1}{3\omega^2 C}$$

*Advantages of Peterson coil grounding:*

1. The Peterson coil grounding is an effective method of clearing both transient faults due to lightning and sustained single line-to-ground faults.

2. There is no tendency of arcing grounds to occur, and the arcs are usually self-extinguishing.

3. Voltage drops to single line-to-ground faults are minimized.

*Disadvantages of Peterson coil grounding:*

1. There is a need for retuning after any network modification.

2. The line should be transposed.

3. There is an increase in corona and radio interference in the event of a double line-to-ground fault.

## WORKED EXAMPLES

### EXAMPLE 17.1

Calculate the reactance of arc suppression coil suitable for a 33-kV, three-phase transmission line having a capacitance to earth of each conductor as 4.5 μF. Assume supply frequency to be 50 Hz.

**Solution**

Supply frequency, $f = 50$ Hz.

Line to earth capacitance, $C = 4.5$ μF $= 4.5 \times 10^{-6}$ F.

For Peterson coil grounding, reactance $X_L$ of the arc suppression coil should be equal to $X_C/3$, where $X_C$ is line to earth capacitive reactance. Therefore, reactance of arc suppression coil is

$$X_L = \frac{X_C}{3} = \frac{1}{3\omega C} = \frac{1}{3 \times 2\pi f \times C}$$

$$= \frac{1}{3 \times 2\pi \times 50 \times 4.5 \times 10^{-6}} = 235.8\ \Omega$$

### EXAMPLE 17.2

A 66-kV, three-phase, 50-Hz, 100-km transmission line has a capacitance to earth of 0.03 μF/km per phase. Calculate the inductance and kVA rating of the Peterson coil used for earthing the above system.

**Solution**

Supply frequency, $f = 50$ Hz.

Capacitance of each line to earth, $C = 100 \times 0.03 = 3 \times 10^{-6}$ F.

Required inductance of Peterson coil is

$$L = \frac{1}{3\omega^2 C}$$

$$= \frac{1}{3 \times (2\pi \times 50)^2 \times 3 \times 10^{-6}} = 1.12\ \text{H}$$

Current through Peterson coil is

$$I_F = \frac{V_{ph}}{X_L} = \frac{66 \times 10^3/\sqrt{3}}{2\pi \times 50 \times 1.12} = 108.296\ \text{A}$$

Voltage across Peterson coil is

$$V_{ph} = \frac{V_L}{\sqrt{3}} = \frac{66 \times 10^3}{\sqrt{3}} \, V$$

Therefore, rating of Peterson coil is

$$V_{ph} \times I_F = \frac{66 \times 10^3}{\sqrt{3}} \times 108.296 \times \frac{1}{1000} = 4126.63 \, kVA$$

### EXAMPLE 17.3

A 33-kV, three-phase, 50-Hz transmission line of 100 km long consists of three conductors of effective diameter 30 mm arranged in a vertical plane with 3 m spacing and regularly transposed. Find the inductance and kVA rating of the arc suppression coil in the system.

**Solution**

Radius of conductor, $r = 30/2 = 15$ mm $= 0.015$ m
  Conductor spacing, $d = 3$ m

∴ Capacitance between phase and neutral or earth

$$= \frac{2\pi\varepsilon_0}{\log_e(d/r)} \, F/m = \frac{2\pi \times 8.854 \times 10^{-12}}{\log_e(3/0.015)} = 10.53 \times 10^{-12} \, F/m$$
$$= 10.53 \times 10^{-9} \, F/km$$

Therefore, capacitance $C$ between phase and earth for 100 km line is

$$C = 100 \times 10.53 \times 10^{-9} = 10.53 \times 10^{-7} \, F$$

The required inductance $L$ of the arc suppression coil is

$$L = \frac{1}{3\omega^2 C} = \frac{1}{3 \times (2\pi \times 50)^2 \times 10.53 \times 10^{-7}} = 3.207 \, H$$

Current through the coil,

$$I_F = \frac{V_{ph}}{X_L} = \frac{33 \times 10^3 / \sqrt{3}}{2\pi \times 50 \times 3.207} = 18.91 \text{ A}$$

$$\therefore \text{ Rating of the coil } = V_{ph} \times I_F = \frac{33}{\sqrt{3}} \times 18.91 = 360.29 \text{ kVA}$$

## EXERCISES

1. What do you mean by grounding or earthing? Explain it with an example.

2. Describe ungrounded or isolated neutral system. What are its disadvantages?

3. What do you mean by equipment grounding?

4. Illustrate the need of equipment grounding.

5. What is neutral grounding?

6. What are the advantages of neutral grounding?

7. What is solid grounding? What are its advantages?

8. What are the disadvantages of solid grounding?

9. What is resistance grounding? What are its advantages and disadvantages?

10. Describe arc suppression coil grounding.

11. What is resonant grounding?

# CHAPTER 18

# Substation

## 18.1 INTRODUCTION

The modern electrical power system is generated, transmitted, and distributed in the form of alternating current. The electric power is produced at the power stations which are located in suitable places, generally quite away from the consumers. It is delivered to the consumers through a large network of transmission and distribution. At many places in the line of the power system, it may be desirable and necessary to change some characteristic (e.g., voltage, AC to DC, frequency, power factor [pf], etc.) of electric supply. For that, in between the power station and ultimate consumer, a number of transformations and switching stations have to be created. These are generally known as substation.

## 18.2 SUBSTATION

The assembly of apparatus used to change some characteristic (e.g., voltage, AC to DC, frequency, pf, etc.) of electric supply is called a substation.

Substations are important part of power system. The continuity of supply depends to a considerable extent upon the successful operation of substations. It is, therefore, essential to exercise utmost care while designing and building a substation. The following are the important points which must be kept in view while laying out a substation:

1. It should be located at a proper site. As far as possible, it should be located at the center of gravity of load.

2. It should provide safe and reliable arrangement. For safety, consideration must be given to the maintenance of regulation clearances,

facilities for carrying out repairs and maintenance, abnormal occurrences such as possibility of explosion or fire, etc. For reliability, consideration must be given for good design and construction, the provision of suitable protective gear, etc.

3. It should be easily operated and maintained.

4. It should involve minimum capital cost.

## 18.3 CLASSIFICATION OF SUBSTATIONS

There are several ways of classifying substations. However, the two most important ways of classifying them are according to (1) service requirement and (2) constructional features.

### 18.3.1 According to Service Requirement

A substation may be called upon to change voltage level or improve power factor or convert AC power into DC power, etc. According to the service requirement, substations may be classified as follows:

1. *Transformer substations.* Those substations which change the voltage level of electric supply are called transformer substations. These substations receive power at some voltage and deliver it at some other voltage. Obviously, transformer will be the main component in such substations. Most of the substations in the power system are of this type.

2. *Switching substations.* These substations do not change the voltage level, that is, incoming and outgoing lines have the same voltage. However, they simply perform the switching operations of power lines.

3. *Power factor correction substations.* Those substations which improve the power factor of the system are called power factor correction substations. Such substations are generally located at the receiving end of transmission lines. These substations generally use synchronous condensers as the power factor improvement equipment.

4. *Frequency changer substations.* Those substations which change the supply frequency are known as frequency changer substations. Such a frequency change may be required for industrial utilization.

5. *Converting substations.* Those substations which change AC power into DC power are called converting substations. These substations

receive AC power and convert it into DC power with suitable apparatus (e.g., ignitron) to supply for such purposes as traction, electroplating, electric welding, etc.

6. *Industrial substations.* Those substations which supply power to individual industrial concerns are known as industrial substations.

## 18.3.2 According to Constructional Features

A substation has many components (e.g., circuit breakers, switches, fuses, instruments, etc.) which must be housed properly to ensure continuous and reliable service. According to constructional features, the substations are classified as follows:

1. *Indoor substations.* For voltages up to 11 kV, the equipment of the substation is installed indoors because of economic considerations. However, when the atmosphere is contaminated with impurities, these substations can be erected for voltages up to 66 kV. According to construction, indoor distribution transformer substations and high voltage switchboards are further subdivided into following categories:

   i. *Substation integrally built type.* In such a substation the apparatus is installed on site. The structures are constructed of concrete or brick.

   ii. *Substation of the composite built-up type.* Here, the assemblies and parts are factory and workshop prefabricated, but are assembled on site within a substation switchgear room. The components of substations take the form of metal cabinets or enclosures, each of which contains the equipment of one main connection cell. Within such cabinets or enclosures an oil circuit breaker, a load interrupter switch, and one or more voltage transformers may be mounted.

   iii. *Unit type factory fabricated substations and metal clay switchboards.* These are built in electrical engineering workshops and are shipped to the site of installation fully preassembled. After installation of the substations switchboard-only connections to the incoming and outgoing power circuits are required.

2. *Outdoor substations.* For voltages beyond 66 kV, equipment is invariably installed outdoor. It is because for such voltages, the clearances

between conductors and the space required for switches, circuit breakers, and other equipments become so great that it is not economical to install the equipment indoor.

3. *Underground substation.* In thickly populated cities, there is scarcity of land as well as the prices of land are very high. This has led to the development of underground substation. In such substations, the equipment is placed underground. Figure 18.1 shows a typical underground substation.

4. *Pole-mounted substation.* It is a distribution substation placed overhead on a pole. It is the cheapest form of substation as it does not involve any building work. Figure 18.2a shows the layout of pole-mounted substation, whereas Figure 18.2b shows the schematic connections. The transformer and other equipments are mounted on H-type pole (or four-pole structure).

The 11 kV line is connected to the transformer (11 kV/400 V) through gang isolator and fuses. The lightning arresters are installed on the high-tension (HT) side to protect the substation from lightning strokes. The transformer steps down the voltage to 400-V, three-phase, four-wire supply. The voltage between any two lines is 400 V, whereas the voltage between any line and neutral is 230 V. The oil circuit breaker (OCB) installed on the low-tension side automatically isolates the transformer

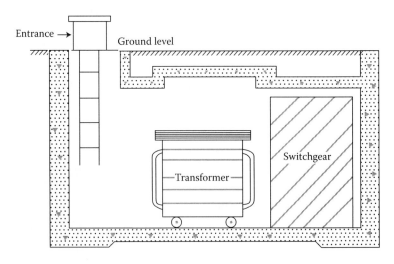

FIGURE 18.1   A typical underground substation.

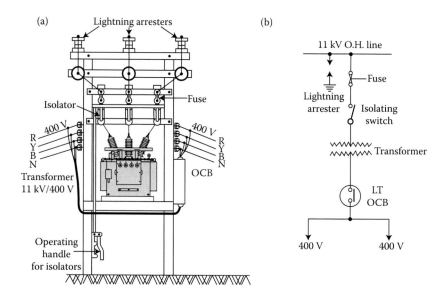

FIGURE 18.2    (a) Layout of pole-mounted substation. (b) Schematic connections of of pole-mounted substation.

from the consumers in the event of any fault. The pole-mounted substations are generally used for transformer capacity up to 200 kVA. The following points may be noted about pole-mounted substations:

- There should be periodical checkup of the dielectric strength of oil in the transformer and OCB.

- In case of repair of transformer or OCB, both gang isolator and OCB should be shut off.

## 18.4 COMPARISON BETWEEN OUTDOOR AND INDOOR SUBSTATIONS

The comparison between outdoor and indoor substations is given in the following table:

| S. No. | Particular | Outdoor Substation | Indoor Substation |
| --- | --- | --- | --- |
| 1 | Space required | More | Less |
| 2 | Time required for extension | Less | More |
| 3 | Future extension | Easy | Difficult |

| 4 | Fault location | Easier because the equipment is in full view | Difficult due to closed equipment |
| 5 | Capital cost | Low | High |
| 6 | Operation | Difficult | Easier |
| 7 | Possibility of fault escalation | Less due to greater clearances can be provided | More |

## 18.5 TRANSFORMER SUBSTATIONS

The majority of the substations in the power system are concerned with the changing of voltage level of electric supply. These are known as transformer substations because transformer is the main component employed to change the voltage level. Depending upon the purpose served, transformer substations may be classified into

1. Step-up substation

2. Primary grid substation

3. Secondary substation

4. Distribution substation

Figure 18.3 shows the block diagram of a typical electric supply system, indicating the position of above types of substations. It may be noted that it is not necessary that all electric supply schemes include all the stages shown in the figure. For example, in a certain supply scheme there may not be secondary substations and in another case, the scheme may be so small that there are only distribution substations.

1. *Step-up substation.* The generation voltage (11 kV in this case) is stepped up to high voltage (220 kV) to affect economy in transmission of electric power. The substations which accomplish this job are called step-up substations. These are generally located in the power houses and are of outdoor type.

2. *Primary grid substation.* From the step-up substation, electric power at 220 kV is transmitted by three-phase, three-wire overhead system to the outskirts of the city. Here, electric power is received by the primary grid substation which reduces the voltage level to 66 kV for secondary transmission. The primary grid substation is generally of outdoor type.

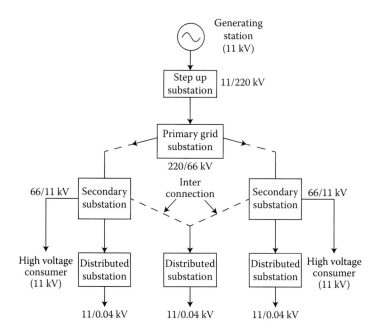

FIGURE 18.3    Block diagram of a typical electric supply system.

3. *Secondary substation.* From the primary grid substation, electric power is transmitted at 66 kV by three-phase, three-wire system to various secondary substations located at the strategic points in the city. At a secondary substation, the voltage is further stepped down to 11 kV. The 11 kV lines run along the important road sides of the city. It may be noted that big consumers (having demand more than 50 kW) are generally supplied power at 11 kV for further handling with their own substations. The secondary substations are also generally of outdoor type.

4. *Distribution substation.* The electric power from 11 kV lines is delivered to distribution substations. These substations are located near the consumer's localities and step down the voltage to 400 V, three phases, four wires for supplying to the consumers. The voltage between any two phases is 400 V and between any phase and neutral it is 230 V. The single-phase residential lighting load is connected between any one phase and neutral, whereas three-phase, 400-V motor load is connected across three-phase lines directly. It may be worthwhile to mention here that majority of the distribution substations are of pole-mounted type.

## 18.6 EQUIPMENT IN A TRANSFORMER SUBSTATION

The equipment required for a transformer substation depends upon the type of substation, service requirement, and the degree of protection desired. However, in general, a transformer substation has the following main equipment:

1. *Bus bars*. When a number of lines operating at the same voltage have to be directly connected electrically, bus bars are used as the common electrical component. Bus bars are copper or aluminum bars (generally of rectangular cross-section) and operate at constant voltage. The incoming and outgoing lines in a substation are connected to the bus bars. The most commonly used bus-bar arrangements in substations are

   a. Single bus-bar arrangement

   b. Single bus-bar system with sectionalization

   c. Double bus-bar arrangement

2. *Insulators*. The insulators serve two purposes. They support the conductors (or bus bars) and confine the current to the conductors. The most commonly used material for the manufacture of insulators is porcelain. There are several types of insulators (e.g., pin type, suspension type, postinsulator, etc.), and their use in the substation will depend upon the service requirement. For example, postinsulator is used for bus bars. A postinsulator consists of a porcelain body, cast iron cap, and flanged cast iron base. The hole in the cap is threaded so that bus bars can be directly bolted to the cap.

3. *Isolating switches*. In substations, it is often desired to disconnect a part of the system for general maintenance and repairs. This is accomplished by an isolating switch or isolator. An isolator is essentially a knife switch and is designed to open a circuit under no load. In other words, isolator switches are operated only when the lines in which they are connected carry no current.

   Figure 18.4 shows the use of isolators in a typical substation. The entire substation has been divided into five sections. Each section can be disconnected with the help of isolators for repair and maintenance. For instance, if it is desired to repair section II, the procedure of disconnecting this section will be as follows. First of all,

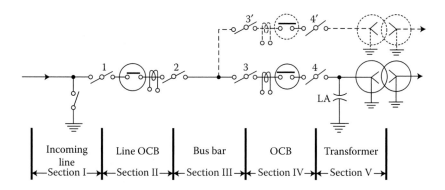

FIGURE 18.4   Use of isolators in a typical substation.

open the circuit breaker in this section and then open the isolators 1 and 2. This procedure will disconnect section II for repairs. After the repairs has been done, close the isolators 1 and 2 first and then the circuit breaker.

4. *Circuit breaker.* A circuit breaker is an equipment which can open or close a circuit under normal as well as fault conditions. It is so designed that it can be operated manually (or by remote control) under normal conditions and automatically under fault conditions.

5. *Power transformers.* A power transformer is used in a substation to step-up or step-down the voltage. Except at the power station, all the subsequent substations use step-down transformers to gradually reduce the voltage of electric supply and finally deliver it at utilization voltage. The modern practice is to use three-phase transformers in substations; although three single-phase bank of transformers can also be used. The use of three-phase transformer (instead of three single-phase bank of transformers) permits two advantages:

   a. Only one three-phase load-tap changing mechanism can be used.

   b. Its installation is much simpler than the three single-phase transformers.

6. *Instrument transformers.* The lines in substations operate at high voltages and carry current of thousands of amperes. The measuring instruments and protective devices are designed for low voltage (generally 110 V) and current (about 5 A). Therefore, they will not work

satisfactorily if mounted directly on the power lines. This difficulty is overcome by installing instrument transformers on the power lines. The function of these instrument transformers is to transfer voltage or current in the power lines to values which are convenient for the operation of measuring instruments and relays. Two types of instrument transformers are as follows:

a. *Current transformer (CT).* A CT is essentially a step-up transformer which steps down the current to a known ratio. The primary of this transformer consists of one or more turns of thick wire connected in series with the line. The secondary consists of a large number of turns of fine wire and provides for the measuring instruments and relays a current which is a constant fraction of the current in the line. Suppose a CT rated at 100/5 A is connected in the line to measure current. If the current in the line is 100 A, then current in the secondary will be 5 A. Similarly, if current in the line is 50 A, then secondary of CT will have a current of 2.5 A. Thus the CT under consideration will step down the line current by a factor of 20.

b. *Potential transformer.* It is essentially a step down transformer and steps down the voltage to a known ratio. The primary of this transformer consists of a large number of turns of fine wire connected across the line. The secondary winding consists of a few turns and provides for measuring instruments and relays a voltage which is a known fraction of the line voltage. Suppose a potential transformer rated at 66 kV/110 V is connected to a power line. If line voltage is 66 kV, then voltage across the secondary will be 110 V.

7. *Metering and indicating instruments.* There are several metering and indicating instruments (e.g., ammeters, voltmeters, energy meters, etc.) installed in a substation to maintain watch over the circuit quantities. The instrument transformers are invariably used with them for satisfactory operation.

8. *Miscellaneous equipment.* In addition to above, there may be following equipments in a substation:

a. Fuses

b. Carrier-current equipment

c. Substation auxiliary supplies

## 18.7 BUS-BAR ARRANGEMENTS IN SUBSTATIONS

Bus bars are the important components in a substation. There are several bus-bar arrangements that can be used in a substation. The choice of a particular arrangement depends upon various factors such as system voltage, position of substation, degree of reliability, cost, etc. The following are the important bus-bar arrangements used in substations.

### 18.7.1 Single Bus-Bar System

As the name suggests, it consists of a single bus bar, and all the incoming and outgoing lines are connected to it. The chief advantages of this type of arrangement are low initial cost, less maintenance, and simple operation. However, the principal disadvantage of single bus-bar system is that if repair is to be done on the bus bar or a fault occurs on the bus, there is a complete interruption of the supply. This arrangement is not used for voltages exceeding 33 kV. The indoor 11 kV substations often use single bus-bar arrangement.

Figure 18.5 shows single bus-bar arrangement in a substation. There are two 11-kV incoming lines are connected to the bus bar through circuit breakers and isolators. The two 400-V outgoing lines are connected to the bus bars through transformers (11 kV/400 V) and circuit breakers.

### 18.7.2 Single Bus-Bar System with Sectionalization

In this arrangement, the single bus bar is divided into sections, and load is equally distributed on all the sections. Any two sections of the bus bar are

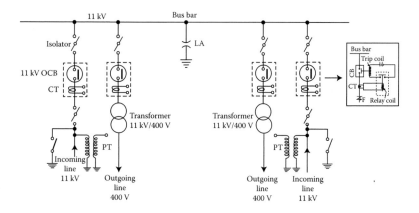

FIGURE 18.5    Single bus-bar arrangement in a substation.

connected by a circuit breaker and isolators. Two principal advantages are claimed for this arrangement. Firstly, if a fault occurs on any section of the bus, that section can be isolated without affecting the supply from other sections. Secondly, repairs and maintenance of any section of the bus bar can be carried out by de-energizing that section only, eliminating the possibility of complete shutdown. This arrangement is used for voltages up to 33 kV. Bus bar with sectionalization where the bus has been divided into two sections.

Figure 18.6 shows bus bar with sectionalization where the bus has been divided into two sections. There are two 33-kV incoming lines connected to sections I and II as shown through circuit breaker and isolators. Each 11 kV outgoing line is connected to one section through transformer (33/11 kV) and circuit breaker. It is easy to see that each bus section behaves as a separate bus bar.

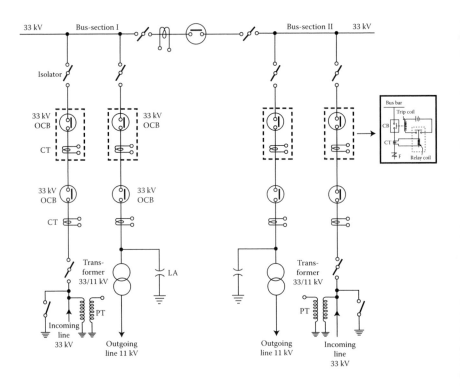

FIGURE 18.6  Bus bar with sectionalization.

### 18.7.3 Duplicate Bus-Bar System

This system consists of two bus bars, a "main" bus bar and a "spare" bus bar. Each bus bar has the capacity to take up the entire substation load. The incoming and outgoing lines can be connected to either bus bar with the help of a bus-bar coupler that consists of a circuit breaker and isolators. Ordinarily, the incoming and outgoing lines remain connected to the main bus bar. However, in case of repair of main bus bar or fault occurring on it, the continuity of supply to the circuit can be maintained by transferring it to the spare bus bar. For voltages exceeding 33 kV, duplicate bus-bar system is frequently used.

Figure 18.7 shows the arrangement of duplicate bus-bar system in a typical substation. The two 66 kV incoming lines can be connected to either bus bar by a bus-bar coupler. The two 11-kV outgoing lines are connected to the bus bars through transformers (66/11 kV) and circuit breakers.

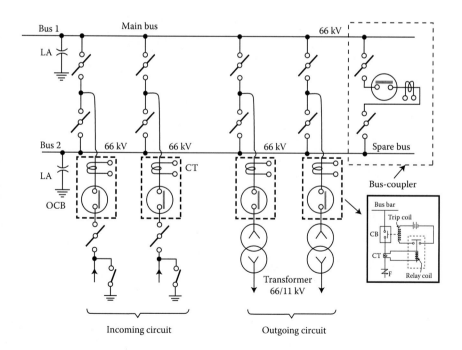

FIGURE 18.7   Duplicate bus-bar system.

## 18.8 KEY DIAGRAM OF 11 kV/400 V INDOOR SUBSTATION

Figure 18.8 shows the key diagram of a typical 11 kV/400 V indoor substation. The key diagram of this substation can be explained as follows:

1. The three-phase, three-wire 11-kV line is tapped and brought to the gang operating switch installed near the substation. The GO switch consists of isolators connected in each phase of the three-phase line.

2. From the GO switch, the 11 kV line is brought to the indoor substation as underground cable. It is fed to the HT side of the transformer (11 kV/400 V) via the 11 kV OCB the transformer steps down the voltage to 400 V, three phases, four wires.

3. The secondary transformer supplies to the bus bars via the main OCB. From the bus bars, 400-V, three-phase, four-wire supply is given to the various consumers via 400 V OCB. The voltage between any two phases 400 V and between any phase and neutral it is 230 V. The single-phase residential load is connected between any one phase and neutral, whereas three-phase, 400-V motor load is connected across three-phase lines directly.

4. The CTs are located at suitable places in the substation circuit and supply for the metering and indicating instruments and relay circuit.

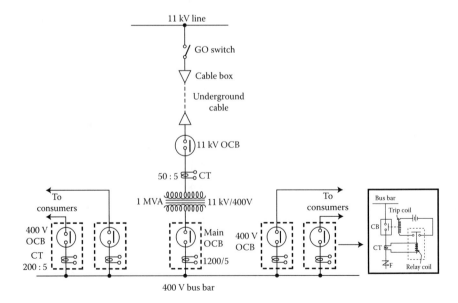

FIGURE 18.8   Key diagram of a typical 11 kV/400 V indoor substation.

## EXERCISES

1. What is a substation? Name the factors that should be taken care of while designing and erecting a substation.

2. Discuss the different ways of classifying the substations.

3. Give the comparison of outdoor and indoor substations.

4. What is a transformer substation? What are the different types of transformer substations? Illustrate your answer with a suitable block diagram.

5. Draw the layout and schematic connection of a pole-mounted substation.

6. Draw the layout of a typical underground substation.

7. Write a short note on the substation equipment.

8. What are the different types of bus-bar arrangements used in substations? Illustrate your answer with suitable diagrams.

9. What are terminal and through substations? What is their purpose in the power system?

10. Draw the key diagram of a typical 11 kV/400 V indoor substation.

11. Why gravel stones are used to pave substation surface?

    The ground of the substation yard is filled with crushed gravel stones. Generally 20 to 25 mm baby gravel stones are used instead big size stones to facilitate movement of persons and equipment in the substation yard. This is because of the following reasons concern to safety from shock:

    1. It provides a high resistance layer or insulation between our foot and the ground. So that the fault currents flow into the ground but not along the ground.

    2. To minimise step potential and touch potential voltages.

    3. It avoids pool of inflammable oil, etc. on the substation ground in case of any spilling of insulation oil from the equipment. This also avoids spreading of fire from one equipment to the other in the substation.

4. It restricts entering of snakes and other reptiles as the surface would be inconvenient to crawl.

5. It avoids growth of plants and weeds in the substation yard to some extent.

# CHAPTER 19

# Power System Stability

## 19.1 INTRODUCTION

The stability of an interconnected power system is its ability to return to normal or stable operation after having been subjected to some form of disturbance. Conversely, instability means a condition denoting loss of synchronism or falling out of step. Stability considerations have been recognized as an essential part of power system planning for a long time. With interconnected systems continually growing in size and extending over vast geographical regions, it is becoming increasingly more difficult to maintain synchronism between various parts of a power system. The stability problem is concerned with the behavior of synchronous machines after a disturbance. Under stable conditions the system stays in synchronism.

Synchronous stability may be divided into two main categories depending upon the magnitude of disturbance.

1. *Steady-state stability.* Steady-state stability refers to the ability of the power system to regain synchronism after small and slow disturbance, such as gradual power changes.

2. *Transient stability.* Transient stability is the ability of the system to regain synchronism after a large disturbance. The large disturbance can occur due to sudden changes in application or removal of large loads, line switching operations, faults on the system, sudden outage of a line, or loss of excitation. Transient stability studies are needed to ensure that the system can withstand the transient conditions following a major disturbance. Frequently, such studies are conducted when new generating and transmitting facilities are planned.

Steady-state stability is subdivided into static stability and dynamic stability.

1. *Static stability.* Static stability refers to inherent stability that prevails without the aid of automatic control devices such as governors and voltage regulators.

2. *Dynamic stability.* Dynamic stability, on the other hand, denotes artificial stability given to an inherently unstable system by automatic control devices. Dynamic stability is concerned with small disturbances lasting for times of the order of 10–30 s with the inclusion of automatic control devices.

Stability studies are helpful for the following purposes:

1. Determination of critical clearing time of circuit breakers (CBs)

2. Investigation of schemes of protective relaying

3. Determination of voltage levels

4. Transfer capability between systems

Analysis of power system stability is complex and nonlinear. Consequently, final designs are generally based on computer simulations. For approximate purposes, simplified calculations are used. Simplified calculations provide a starting point for, and check of, computer simulations. They are also useful in studying the factors that influence the power system stability. Invariably stability studies of power systems are carried out on a digital computer. In the following, special cases to illustrate certain principles and basic concepts are presented.

## 19.2 STABILITY LIMITS AND POWER TRANSMISSION CAPABILITY

The stability limit is the maximum power that can be transferred in a network between sources and loads without loss of synchronism. The steady-state limit is the maximum power that can be transferred without the system becoming unstable when the load is increased gradually under steady-state conditions. Transient limit is the maximum power that can be transferred without the system becoming unstable when a sudden or large disturbance occurs.

The system experiences a shock by sudden and large power changes and violent fluctuations of voltage occur. Consequently, individual machines or group of machines may go out of step. The rapidity of application of a large disturbances is responsible for the loss of stability, otherwise it may be possible to maintain stability if the same large load is applied gradually. Thus, the transient stability limit is lower than the steady-state limit.

### 19.2.1 Power Transmission Capability

The power transmission capability of a line is limited by the thermal loading limit and the stability limit. The real power loss increases the conductor temperature. This will increase the sag of the conductors between the transmission towers. The thermal limit is specified by the current-carrying capacity of the conductor and is available in the manufacturer's data.

Let $I_{\text{thermal}}$ be the current-carrying capacity, $S_{\text{thermal}}$, the thermal loading of the line, $V_p$, the rated phase voltage, and $S_{\text{thermal}}$, the $3V_p I_{\text{thermal}}$.

## 19.3 INFINITE BUS

In a power system, normally more than two generators operate in parallel. The machines may be located at different places. A group of machines located at one place may be treated as a single large machine. Also, the machines not connected to the same bus but separated by lines of low reactance may be grouped into one large machine. The operation of one machine connected in parallel with such a large system comprising many other machines is of great interest. The capacity of the system is so large that its voltage and frequency may be taken constant. The connection or disconnection of a single small machine on such a system would not affect the magnitude and phase of the voltage and frequency. Such a system of constant voltage and constant frequency regardless of the load is called infinite bus-bar system or simply infinite bus. Physically it is not possible to have a perfect infinite bus. An infinite bus is an ideal voltage source.

## 19.4 SYNCHRONOUS GENERATOR CONNECTED TO AN INFINITE BUS

Consider a simple system consisting of a synchronous generator connected to an infinite bus through a network presented by the ABCD parameters as shown in Figure 19.1.

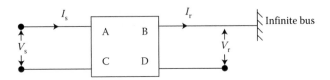

FIGURE 19.1  A synchronous generator connected to an infinite bus through a two-port network.

The sending end and receiving end voltages are assumed as

$$\mathbf{V_S} = V_S \angle\delta, \quad \mathbf{V_r} = V_r \angle 0°$$

We have

$$\mathbf{V_s} = \mathbf{AV_r} + \mathbf{BI_r}$$

$$\mathbf{I_r} = \frac{\mathbf{V_s}}{\mathbf{B}} - \frac{\mathbf{A}}{\mathbf{B}}\mathbf{V_r} = \frac{V_S\angle\delta}{B\angle\beta} - \frac{A\angle\alpha}{B\angle\beta}V_r\angle 0° = \frac{V_S}{B}\angle(\delta-\beta) - \frac{AV_r}{B}\angle(\alpha-\beta)$$

Complex power received at the infinite bus is

$$\mathbf{S_r} = \mathbf{V_r}\mathbf{I_r^*} = P_r + jQ_r$$

where $I_r^*$ is the complex conjugate of $I_r$.

$$\mathbf{I_r^*} = \frac{V_s}{B}\angle(\beta-\delta) - \frac{AV_r}{B}\angle(\beta-\alpha)$$

$$P_r + jQ_r = \mathbf{V_r}\mathbf{I_r^*} = \frac{V_sV_r}{B}\angle(\beta-\delta) - \frac{AV_r^2}{B}\angle(\beta-\alpha)$$
$$= \frac{V_sV_r}{B}[\cos(\beta-\delta) + j\sin(\beta-\delta)] - \frac{AV_r^2}{B}[\cos(\beta-\alpha) + js(\beta-\alpha)]$$

Equating real and imaginary part, we get

$$P_r = \frac{V_sV_r}{B}\cos(\beta-\delta) - \frac{AV_r^2}{B}\cos(\beta-\alpha)$$

$$Q_r = \frac{V_sV_r}{B}\sin(\beta-\delta) - \frac{AV_r^2}{B}\sin(\beta-\alpha)$$

The power received is a maximum when $\delta = \beta$. Therefore,

$$P_{r\,max} = \frac{V_s V_r}{B} - \frac{A V_r^2}{B}\cos(\beta - \alpha)$$

## 19.5 POWER–ANGLE CURVE

Figure 19.2 shows a synchronous machine connected to an infinite bus through a transmission line of reactance $X_1$. Let us assume that line resistance and capacitances are neglected.

Let $\mathbf{V} = V\angle 0°$ be the voltage of infinite bus, $\mathbf{E} = E\angle\delta$, the voltage behind the direct axis synchronous reactance of the machine, and $X_d$, the synchronous/transient reactance of the machine.

The complex power delivered by the generator to the system is

$$\mathbf{S} = \mathbf{V}\mathbf{I}^* = V\left[\frac{E\angle\delta - V\angle 0°}{j(X_d + X_1)}\right]^*$$

$$(19.1)$$

Let

$$X_d + X_1 = X \qquad\qquad (19.2)$$

$$\mathbf{S} = V\left[\frac{E\angle\delta}{X\angle 90°} + j\frac{V}{X}\right]$$

$$\frac{EV}{X}\angle(90° - \delta) - j\frac{V^2}{X} = \frac{EV}{X}\sin\delta + j\frac{EV}{X}\cos\delta - j\frac{V^2}{X}$$

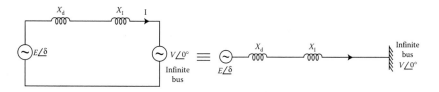

FIGURE 19.2  Synchronous machine connected to an infinite bus through a transmission line of series reactance $X_1$.

$$P_e + jQ_e = \frac{EV}{X}\sin\delta + j\left(\frac{EV}{X}\cos\delta - \frac{V^2}{X}\right) \qquad (19.3)$$

Active power transferred to the system is

$$P_e = \frac{EV}{X}\sin\delta \qquad (19.4)$$

The reactive power transferred to the system is

$$Q_e = \left(\frac{EV}{X}\cos\delta - \frac{V^2}{X}\right) \qquad (19.5)$$

The maximum steady-state power transfer occurs when $\delta = 90°$. From Equation 19.4,

$$P_{e\,max} = \frac{EV}{X}\sin 90° = \frac{EV}{X} \qquad (19.6)$$

$$P_e = P_{e\,max}\sin\delta \qquad (19.7)$$

The graphical representation of power $P_e$ and the load angle $\delta$ is called the power–angle diagram or power–angle curve. Such diagram is widely used in power-system stability studies. A power–angle diagram is shown in Figure 19.3.

Maximum power is transferred when $\delta = 90°$. As $\delta$ is increased beyond 90°, $P_e$ decreases and becomes zero at $\delta = 180°$. Beyond $\delta = 180°$, $P_e$ becomes negative which implies that the power flow direction is reversed and the power is supplied from the infinite bus to the generator, the value of $P_{e\,max}$ is often called the pull-out power. It is also called the steady-state limit.

The total reactance $X$ between two voltage sources $V$ and $E$ is called the transfer reactance. It is seen that the maximum power limit is inversely proportional to the transfer reactance. Equation 19.4 is valid for both steady-state and transient conditions. For steady-state conditions, we use synchronous reactance and we take $E$ as the electromotive force behind synchronous reactance. For transient conditions, the transient reactance $X_d'$ is used and $E$ is taken as the electromotive force behind transient reactance.

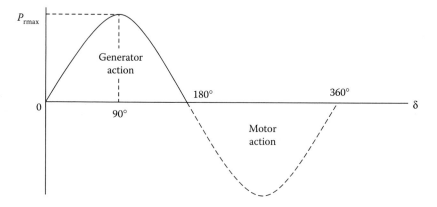

FIGURE 19.3  Power–angle curve.

## 19.6 POWER–ANGLE RELATIONS FOR GENERAL NETWORK CONFIGURATION

In general, active power flow from bus $i$ to bus $j$ of an AC network when resistances are neglected is given by

$$P_{ij} = \frac{V_i V_j}{X_{ij}} \sin \delta_{ij} \tag{19.8}$$

where $P_{ij}$ is the active power flow from bus $i$ to bus $j$, $V_i$ is the voltage at bus $i$, $V_j$ is the voltage at bus $j$, $\delta_{ij}$ is the angle between bus $i$ and bus $j$ with bus $i$ taken as reference, and $X_{ij}$ is the equivalent transfer reactance between buses $i$ and $j$.

The reactive power flow is given by

$$Q_{ij} = \frac{V_i V_j}{X_{ij}} \cos \delta_{ij} = \frac{V_{ij}^2}{X_{ij}} \tag{19.9}$$

where $Q_{ij}$ is the reactive power flow from bus $i$ to bus $j$.

Let us determine the power–angle relation for the network configuration, as shown in Figure 19.4.

Here $E_G$ is the generator voltage, $V$ is the infinite bus voltage, $X_G$ is the generator reactance, $X_s$ is the system reactance, $X_f$ is the fault reactance, and $P_e$ is the machine electrical power output.

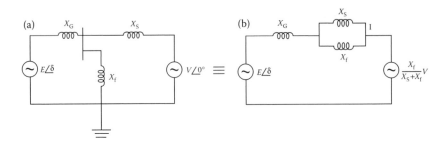

FIGURE 19.4    An equivalent series circuit of T-circuit.

We can replace the T-circuit of Figure 19.4a by a series circuit, as shown in Figure 19.4b.

From Equation 19.8,

$$P_e = \frac{E((X_f)/(X_s + X_f)V)}{X_G + ((X_s X_f)/(X_s + X_f))}\sin\delta = \frac{EV}{X_G((X_s + X_f)/X_f) + X_s}$$

$$\sin\delta = \frac{EV}{X_G + X_s + ((X_G X_s)/X_f)}\sin\delta \qquad (19.10)$$

Equation 19.10 gives the power–angle relation for the network configuration, as shown in Figure 19.4a.

## 19.7  STEADY-STATE STABILITY CRITERION

The rate $dP/d\delta$, that is, the differential power increase obtained per differential load angle increase is called the synchronizing power coefficient or electrical stiffness of the synchronous machine. It is taken as the measure of the stability of the system. The direct-state synchronous stability criterion for a simple system is $dP/d\delta > 0$, that is, the synchronizing coefficient is positive. The steady-state stability limit is reached when $dP/d\delta = 0$ and if $dP/d\delta < 0$, the system is unstable.

The criterion of stability holds only under conditions satisfying the following assumptions:

1. Generator are represented by constant impedances in series with the no-load voltages.

2. The mechanical power input is constant.

3. Damping is negligible.

4. Load angle variation are small.

5. Speed variations are negligible.

When the effect of inertia of machines, governor action, and automatic voltage regulators are considered the problem become more complex. The criterion $dP/d\delta = 0$ alone gives a low fault which is safe.

## 19.8 TRANSIENT STABILITY

The transient stability is the ability of system to maintain synchronous operation and to reach a stable state or the one close to it after a large disturbance. The following simplifying assumptions are made in the study of transient stability:

1. System resistance may be reduced to an equivalent two-machine system.

2. Each machine has cylindrical rotor. The direct-axis reactance $(X_d)$ is equal to the quadrature-axis reactance $(X_q)$.

3. The system may be reduced to an equivalent two-machine system.

4. Each machine may be assumed to supply an infinite bus.

5. Direct axis transient reactance $(X_d)$ is used for machine representation.

6. The shaft input power may be assume constant for few seconds after occurrence of a disturbance. This assumption may be valid on the grounds that the mechanical system involving governors, steam valves, etc., is relatively sluggish in operation as compared to rapidly charging electrical quantities. With fast action valves, the assumption of constant input will not be true.

The problem of stability removes around the determination of whether or not the torque angle $\delta$ will stabilize after a sudden disturbance. In case $\delta$ continues to increase after a disturbance, the machine will loss synchronism.

In a synchronous generator, the input is the mechanical or shaft torque and the output is the electromagnetic torque. Both these torques are assumed positive in the following discussion. For a synchronous motor, the input is the electromagnetic torque and the output is the shaft torque.

Based upon the sign conventions adopted for synchronous generators, the value of the shaft torque and electromagnetic torque are taken as negative for motor action.

Let $T_e$ be the electromagnetic torque and $T_s$, the shaft torque. If the losses are neglected, the difference between the shaft torque and the electromagnetic torque is equal to the accelerating or deceleration torque. For a generator, when $T_s > T_e$, then $T_a$ is positive and the rotor accelerates. In case of synchronous motor $T_a$ is positive only when $T_e > T_s$, since $T_s$ and $T_e$ are both negative.

## 19.9 SWING EQUATION

The behavior of a synchronous machine during transient is described by the swing equation. Let $\theta$ be the angular position of the rotor at any instant $t$. However, $\theta$ is continuously changing with time. It is convenient to measure $\theta$ with respect to reference axis that is rotating at synchronous speed. If $\delta$ is the angular displacement of the rotor in electrical degrees from the synchronously rotating reference axis and $\omega_s$ the synchronous speed in electrical radians, then $\theta$ can be express as the sum of (1) time varying angle $\omega_s t$ on the rotating reference axis, plus (2) the torque angle $\delta$ of the rotor with respect to the rotating reference axis. In other word,

$$\theta = \omega_s t + \delta \text{ electrical radians} \tag{19.11}$$

Differentiating Equation 19.11 with respect to $t$, we get

$$\frac{d\theta}{dt} = \omega_s + \frac{d\delta}{dt} \tag{19.12}$$

Differentiation of Equation 19.12 gives

$$\frac{d^2\theta}{dt^2} = \frac{d^2\delta}{dt^2} \tag{19.13}$$

Angular acceleration of rotor is

$$\alpha = \frac{d^2\theta}{dt^2} = \frac{d^2\delta}{dt^2} \text{ elec.rad/s}^2 \tag{19.14}$$

If damping is neglected the accelerating torque, a in a synchronous generator is equal to the difference of input mechanical or shaft toque $T_s$ and the output electromagnetic (electrodynamics) torque $T_e$. That is,

$$T_a = T_s - T_e \tag{19.15}$$

Let $\omega$ be the synchronous speed of the rotor, $J$, the moment of inertia of the rotor, $M$, the angular momentum of the rotor, $P_s$, the mechanical power input, $P_e$, the electrical power input, and $P_a$, the accelerating power. Now

$$M = J\omega \tag{19.16}$$

Multiplying both the sides of Equation 19.15 by $\omega$, we get

$$\omega T_a = \omega T_s - \omega T_e$$

$$P_a = P_s - P_e$$

But

$$J\frac{d2\theta}{dt2} = T_a, \quad J\frac{d2\delta}{dt2} = T_a$$

$$\omega J\frac{d2\delta}{dt2} = \omega T_a$$

$$M\frac{d2\delta}{dt2} = P_a = P_s - P_e \tag{19.17}$$

Equation 19.17 gives the relation between the accelerating power and angular acceleration. It is called the swing equation. It is a nonlinear differential equation of the second order. With this differential equation, we can discuss stability in a quantitative way, because it describes swings in the power angle $\delta$ during transient.

## 19.10 SWING CURVES

A graph of $\delta$ (usually in electrical radians) versus time in seconds is called the swing equation. Swing curves (Figure 19.5) provide information regarding stability. They show any tendency of $\delta$ to oscillate and/or increase beyond the point of return. If $\delta$ increases continuously with time, the system is unstable. While if starts decreasing after reaching a maximum value, it is inferred that the system will remain stable.

Swing curves are useful in determining the adequacy of relay protection on power system with regard to the clearing of faults before one or more machines become unstable and fall out of synchronism. The critical clearing time is found to specify the correct speed of the CB.

The solution of swing equation involves elliptic integrals. Step-by-step method may be used for numerical solution of swing equation. At present digital computer is used for solving swing equation.

## 19.11 *M* AND *H* CONSTANTS

The transient conditions of synchronous machine depend, in part, on the mechanical constant of the rotor and load or prime mover.

Let $\omega$ be the synchronous speed of the rotor in rad/s, $m$, the mass of the rotor in kilogram, $r$, the radius of gyration in meter, $j$, the moment of inertia of the rotor in kg m², $M$, the angular momentum of the rotor in Js/rad, $W$, the kinetic energy of the rotor in J, $F$, the system frequency in

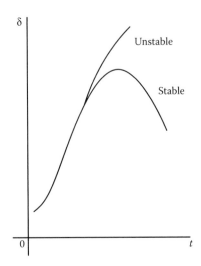

FIGURE 19.5 Swing curves.

Hz, $T$, the torque in N m, $P$, the power in watts, and $\alpha$, the angular acceleration of the rotor.

### 19.11.1 $M$ Constant

Now, $j = mr^2$

$$W = \frac{1}{2}J\omega^2$$

$$M = J\omega = \frac{2W}{\omega}$$

$$\omega = 2\pi f \text{ rad/s} = 360 f \text{ elec.deg/s}$$

$$T = J\alpha$$

$$P = \omega T = \omega J\alpha = M\alpha$$

$$M = \frac{P}{\alpha}$$

Thus, $M$ constant may be defined as the power in $MW$ required to producer unit angular acceleration.

### 19.11.2 $H$ Constant or Per-Unit Inertia Constant

Another constant $H$, called the unit per-unit inertia constant, is more frequently used by the manufactures. The per-unit inertia constant $H$ is defined as the kinetic energy stored in the parts of the machine at synchronous speed per-unit megavolt amperes (MVA) of the machine.
    Thus

$$H = \frac{\text{Kinetic energy in MJ at rated speed}}{\text{Machine rating in MVA}}$$

It is expressed in MJ/MVA.
    If $W$ is the stored energy in megajoles. MJ and $S$ is the rating of the machine in MVA.

Then

$$H = \frac{W}{S} = \frac{\omega M}{2S} = \frac{2\pi f M}{2S} = \frac{\pi f M}{S} \qquad (19.18)$$

$$M = \frac{HS}{\pi f} \text{ MJs/elec.radian} \qquad (19.19)$$

$$M = \frac{HS}{180 f} \text{ MJs/elec. degree} \qquad (19.20)$$

The value of angular momentum $M$ varies over a wide range of MVA for a given type of machine and prime mover, but the value of $H$ is fairly constant. Hence $H$ is more convenient use, typically value are

- Cylindrical-rotor alternator 4–10

- Salient-pole alternator 2–4

- Synchronous compensators 1–2

- Salient-pole synchronous motors 0.5–2

The swing equation can be written as

$$\frac{HS}{180 f} \frac{d^2\delta}{dt^2} = P_s - P_e \qquad (19.21)$$

By combining Equations 19.20 and 19.21 and divided by $S$, we obtain the per-unit swing as

$$\frac{HS}{180 f} \frac{d^2\delta}{dt^2} = P_{s\,pu} - P_{e\,pu} = P_{a\,pu} \qquad (19.22)$$

### 19.11.3 $H$ Constant on a Common Base

An inertia constant $H_{mach}$ based on a machine, own MVA rating may be converted to value $H_{syst}$ relative to the system base $S_{syst}$ with the formula

$$H_{syst} = H_{mach} \frac{S_{mach}}{S_{syst}}$$

A convenient system base value is 100 MVA.

## 19.12 EQUIVALENT SYSTEM

Suppose that a number of generators are connected in parallel to the same bus bars.

Let $S_1, S_2, ..., S_n$ be the MVA rating of individual machines, $S_e$, the MVA rating of the equivalent machine, $S_b$, the base MVA, $H_1, H_2, ..., H_n$, the inertia constant of individual machine, $H_e$, the inertia constant of a single equivalent machine, and $S$, the total rating of the machine.

Energy stored by the equivalent machine = Sum of the energies stored by individual machine,

$$W = W_1 + W_2 + \cdots + W_n \tag{19.23}$$

$$S_e H_e = S_1 H_1 + S_2 H_2 + \cdots + S_n H_n \tag{19.24}$$

$$S_e = S_1 + S_2 + \cdots + S_n \tag{19.25}$$

If the base MVA is equal to the combined MVA rating of the individual machine, that is, $S_b = S_p$, Equation 19.24 becomes

$$H_e = H_1 \left( \frac{S_1}{S_b} \right) + H_2 \left( \frac{S_2}{S_b} \right) + \cdots + H_n \left( \frac{S_n}{S_b} \right)$$

Thus the equivalent inertia constant is the sum of the individual constant, when this is referred to the total rating of the machine.

If the machine is identical,

$$S_1 = S_2 = \cdots = S_n = S \text{ (say)}$$

$$H_1 = H_2 = \cdots = H_n = H \text{ (say)}$$

$$S_b = S_e = nS$$

$$H_e = \frac{nHS}{nS} = H$$

Thus the equivalent $H$ constant of several identical machine operating in parallel is the same as that of any one of the machine.

## 19.13 EQUIVALENT $M$ CONSTANT OF TWO MACHINES

Two synchronous machine connected by a reactance can be replaced by one equivalent machine connected by the reactance to an infinite bus bar. Let suffix 1 and suffix 2 be used for the two machines.

For one machine connected to infinite bus bar,

$$M\frac{d^2\delta}{dt^2} = P_s - P_e \tag{19.26}$$

$$M_1\frac{d^2\delta_1}{dt^2} = P_{s1} - P_{e1} \tag{19.27}$$

$$M_2\frac{d^2\delta_2}{dt^2} = P_{s2} - P_{e2} \tag{19.28}$$

Let $\delta$ be the relative angle between the rotors of the two machine,

$$\delta = \delta_1 - \delta_2 \tag{19.29}$$

$$\frac{d^2\delta}{dt^2} = \frac{d^2\delta_1}{dt^2} - \frac{d^2\delta_2}{dt^2} = \frac{1}{M_1}(P_{s1} - P_{e1}) - \frac{1}{M_2}(P_{s2} - P_{e2})$$

Multiplying both sides of the above equation by $M_1M_2/(M_1+M_2)$

$$\frac{M_1M_2}{M_1+M_2} \cdot \frac{d^2\delta}{dt^2} = \frac{M_2}{M_1+M_2}(P_{s1}-P_{e1}) - \frac{M_1}{M_1+M_2}(P_{s2}-P_{e2})$$

$$= \frac{M_2P_{s1}-M_1P_{s2}}{M_1+M_2} - \frac{M_2P_{e1}-M_1P_{e2}}{M_1+M_2} \tag{19.30}$$

Equation 19.30 can be represented as

$$M' \frac{d^2\delta}{dt^2} = P'_s - P'_e \tag{19.31}$$

$$M' = \frac{M_1 M_2}{M_1 + M_2} \tag{19.32}$$

$$P'_s = \frac{M_2 P_{s1} - M_1 P_{s2}}{M_1 + M_2} \tag{19.33}$$

$$P'_e = \frac{M_2 P_{e1} - M_1 P_{e2}}{M_1 + M_2} \tag{19.34}$$

It is seen that Equation 19.31 is similar to Equation 19.26. Thus two interconnected machine can be represented as a single source. The quantities $M'$, $P'_s$, and $P_e'$ represent the equivalent values of the inertia constant, the input at the shaft and the electromagnetic output respective. The load angle $\delta$ of the equivalent machine is given by Equation 19.29.

### 19.13.1 Multi-Machine Systems

The swing equation of a machine is given by

$$M \frac{d^2\delta}{dt^2} = P_s - P_e$$

In a system with $n$ machine, the rotor of each machine will respond in accordance with Equation 19.26 as

$$M_1 \frac{d^2\delta_1}{dt^2} = P_{s1} - P_{e1}$$

$$M_2 \frac{d^2\delta_2}{dt^2} = P_{s2} - P_{e2}$$

$$M_n \frac{d^2\delta_n}{dt^2} = P_{sn} - P_{en}$$

It is seen that there is a separate swing equation for each machine. Consequently, it is the relative displacement between power angles of the machine that is essential in determining the system stability. In a multi-machine system, a single machine is usually chosen as a reference and the rotor swing (changes in power angles) of the remaining machines are determined relative to the reference machine. Stability is maintained if the machine rotors return to a value operating state relative to each other.

## 19.14 EQUAL-AREA CRITERION OF STABILITY

Equal area criterion may be used to assess the transient stability of a two-machine system or one machine connected to an infinite bus without actually solving the swing equation. Consider a loss free synchronous generator supplying an infinite bus through a purely reactive transmission line of reactance $X_1$ as shown in Figure 19.2. We know

$$P_e = P_{e\,max} \sin\delta \tag{19.35}$$

The power–angle curve is shown in Figure 19.6. Suppose that initially the mechanical input (shaft power) is $P_{s0}$ at load angle $\delta_0$. It is represented by point a on the power–angle curve. Let the mechanical input power suddenly increase to $P_{s1}$. With the sudden increase of shaft power, there is momentarily more shaft input than electrical output. The increase in

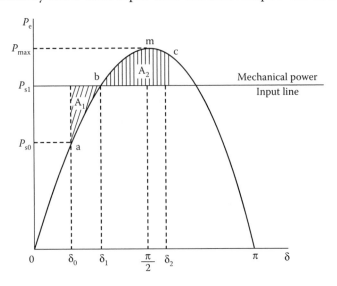

FIGURE 19.6   Power–angle curve.

power $(P_{s1} - P_{s0})$ accelerates the rotor so that it is advanced with respect to the initial position with the result that load angle is increased. Let this new load angle be $\delta_1$ corresponding to $P_{s1}$. Since the rotor is in acceleration and running slightly above synchronous speed, the load angle goes on increasing overshooting point b. When the load angle is more than $\delta_1$, the rotor retards since the power transferred to the bus bar is greater than input power $P_{s1}$. The rotor decelerates until it reaches some maximum point c, where it is again running at synchronous speed. The rotor swings start in the reverse direction. The load angle goes on decreasing until it is equal to $\delta_0$ where again the rotor is running at synchronous speed. The cycle is repeated. The rotor will oscillate for sometime about b, before finally coming to rest at b.

Here, we have made an important assumption that the first swing or oscillation of the rotor does not make the system unstable. In practice, the system is more likely to be stable during subsequent swings, particularly if it is stable for the subsequent steady-state condition. This assumption may be justified by the fact that the losses of the system progressively damp the amplitude of the swing.

Consider the swing equation

$$\frac{d^2\delta}{dt^2} = \frac{P_a}{M} \tag{19.36}$$

Multiplying both the sides by $2(d\delta/dt)$:

$$2\frac{d\delta}{dt}\left(\frac{d^2\delta}{dt^2}\right) = 2\frac{P_a}{M} \cdot \frac{d\delta}{dt}$$

$$\frac{d}{dt}\left(\frac{d\delta}{dt}\right)^2 = 2\frac{P_a}{M} \cdot \frac{d\delta}{dt} \tag{19.37}$$

The time rate of change of load angle $d\delta/dt$ is the speed of the machine with respect to the synchronously revolving reference frame. For the stability, this speed must become zero at sometime after disturbance. That is, $d\delta/dt = 0$.

Since the condition $(d\delta/dt) = 0$ implies synchronous running, Equation 19.37 is integrated between the limits of swinging of $\delta$, that is, from

$\delta = \delta_0$ where $(d\delta/dt) = 0$ to $\delta = \delta_2$ where again $(d\delta/dt) = 0$. On integration, Equation 19.37 gives

$$\left(\frac{d\delta}{dt}\right)^2 = \frac{2}{M}\int_{\delta_0}^{\delta_2} P_a \, d\delta$$

For stability,

$$\frac{d\delta}{dt} = 0$$

$$\therefore \quad \int_{\delta_0}^{\delta_2} P_a = 0 \tag{19.38}$$

$$\int_{\delta_0}^{\delta_2} P_a \, d\delta + \int_{\delta_1}^{\delta_2} P_a \, d\delta = 0$$

$$\int_{\delta_0}^{\delta_2} P_a \, d\delta = -\int_{\delta_1}^{\delta_2} P_a \, d\delta \tag{19.39}$$

$$A_1 = -A_2 \tag{19.40}$$

where

$$A_1 = \int_{\delta_0}^{\delta_1} P_a \, d\delta = \text{Positive or accelerating area}$$

= The amount of work done on the rotor to move it from point a to point b in increasing the kinetic energy of the rotor

$$A_2 = \int_{\delta_1}^{\delta_2} P_a \, d\delta = \text{Negative or decelerating area.}$$

= The amount of work done on the rotor to move it from point b to point c when the rotor returns its energy to the circuit

Since the positive or accelerating area $A_1$ is equal to the negative or decelerating area $A_2$, it is called the equal-area criterion of stability. Thus, for stability area $A_1$ is equal to area $A_2$.

The equal area criterion of stability provides the following information:

1. It is an easy means of finding the maximum angle of swing.

2. An estimate of whether synchronism will be maintained.

3. The maximum amount of disturbance that can be allowed without losing synchronism.

The equal area criterion is applicable only to a two-machine system or one machine connected to infinite bus. It is not applicable to multi-machine system.

### 19.14.1 Application to Sudden Increase in Mechanical Power Input

The equal-area criterion is used to determine the maximum additional power $P_s$ can be applied for stability to be maintained. With a sudden change in power input, the stability is maintained only if area $A_2$ is at least equal to $A_1$ can be located above $P_s$. If area $A_2$ is less than area $A_1$, the accelerating momentum can never be overcome. For the system to remain stable, it is possible to find angle $\delta_2$ such that $A_2 = A_1$. As $P_{s1}$ is increased a limiting condition is finally reached when $A_1$ equals the area above the $P_{s1}$ line as shown in Figure 19.7. The limit of stability occurs when $\delta_{max}$ is at the intersection of line $P_s$ and the power–angle curve for $90° < \delta < 180°$, as shown in Figure 19.7.

Under this condition $\delta_2$ acquires the maximum value $\delta_{max}$ such that

$$\delta_2 = \delta_{max} = 180° - \delta_1$$

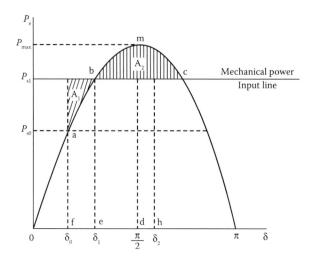

FIGURE 19.7 Limiting case of transient stability with mechanical input suddenly increased.

Any further increase in $P_{s1}$ means that the area available for $A_2$ is less than the area $A_1$, so that excess kinetic energy causes $\delta$ to increase beyond point c and the retarding power changes over to accelerating power with the system consequently become unstable.

It may also be noted from Figure 19.7 that the system will remain stable even though the rotor may oscillate beyond $\delta = 90°$ so long as equal area criterion is met. The condition $\delta = 90°$ is meant for use in steady-state stability only, and does not apply to the transient stability.

Applying the equal-area criterion to Figure 19.7, we have

$$\text{Area } A_1 = \text{Area } A_2$$

$$\text{Area agb} = \text{Area bmc}$$

Area $A_1$ = Area agb

= Area of the rectangle gbef − Area abef under the sine curve

$$= P_{s1}(\delta_1 - \delta_0) - \int_{\delta_0}^{\delta_1} P_m \sin\delta\, d\delta$$

$$= P_{s1}(\delta_1 - \delta_0) + P_m(\cos\delta_1 - \cos\delta_0)$$

Area $A_2$ = Area bmc

= Area bmche under the sine curve − Area of rectangle bche

$$= \int_{\delta_1}^{\delta_2} P_m \sin\delta \, d\delta - P_{s1} (\delta_2 - \delta_1)$$

$$= P_m(\cos\delta_1 - \cos\delta_2) - P_{s1} (\delta_2 - \delta_1)$$

By equal area criterion,

Area $A_1$ = Area $A_2$

$$P_{s1}(\delta_1 - \delta_0) + P_m(\cos\delta_1 - \cos\delta_0) = P_m(\cos\delta_1 - \cos\delta_2) - P_{s1} (\delta_2 - \delta_1)$$

or

$$P_{s1}(\delta_2 - \delta_0) = P_m(\cos\delta_0 - \cos\delta_2)$$

Also,

$$\delta_2 = \delta_{max}$$

and

$$P_{s1} = P_m \sin\delta_{max}$$

At point c of the sine curve.

Substitution of these values in the above equation gives

$$(\delta_{max} - \delta_0)\sin\delta_{max} + \cos\delta_{max} = \cos\delta_0$$

The above nonlinear algebraic equation can be solved by trial and error method for $\delta_{max}$. Once $\delta_{max}$ is obtained, the maximum permissible power or the transient stability limit is found from

$$P_{s1} = P_m \sin\delta_1$$

where

$$\delta_1 = \pi - \delta_{max}$$

## 19.14.2 One of the Parallel Lines Suddenly Switched Off

Consider a system (Figure 19.8) consisting of a synchronous generator feeding an infinite bus through a double-circuit line. The two circuits are operating in parallel. If one of the circuits is switched off suddenly, the system may become unstable in spite of the fact that the load could be supplied over by the other circuit under steady-state conditions.

We shall use the equal-area criterion to study the transient stability of the system when one of the lines is switched off. When both the lines are operating in parallel; the power transfer is given by

$$P_{el} = \frac{EV}{X_A} \sin \delta$$

where $X_A$ is the transfer reactance when both the lines are operating in parallel.

$$X_A = X_d' + X_1 \| X_2 = X_d' + \frac{X_1 X_2}{X_1 + X_2}$$

$$P_{el} = \frac{EV}{X_d' + (X_1 X_2 / (X_1 + X_2))} \sin \delta$$

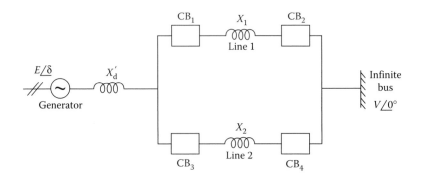

FIGURE 19.8 A synchronous generator feeding an infinite bus through a double-circuit line.

or

$$P_{e1} = P_{max1} \sin \delta \qquad (19.41)$$

where

$$P_{max1} = \frac{EV}{X_d' + (X_1 X_2 / (X_1 + X_2))}$$

The power–angle curve given by Equation 19.41 is shown in Figure 19.9 as curve A.

Let the mechanical power input to the generator be $P_s$ corresponding to a point a on the power–angle curve A of the two lines in parallel. At point a, load angle is $\delta_0$ and the input power $P_s$ at the shaft is equal to the output power $P_e$ of the generator.

When power $P_s$ is being transferred at an angle $\delta_0$, suppose that the line 1 is suddenly switched off by opening the $CB_1$ and $CB_2$. The power transfer is given by

$$P_{e2} = \frac{EV}{X_B} \sin \delta$$

where $X_B$ is the transfer reactance when only line 2 is in operation.

$$X_B = X_d' + X_2$$

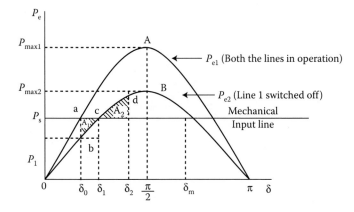

FIGURE 19.9   Power–angle curves. Curve A for two lines in parallel and curve B for line 2.

$$P_{e2} = \frac{EV}{X'_d + X_2} \sin\delta$$

$$P_{e2} = P_{max2} \sin\delta \qquad (19.42)$$

where

$$P_{max2} = \frac{EV}{X'_d + X_2}$$

The power–angle curve given by Equation 19.42 is shown as curve B in Figure 19.9. Immediately on switching off line 1, the load angle $\delta_0$ cannot change instantaneously due to rotor inertia. Since the load angle is still $\delta_0$, the output power has reduced to $P_1$. The operating shifts to point b on the new operating curve B. This sudden change in generator output is not immediately detected by the governor of the prime mover. Thus the shaft power $P_s$ is not changed. Now, the power input to the generator from the shaft is $P_s$ and power output of the generator in the line 2 is $P_1$. The power output $P_1$ is less than the power input $P_s$. The difference $(P_s - P_1)$ acceler-ates the rotor. With a slight increase in speed, the load angle $\delta_0$ increases. The operating points move along the curve B from b to c. At point c, $\delta = \delta_1$ and the accelerating power is zero, but the rotor is running slightly above synchronous speed. It will, therefore, continue to advance up to $\delta_2$, due to rotor inertia. However, in the region between $\delta_1$ and $\delta_2$, the electrical out-put $P_e$ is more than the shaft input $P_s$, with the result the rotor slows down. At point d, the rotor relative speed (wrt the synchronous speed) becomes zero. When $\delta = \delta_2$, the area $A_1$ is equal to the area $A_2$, and the rotor starts swinging back toward $\delta_1$. The rotor will oscillate about c and the oscil-lations go on diminishing due to damping. The operating point finally comes at c, where the input power is equal to the output power again, that is, $P_{e2} = P_{max2} \sin\delta_1$ as shown in Figure 19.9.

If the initial input power $P_s$ is increased (so that the input line $P_s$ is shifted upwards), a limit is reached beyond which the retarding area $A_2$ cannot be equal to the accelerating area $A_1$. If $A_2 < A_1$, the rotor will over-shoot past $\delta_m$ and the machine will lose synchronism. Thus, the maximum value that $\delta_2$ can attain without loss of stability is $\delta_m$ and is given by

$$\delta_2 = \delta_m = (\pi - \delta_1) \text{ elec.radians}$$

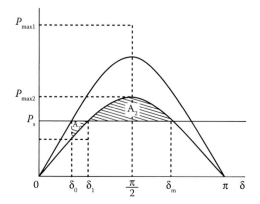

FIGURE 19.10   Transient stability limit for the system in Figure 19.8.

This is shown in Figure 19.10. The value of corresponding $P_s$ to this condition is called the transient stability limit.

## 19.15  SYSTEM FAULT AND SUBSEQUENT CIRCUIT ISOLATION

Consider a synchronous generator supplying power to an infinite bus through a double-circuit line as shown in Figure 19.11. Suppose that some type of fault (say line-to-ground fault) occurs in the middle of line 2. Let us further assume that the fault is not sustained but it is cleared after sometime by opening of the CBs at both the ends of the faulted line. The fault produces a transient change which may render the system unstable. However, if the CBs clear the fault in time, it is possible to maintain stability. The maximum value of time allowed for protective gear to operate without loss of stability is called the critical clearing time. The torque angle corresponding to this time is called critical clearing angle.

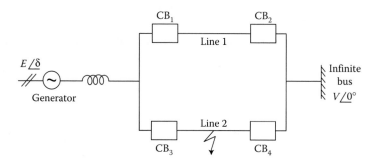

FIGURE 19.11   Line-to-ground fault in the middle of line 2.

The conditions of operation are shown by three power angles—curves A, B, and C. Curve A is called the prefault curve. It represents the prefault condition when both the lines are in healthy condition. Curve B is called the fault-duration curve. It represents the condition of the system during fault when one healthy line and one faulted line are in the circuit. Curve C is called the post-fault curve. This curve represents the condition of the system after the faulted line has been switched out and only one healthy line is in the circuit. The three power–angle curves are drawn from the following equations.

For prefault condition,

$$P_{e1} = \frac{EV}{X_A} \sin \delta = P_{max\,1} \sin \delta$$

For the condition during fault,

$$P_{e2} = \frac{EV}{X_B} \sin \delta = P_{max\,2} \sin \delta$$

For the post-fault condition,

$$P_{e3} = \frac{EV}{X_C} \sin \delta = P_{max\,3} \sin \delta$$

where $X_A$ is the transfer reactance prior to fault, $X_B$ is the transfer reactance during fault, and $X_C$ is the transfer reactance for the post-fault condition.

The three power–angle curves are shown in Figure 19.12. The input line is given by $P_e = P_s$. Before the occurrence of the fault, the system was operating at a point a on the prefault curve A. The initial load angle $\delta_0$ is obtained by intersection of the input line $P_e = P_s$ and the prefault curve A. When the fault occurs, the transfer reactance is changed and the power output is reduced. The operating point shifts from point a to point b corresponding to load angle $\delta_0$ on the fault-duration curve B. At this point, the input $P_s$ is greater than the electrical output, with the result that the rotor accelerates and the operating point moves to point c. At point c, the $CB_3$ and $CB_4$ at the two ends of the faulted line 2 open and the fault is cleared. Thus, only

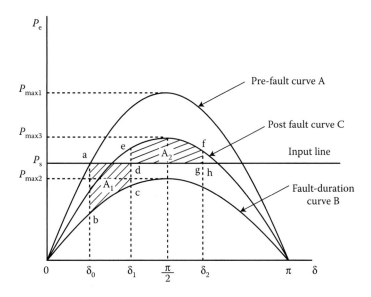

FIGURE 19.12    Power–angle curves when fault occurs on one line.

one healthy line remains in the circuit. The transfer reactance changes to a new value due to the opening of the CBs. The operating point moves to point e on the post-fault curve C. The load angle δ goes on increasing due to the inertia of the rotor. Now, the output power is greater than the input power, and the rotor starts retarding till the point f is reached. At the point f, the angle δ is $δ_2$ and the speed of the rotor with respect to the synchronous speed becomes zero. The extent of overshoot, that is, the value of $δ_2$ can be determined by equating the area s defg and abcd. If the area included between the curve C and the line $P_e = P_s$, bounded by $δ = δ_1$ is less than area abcd, the machine will lose synchronism after the operation of the CBs. The system will be stable if the retardation area $A_2$ (area defg) is equal to the acceleration area $A_1$ (area abcd). It is to be noted that the acceleration area $A_1$ depends upon the clearing angle $δ_1$. When $δ_1$ increases, area $A_1$ also allowable value of δ for stability. Such a condition is shown in Figure 19.13a, where $δ_1 = δ_c$. The angle $δ_c$ is called the critical clearing angle. For the system to be stable, the clearing angle should be less than the critical angle. If the actual clearing angle is greater than the critical clearing angle, the system becomes unstable. Thus, more rapidly the fault is cleared, the smaller will the accelerating area be, and the greater the chance of stable operation being restored.

For simplicity, let us take $P_{max1} = P_{m1}$, $P_{max2} = P_{m2}$, and $P_{max3} = P_{m3}$.

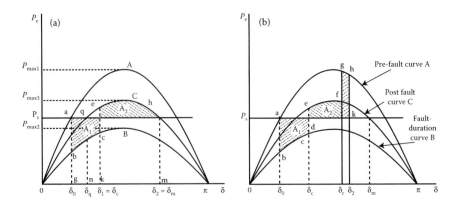

FIGURE 19.13 (a) Determination of critical clearing angle. (b) Equal area criterion for system fault, circuit isolation, and reclosing.

For transient stability limit, area abcd = area defh.

Area abcd = Area adkg − Area bckg

$$= P_s(\delta_c - \delta_0) - \int_{\delta_0}^{\delta_c} P_{m2} \sin \delta \, d\delta$$

$$= P_s(\delta_C - \delta_0) - [-P_{m2}\cos\delta]_{\delta_0}^{\delta_C} = P_S(\delta_C - \delta_0) + P_{m2}\cos\delta_C - P_{m2}\cos\delta_0$$

Area defh = Area kdefhm − Area dhmk

$$= \int_{\delta_C}^{\delta_m} P_{m3} \sin\delta \, d\delta - P_s(\delta_m - \delta_C)$$

$$= [-P_{m3}\cos\delta]_{\delta_C}^{\delta_m} - P_S(\delta_m - \delta_C) = P_{m3}\cos\delta_c - P_{m3}\cos\delta_m - P_S (\delta_m - \delta_C)$$

Equating the two areas, we get

$$P_s(\delta_C - \delta_0) + P_{m2}\cos\delta_C - P_{m2}\cos\delta_0 = P_{m3}\cos\delta_c - P_{m3}\cos\delta_m - P_s(\delta_m - \delta_C)$$

(19.43)

Also, $P_s = P_{m1} \sin \delta_0$ at point a on curve A.
$P_s = P_{m3} \sin \delta_q$ at point q on curve C

$$\delta_m = \pi - \delta_q = \pi - \sin^{-1}\left(\frac{P_s}{P_{m3}}\right) \text{radians}$$

From Equation 19.43,

$$\cos\delta_c = \frac{P_S(\delta_m - \delta_0) - P_{m2}\cos\delta_0 + P_{m3}\cos\delta_m}{P_{m3} - P_{m2}} \quad (19.44)$$

This Equation 19.44 can be used to determine critical clearing angle. The angles in this equation are in radians. If the angles are in degrees, then the Equation 19.44 becomes

$$\cos\delta_c = \frac{(\pi/180)P_S(\delta_m - \delta_0) - P_{m2}\cos\delta_0 + P_{m3}\cos\delta_m}{P_{m3} - P_{m2}} \quad (19.45)$$

### 19.15.1 System Fault, Circuit Isolation, and Reclosing

Most of the faults on the system are of transient nature. Automatic quick reclosing CBs are used with transmission lines. When a fault occurs, the fault line is disconnected. After an interval, the CBs of faulted line are reclosed automatically. The input is $P_s$ and the initial angle is $\delta_0$ when a fault occurs, the operation shifts to be point b on the fault duration curve B as shown in Figure 19.13b. When the load angle is $\delta_c$, the faulted line is isolated and the operation shifts to the post-fault curve C. When the load angle is $\delta_r$, the CBs reclose and the operation shifts to the prefault curve A.

For stable operation, the accelerating area $A_1$ (=area abcd) should be equal to the decelerating area $A_2$ (=area defghk). The maximum angle to which the rotor angle swings is $\delta_2$. It is less than $\delta_m$ (i.e., the maximum permissible rotor swing if stability is to be maintained.

## 19.16 METHODS OF IMPROVING STABILITY

Figure 19.13 shows that when the maximum power limit various power–angle curves is raised, the accelerating area decreases $n$ decelerating area increases for a given clearing angle. Consequently $\delta_0$ is decreased and $\delta_m$ is increased. This means that by increasing $P_{max}$, the rotor can swing through a larger angle from its original position before it reaches a critical clearing angle. Thus, raising the value of $P_{max}$, the rotor can swing through a larger angle from its original position before it reaches a critical clearing angle. Thus raising the value of $P_{max}$ increases the critical clearing time and improves stability.

The steady-state power limit is given by

$$P_{max} = \frac{EV}{X}$$

It can be seen from this expression that $P_{max}$ can be increased by increasing either $V$ or $E$, or both and reducing the transfer reactance. The following methods are available for reducing the transfer reactance:

1. *Use of double-circuit lines.* The impedance of a double-circuit line is less than that of a single-circuit line. A double-circuit line doubles the transmission capability. An additional advantage is that the continuity of supply is maintained over one line with reduce capacity when the other line is out of service for maintenance or repair. But the provision of additional line can hardly be justified by stability consideration alone.

2. *Use of bundled conductors.* Bundling of conductors reduces to a considerable extent the line reactance and so increases the power limit of the line.

3. *Series compensation of the lines.* The inductive reactance of a line can be reduced by connecting static capacitors in series with the line.

It is to be noted that any measure to increase the steady-state limit $P_{max}$ will improve the transient stability limit. The use of generators of high inertia and low reactance improves the transient stability, but generators with these characteristics are costly. In practice, only those methods are used which are economical.

## WORKED EXAMPLES

### EXAMPLE 19.1

A round rotor generator with internal voltage $E_1 = 2$ pu, $X = 1.1$ pu is connected to a round rotor synchronous motor with internal voltage $E_1 = 1.3$ pu, $X = 1.2$ pu. The reactance of the line connecting generator to the motor is 0.5 pu when the generator supplies 0.5 pu power (Figure 19.14), what will be the rotor angle difference between the machines?

### Solution

Let $\delta_1$ and $\delta_2$ be the load angle of the generator and motor, respectively.

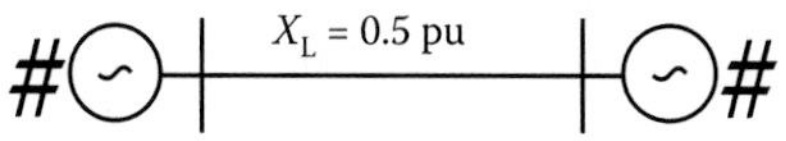

$$X_L = 0.5 \text{ pu}$$

#○| —————————— |○#

FIGURE 19.14   A round rotor generator connected with a round rotor synchronous motor.

Active power transferred, $P_e = \dfrac{EV}{X}\sin\delta = \dfrac{E_1 E_2}{X}\sin(\delta_1 - \delta_2)$

or

$$0.5 = \frac{2 \times 1.3}{(1.1 + 0.5 + 1.2)}\sin(\delta_1 - \delta_2)$$

or

$$(\delta_1 - \delta_2) = 32.58°$$

∴ Rotor angle difference between the machines is 32.58°.

## EXAMPLE 19.2

A generator with constant 1 pu terminal voltage supplies power through a step up transformer of 0.12 pu reactance and a double circuit line to an infinite bus bar as shown in Figure 19.15. Neglecting resistance and susceptances of system, SSSPL (steady state stability power limit) of the system is 6.25 pu. If one of the double circuit is tripped then what will be the resulting SSSPL in pu?

## Solution

Reactance of the transformer $(X_1) = 0.12_{pu}$

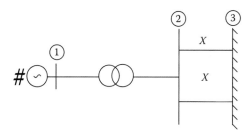

FIGURE 19.15   A double circuit line to an infinite bus bar.

Reactances of the double circuit line $X_2$ and $X_3$ are

$$\therefore X_2 = X_3 = X'$$

Steady-state stability limit of the system, SSSPL $= P_{e\,max} = (EV/X)$

or

$$6.25 = \frac{1 \times 1}{X_1 + (X' \parallel X')}$$

or

$$6.25 = \frac{1 \times 1}{0.12 + (X' \parallel X')}$$

or

$$6.25 = \frac{1}{0.12 + (X'/2)}$$

or

$$X' = 0.08_{pu}$$

If one of the double circuit is tripped,

$$\therefore P_e = \frac{EV}{X} = \frac{1 \times 1}{0.12 + X'} = \frac{1}{0.12 + 0.08} = 5_{pu}$$

### EXAMPLE 19.3

The power angle characteristics of a machine-infinite bus system are $P_e = 2 \sin \delta_{pu}$. It is operating at $\delta = 30°$. Determine the synchronizing power coefficient at the operating point.

### Solution

Synchronizing power coefficient,

$$\frac{dP_e}{d\delta} = \frac{d}{d\delta}(2 \sin \delta)$$
$$= 2 \cos \delta$$
$$= 2 \cos 30°$$
$$= \sqrt{3}_{pu}$$

## EXAMPLE 19.4

A synchronous generator is connected to an 11-kV infinite bus through a transmission line. The reactances of generator and transmission line are 1.2 and 0.8 Ω, respectively. The terminal voltage of synchronous generator is 15 kV. If the generator delivers 70 MW power to infinite bus, then what will be the load angle?

**Solution**

Active electrical power transmitted

$$P_e = \frac{EV}{X} \sin \delta$$

or

$$70 = \frac{11 \times 15}{2} \sin \delta$$

or

$$\sin \delta = \frac{140}{165}$$

or

$$\delta = 58.04°$$

## EXAMPLE 19.5

A 500-MW, 21-kV, three-phase, two-pole synchronous generator having a rated power factor=0.9 has moment of inertia $27.5 \times 10^3$ Nm. Determine inertia constant $H$.

**Solution:**

$$\text{Angular velocity,} \quad \omega = \frac{2\pi N}{60}$$
$$= \frac{2\pi \times \left((120 \times 50)/2\right)}{60}$$
$$= 314.15 \text{ rad/s}$$

$\therefore$ Kinetic energy $= \dfrac{1}{2}I\omega^2 = \dfrac{1}{2} \times (27.5 \times 10^3) \times (314.15)^2 = 1357$ MJ

Let machine rating in MVA is $S$

$$\therefore \text{ Inertia constant, } H = \frac{\text{Kinetic energy}}{S}$$

$$= \frac{1357}{(P/\cos\phi)}$$

$$= \frac{1357}{(500/0.9)}$$

$$= 2.44 \text{ MJ/MVA}$$

### EXAMPLE 19.6

A 50-Hz, four-pole, turbo alternator 20 MVA, 13.2 kV has an inertia constant of $H$ as 9 kW s/kVA. If the shaft input less the rotational losses is 26,800 HP of metric and electrical power developed 16,000 kW. Then determine kinetic energy stored by rotor and accelerating torque.

### Solution

Inertia constant, $H = 9$ kW s/kVA $= 9$ MJ/MVA
  Kinetic energy stored by rotor $= S \times H = 20 \times 9 = 180$ MJ
  Accelerating torque, $T_a$

$$\text{Mechanical power input, } P_s = 26,800\,\text{HP}$$
$$= 26,800 \times 0.736$$
$$= 19,698 \text{ kW}$$

And electrical power developed, $P_e = 16,000$ kW.

$$\text{Accelerating power, } P_a = P_s - P_e = (19,698 - 16,000) \text{ kW}$$
$$= 3698 \text{ kW}$$

Let accelerating torque is $T_a$.
Now, accelerating power, $P_a = (2\pi N T_a)/60$.

or

$$3698 = \frac{2\pi \times ((120 \times 50)/4)}{60} \times T_a$$

$$T_a = 23.541 \text{ Nm}$$

## EXAMPLE 19.7

Inertia constant of a 100-MVA, 50-Hz, four-pole generator is 10 MJ/MVA. If mechanical input is suddenly raised to 75 MW from 50 MW, then determine rotor acceleration.

**Solution**

Initial mechanical input, $P_{s1} = 50 \text{MW} = P_{e1} = $ Initial electrical output

Initial accelerating power, $P_{a1} = P_{s1} - P_{e1} = 0$
Now, final mechanical power, $P_{s2} = 75$ MW
But $P_{e2} = P_{e1} = 50$ MW
Final accelerating power, $P_{a2} = P_{s2} - P_{e2} = 25$ MW

$$\text{Rotor acceleration, } \alpha = \frac{d^2\delta}{dt^2} = \frac{P_a}{M} = \frac{25\,\text{MW}}{SH/nf} = \frac{25}{(100 \times 10)/(180 \times 50)}$$

$$= 225 \text{ elec. degree/s}^2$$

## EXAMPLE 19.8

A synchronous motor of negligible resistance is receiving 25% of power that is capable of receiving from infinite bus. If the motor load is suddenly doubled, determine angle at which the system is stable, after making an oscillation between the two possible swings.

**Solution**

Initially,

Mechanical power input = Electrical power output

or

$$P_{s1} = P_{e1}$$

Now, initial accelerating power, $P_{a1} = P_{s1} - P_{e1} = 0$. Therefore,

$$P_{s1} = P_{e1} = 0.25P_m$$

Now,

$$P_{s2} = 2, \quad P_{s1} = P_{e2}$$

or

$$2P_{s1} = P_{e2}$$

or

$$2 \times (0.25P_m) = P_m \sin\delta_2$$

or

$$\delta_2 = 30°$$

### EXAMPLE 19.9

A synchronous generator having inertia constant 6 MJ/MVA is delivering power of $1_{pu}$ to an infinite bus through a purely reactive network. Suddenly a fault occurs and reduces the output power to 0. The maximum power that could be delivered is $2.5_{pu}$. Determine

    1. Critical clearing angle
    2. Critical clearing time

### Solution

Let $P_s = P_{e1} = 1_{pu}$.

$$P_{e2} = 0, \quad P_{m1} = 2.5_{pu} = P_{m3}$$

$$\cos\delta_C = \frac{P_S(\delta_{max} - \delta_0) + P_{m3}\cos\delta_{max} - P_{m2}\cos\delta_0}{P_{m3} - P_{m2}}$$

Critical clearing angle, $\delta_C = \cos^{-1}\left[\dfrac{P_S(\delta_{max} - \delta_0) + P_{m3}\cos\delta_{max} - P_{m2}\cos\delta_0}{P_{m3} - P_{m2}}\right]$

Now,

$$\delta_0 = \sin^{-1}\left(\frac{P_s}{P_{m1}}\right) = 23.57° = 0.411 \text{ rad}$$

and

$$\delta_{max} = 180° - \sin^{-1}\left(\frac{P_s}{P_{m3}}\right) = 156.43° = 2.73 \text{rad}$$

$$\therefore \delta_C = \cos^{-1}\left[\frac{1 \times (2.73 - 0.411) + 2.5\cos(156.43°)}{2.5}\right]$$

$$= 89.27° = 1.56 \text{ rad}$$

Critical clearing time, $t_c = \left[\frac{2M(\delta_c - \delta_0)}{P_s}\right]^{1/2}$

Now,

$$M = \frac{GH}{\pi f} = \frac{1 \times 6}{\pi \times 50} = 0.0382$$

$$\therefore t_c = \left[\frac{2 \times 0.0382(1.56 - 0.411)}{1}\right]^{1/2}$$

$$= 0.296 \text{ s}$$

## EXERCISES

1. Define the terms

   a. Steady-state stability

   b. Transient stability

   c. Steady-state limit

   d. Transient limit

2. Distinguish between steady-state and transient stability of a power system and discuss the factors on which it depend.

3. Explain briefly the equal-area criterion and how it may be used to study in stability of a two-machine system. List the factors determining the stability limit and indicate how it may be improved.

4. Explain the equal-area criterion for the stability of an alternator supplying infinite bus bar via an inductive interconnector.

5. What is meant by swing curve and how is it determined? What information is supplied by it?

6. Explain the equal-area criterion as applied to the power–angle diagram for assessing the transient stability of a transmission line acting as an interconnector between two constant voltage networks.

# CHAPTER 20

# Load Flows

## 20.1 INTRODUCTION

The flow of active and reactive power is called power flow or load flow. Load flow (or power flow) analysis of the determination of current, voltage, active power, and reactive volt-amperes at various points in a power system operating under normal steady-state or static conditions. Load flow studies are made to plan the best operation and control of the existing system as well as to plan the future expansion to keep space with the load growth. Such studies help in ascertaining the effects of new loads, new generating stations, new lines, and new interconnections before they are installed. The prior information serves to minimize the system loses and to provide a check on the system stability.

The mathematical formulation of load flow problem results in a set of algebraic nonlinear equations. A lot of calculation work is involved in the solution of this equation. Hand computations are very tedious and time consuming. Earlier load flow studies were made by AC network analyzers (analog computers). Digital computers, because of greater flexibility, economy accuracy, and quicker operation, have practically replaced network analyzers for the solution of load flow problems.

## 20.2 BUS CLASSIFICATION

Load flow studies are performed to calculate the magnitude and phase angles of voltages and buses, and also the active power and reactive volt-amperes flow for the given terminal or bus condition.

The following variables associated with each bus or node are

1. Magnitude of the voltage, $|V_i|$

2. Phase angle of the voltage, $\delta_i$

3. Active power, $P_i$

4. Reactive power, $Q_i$

Three types of buses or nodes are identified in a power system network for load flow studies. In each bus, two variables are known (specified) and two are to be determined. The bus classification depends upon the specified variables. The buses are classified as follows:

1. *Load bus or P-Q bus.* A load bus is a bus where active power $P_i$ and reactive power $Q_i$ are specified. Magnitude and phase angle of the bus voltages are to be found.

2. *Generator bus or voltage-controlled bus or PV bus.* A generator bus is a bus where the magnitude of bus voltage $|V_i|$ and the corresponding generated power $P_i$ are known. Reactive power $Q_i$ and power angle $\delta$ are to be obtained.

3. *Stack bus or swing bus or reference bus.* A stack bus is a generator bus where the magnitude and phase angle of bus voltage are specified. Real ($P_i$) and reactive ($Q_i$) power are to be obtained. This bus is first to respond to a changing load condition.

Table 20.1 summarizes the above discussion.

## 20.3 BUS ADMITTANCE MATRIX

Consider a small power system network (Figure 20.1) consisting of two generating stations, three transmission lines, one load, and a static capacitor connected to load bus. We shall assume that the network is symmetrical and operating under balanced condition.

Applying Kirchhoff's law:

$$I_1 = y_{12}(V_1 - V_2) + y_{31}(V_1 - V_3) = (y_{12} + y_{31})V_1 - y_{12}V_2 - y_{31}V_3$$

TABLE 20.1  Bus Types for Power Flow Analysis

| Bus Type | Specification Variables | Unknown Variables |
|---|---|---|
| Reference bus | $|V_i|$, $\delta_i$ | $P_i$, $Q_i$ |
| Generator bus | $P_i$, $|V_i|$ | $Q_i$, $\delta_i$ |
| Load bus | $P_i$, $Q_i$ | $|V_i|$, $\delta_i$ |

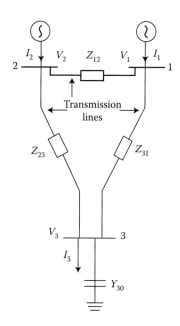

FIGURE 20.1   Power system network for power flow.

$$I_2 = y_{12}(V_2 - V_1) + y_{23}(V_2 - V_3) = -y_{12}V_1 + (y_{12} + y_{31})V_2 - y_{23}V_3$$

$$-I_3 = y_{31}(V_3 - V_1) + y_{23}(V_3 - V_2) + y_{30}V_3 = -y_{31}V_1 - y_{23}V_2 + (y_{31} + y_{23} + y_{30})V_3$$

where $y_{12} = 1/z_{12}$, $y_{23} = 1/z_{23}$, $y_{31} = 1/z_{31}$.

In matrix form,

$$
\begin{bmatrix} I_1 \\ I_2 \\ I_3 \end{bmatrix} = \begin{bmatrix} y_{12} + y_{31} & -y_{12} & -y_{31} \\ -y_{12} & y_{12} + y_{23} & -y_{23} \\ -y_{31} & -y_{23} & y_{31} + y_{23} + y_{30} \end{bmatrix} \begin{bmatrix} V_1 \\ V_2 \\ V_3 \end{bmatrix}
\tag{20.1}
$$

It is to be noted that all injected current are positive and extracted current are negative.

The above equation can be written as

$$
\begin{bmatrix} I_1 \\ I_2 \\ I_3 \end{bmatrix} = \begin{bmatrix} Y_{11} & Y_{12} & Y_{13} \\ Y_{21} & Y_{22} & Y_{23} \\ Y_{31} & Y_{32} & Y_{33} \end{bmatrix} \begin{bmatrix} V_1 \\ V_2 \\ V_3 \end{bmatrix}
\tag{20.2}
$$

$$Y_{11} = y_{12} + y_{31}, \quad Y_{22} = y_{12} + y_{23}, \quad Y_{33} = y_{31} + y_{23} + y_{30}$$
$$Y_{12} = Y_{21} = -y_{12}, \quad Y_{23} = Y_{32} = -y_{23}, \quad Y_{13} = Y_{31} = -y_{31}$$

Here, $Y_{ii}$ ($i = 1, 2, 3, 4,...$) is called driving point admittance or self-admittance of node $i$. It is the algebraic sum of all the admittances terminated at the node.

$Y_{ik}$ is the off diagonal term in the matrix, where $i, k = 1,2,3,...$ is called transfer admittance or mutual admittance. It is the admittance connected between nodes $i$ and $k$ and is equal to the negative sum of all admittance connected directly between these nodes.

Also $Y_{ik} = Y_{ki}$.

For a network with $N$ nodes, obviously,

$$[I] = [Y_{bus}][V] \tag{20.3}$$

where $[Y_{bus}]$ is called bus admittance matrix, $[V]$ and $[I]$ are $N$ element voltage matrix and current matrix, respectively, where

$$[Y_{bus}] = \begin{bmatrix} Y_{11} & Y_{12} & \cdots & Y_{1N} \\ Y_{21} & Y_{22} & \cdots & Y_{2N} \\ \vdots & \vdots & \vdots & \vdots \\ Y_{31} & Y_{32} & \cdots & Y_{3N} \end{bmatrix} \tag{20.4}$$

Obviously at any node $K$,

$$I_K = \sum_{n=1}^{N} Y_{Kn} V_n \tag{20.5}$$

*Advantages:*

1. Data preparation is simple.

2. Its formation and modification are easy.

3. Since the bus admittance matrix is a sparse matrix (i.e., most of its elements are zero), the computer memory requirements are less.

For a large power system, more than 90% of its off-diagonal elements are zero. This is due to the fact that in power system networks, each node (bus) is connected to not more than three nodes in general and an element $Y_{pq}$ exists only if a transmission line links nodes $p$ and $q$.

## 20.4  DEVELOPMENT OF STATIC LOAD FLOW EQUATION

From the nodal current equation, the total current entering the $i$th bus of an $n$ bus system is given by

$$I_i = Y_{i1}V_1 + Y_{i2}V_2 + \cdots + Y_{in}V_n = \sum_{k=1}^{n} Y_{ik}V_k$$

The complex power ($S_i$) injected into the $i$th is given by

$$S_i = P_i + jQ_i = V_i I_i^* = V_i \left[ \sum_{k=1}^{n} Y_{ik}V_k \right]^* \tag{20.6}$$

where

$$V_i = |V_i|\angle\delta_i, \quad V_k = |V_k|\angle\delta_k$$

From Equation 20.6,

$$S_i^* = V_i^* I_i$$

$$P_i - jQ_i = V_i^* \sum_{k=1}^{n} Y_{ik}V_k$$

$$= \sum_{k=1}^{n} |Y_{ik}||V_i||V_k|\angle(\delta_k - \delta_i + \phi_{ik})$$

$$[Y_{ik} = |Y_{ik}|\angle\phi_{ik} = |Y_{ik}|\cos\phi_{ik} + j|Y_{ik}|\sin\phi_{ik}]$$

Separating equation into real and imaginary parts, we have

$$P_i = \sum_{k=1}^{n} |Y_{ik} V_i V_k| \cos|\phi_{ik} + \delta_k - \delta_i|$$

or

$$P_i = |V_i| \sum_{k=1}^{n} |Y_{ik}||V_k| \cos|\phi_{ik} + \delta_k - \delta_i|, \quad i = 1,2,\dots,n \qquad (20.7)$$

$$Q_i = \sum_{k=1}^{n} |Y_{ik} V_i V_k| \sin|\phi_{ik} + \delta_k - \delta_i|$$

or

$$Q_i = |V_i| \sum_{k=1}^{n} |Y_{ik}||V_k| \sin|\phi_{ik} + \delta_k - \delta_i|, \quad i = 1,2,\dots,n \qquad (20.8)$$

Equations 20.7 and 20.8 are called static load flow equations (SLFE). Equation 20.7 gives $n$ real power flow equations. Similarly, Equation 20.8 gives $n$ reactive power flow equation. Thus, Equations 20.7 and 20.8 represent $2n$ power flow equations. At each bus, we have four variables $P_i$, $Q_i$, $V_i$, and $\delta_i$ resulting in total of $4n$ variables. In order to find a solution, it is necessary to specify two variables at each bus. Thus, the number of unknown variables is reduced to $2n$. The solution of these remaining $2n$ variables is done by numerical methods because Equations 20.7 and 20.8 are nonlinear.

No exact analytical solution of nonlinear equation is possible. These equations may be solved by iterative techniques that employ successive approximations eventually converging upon a solution. Before the advent of digital computers, these trial and error techniques were tedious and time consuming. However, today these methods find widespread applications for solving load flow problems. The iteration procedure involves an initial assumed value for each of the unknown independent variable. These numerical values are substituted in the original equation to obtain a new set of corrected values of these independent variables. The second set is used to find the third corrected set. The process is repeated. Each calculation of a new set of variables is called iteration. The iteration is continued until the unknown values converge within required limits.

## 20.5  GAUSS–SEIDEL ITERATIVE TECHNIQUE

This is of the simplest iterative methods and has been in use since early days of digital computer methods of analysis. It has the following advantages:

1. It is very simple.

2. Computing costs are less.

3. Sometimes it is used to find the initial solution for the other iterative procedures.

To illustrate the Gauss–Seidel (GS) method, let us show the following example. Say

$$f(x) = x^2 - 3x + 2 = 0$$

GS iteration is to be applied here. As

$$x^2 - 3x + 2 = 0$$

$$\therefore x = \frac{1}{3}x^2 + \frac{2}{3}$$

*Step 1.* Let the initial guess be such that

$$x^{(0)} = 1.5$$

$$\therefore x^{(1)} = \frac{1}{3} \times (1.5)^2 + \frac{2}{3} = 1.4167$$

*Step 2.* In the next step,

$$x^{(r+1)} = \frac{1}{3}(x^1)^2 + \frac{2}{3}$$

Here,

$$x^{(r+1)} = 1.3356$$

*Step 3.*

$$x^{(r+1)+1} = \frac{1}{3}(x^{r+1})^2 + \frac{2}{3} = \frac{1}{3}(1.3356)^2 + \frac{2}{3} = 1.261$$

The iterative process is ended when $|x^{r+1}| - |x^r| < \varepsilon$, where $\varepsilon$ is the tolerance.

## 20.6 GS METHOD OF SOLUTION OF LOAD FLOW EQUATION USING $Y_{BUS}$

From the nodal current equations, the total current entering the $k$th bus of an $n$-bus system is given by

$$I_k = Y_{k1}V_1 + Y_{k2}V_2 + \cdots + Y_{kn}V_n = \sum_{i=1}^{n} Y_{ki}V_i \qquad (20.9)$$

The complex power injected into the $k$th bus is

$$S_k = P_k + jQ_k = V_k I_k^* \qquad (20.10)$$

The complex conjugate of Equation 20.10 gives

$$S_k^* = P_k - jQ_k = V_k^* I_k \qquad (20.11)$$

$$I_k = \frac{1}{V_k^*}(P_k - jQ_k) \qquad (20.12)$$

Elimination of $I_k$ from Equations 20.9 and 20.12 gives

$$Y_{k1}V_1 + Y_{k2}V_2 + \cdots + Y_{kk}V_k + \cdots + Y_{kn}V_n = \frac{1}{V_k^*}(P_k - jQ_k) \qquad (20.13)$$

$$V_k = \frac{1}{Y_{kk}} \left[ \frac{P_k - jQ_k}{V_k^*} - \sum_{\substack{i=1 \\ i \neq k}}^{n} Y_{ki} V_i \right] \tag{20.14}$$

Equation 20.14 is the heart of iterative algorithm.
At bus 2,

$$V_2 = \frac{1}{Y_{22}} \left[ \frac{P_2 - jQ_2}{V_2^*} - Y_{21} V_1 - Y_{23} V_3 - \cdots - Y_{2n} V_n \right] \tag{20.15}$$

At bus 3,

$$V_3 = \frac{1}{Y_{33}} \left[ \frac{P_3 - jQ_3}{V_3^*} - Y_{31} V_1 - Y_{32} V_2 - Y_{34} V_4 \cdots - Y_{3n} V_n \right] \tag{20.16}$$

For $k$th bus, the voltage at the $(r + 1)$th iteration is given by

$$V_k^{(r+1)} = \frac{1}{Y_{kk}} \left[ \frac{P_k - jQ_k}{\left(V_k^{(r)}\right)^*} - \sum_{i=1}^{k-1} Y_{ki} V_i^{(r+1)} - \sum_{i=k+1}^{n} Y_{ki} V_i^{(r)} \right] \tag{20.17}$$

In the above equation, the quantities $P_k$, $Q_k$, $Y_{kk}$, and $Y_{ki}$ are known and do not vary during the iteration cycle.

## 20.7  NEWTON–RAPHSON METHOD FOR LOAD FLOW SOLUTION

Let us consider two functions with two variables $x_1$ and $x_2$ such that

$$f_1(x_1, x_2) = c_1 \tag{20.18}$$

$$f_2(x_1, x_2) = c_2 \tag{20.19}$$

where $c_1$ and $c_2$ being constants.

Let $x_1^{(0)}$ and $x_2^{(0)}$ be the initial estimates for solutions of Equations 20.18 and 20.19. Let $\Delta x_1^{(0)}$ and $\Delta x_2^{(0)}$ be the values by which the initial estimates differ the exact solution. Thus,

$$f_1\left[\left(x_1^{(0)} + \Delta x_1^{(0)}\right),\left(x_2^{(0)} + \Delta x_2^{(0)}\right)\right] = c_1 \tag{20.20}$$

$$f_2\left[\left(x_1^{(0)} + \Delta x_1^{(0)}\right),\left(x_2^{(0)} + \Delta x_2^{(0)}\right)\right] = c_2 \tag{20.21}$$

Expanding the left-hand side of each of these equations in the form of a Taylor's series, we obtain

$$f_1\left(x_1^{(0)}, x_2^{(0)}\right) + \Delta x_1^{(0)} \left.\frac{\delta f_1}{\delta x_1}\right|_{x_1^{(0)}} + \Delta x_2^{(0)} \left.\frac{\delta f_1}{\delta x_2}\right|_{x_2^{(0)}} = c_1 \tag{20.22}$$

$$f_2\left(x_1^{(0)}, x_2^{(0)}\right) + \Delta x_1^{(0)} \left.\frac{\delta f_2}{\delta x_1}\right|_{x_1^{(0)}} + \Delta x_2^{(0)} \left.\frac{\delta f_2}{\delta x_2}\right|_{x_2^{(0)}} = c_2 \tag{20.23}$$

or

$$\begin{bmatrix} c_1 - f_1\left(x_1^{(0)}, x_2^{(0)}\right) \\ c_2 - f_2\left(x_1^{(0)}, x_2^{(0)}\right) \end{bmatrix} = \begin{bmatrix} \dfrac{\delta f_1}{\delta x_1} & \dfrac{\delta f_1}{\delta x_2} \\ \dfrac{\delta f_2}{\delta x_1} & \dfrac{\delta f_2}{\delta x_2} \end{bmatrix}_{x_1^{(0)}, x_2^{(0)}} \begin{bmatrix} \Delta x_1^{(0)} \\ \Delta x_2^{(0)} \end{bmatrix} \tag{20.24}$$

or

$$\begin{bmatrix} \Delta c_1^{(0)} \\ \Delta c_2^{(0)} \end{bmatrix} = j(0) \begin{bmatrix} \Delta x_1^{(0)} \\ \Delta x_2^{(0)} \end{bmatrix} \tag{20.25}$$

where

$$\begin{bmatrix} \Delta c_1^{(0)} \\ \Delta c_2^{(0)} \end{bmatrix} = \begin{bmatrix} c_1 - f_1\left(x_1^{(0)}, x_2^{(0)}\right) \\ c_2 - f_2\left(x_1^{(0)}, x_2^{(0)}\right) \end{bmatrix}$$

$$j(0) = \begin{bmatrix} \dfrac{\delta f_1}{\delta x_1} & \dfrac{\delta f_1}{\delta x_2} \\ \dfrac{\delta f_2}{\delta x_1} & \dfrac{\delta f_2}{\delta x_2} \end{bmatrix}$$

Matrix $j(0)$ is called "Jacobian" and $\Delta c_1^{(0)}$ and $\Delta c_2^{(0)}$ are the differences.
Solution of Equation 20.25 gives $\Delta x_1^{(0)}$ and $\Delta x_2^{(0)}$.
In the next step, a better estimate of the solution is made when

$$x_1^{(1)} = x_1^{(0)} + \Delta x_1^{(0)}$$

$$x_2^{(1)} = x_2^{(0)} + \Delta x_2^{(0)}$$

The whole process is then repeated, and the iterations continued until $\Delta x_1$ and $\Delta x_2$ become very much smaller than predetermined value.
In the two-bus power system model,

$$P = f_1(\delta, V) = \frac{EV}{X} \sin\delta \tag{20.26}$$

$$Q = f_2(\delta, V) = \frac{EV}{X} \cos\delta - \frac{V^2}{X} \tag{20.27}$$

Thus, Equation 20.25 becomes

$$\begin{bmatrix} \Delta P^{(0)} \\ \Delta Q^{(0)} \end{bmatrix} = \begin{bmatrix} \dfrac{\delta f_1}{\delta\delta} & \dfrac{\delta f_1}{\delta V} \\ \dfrac{\delta f_2}{\delta\delta} & \dfrac{\delta f_2}{\delta V} \end{bmatrix} \begin{bmatrix} \Delta\delta^{(0)} \\ \Delta V^{(0)} \end{bmatrix} \tag{20.28}$$

$$\begin{bmatrix} \Delta\delta^{(0)} \\ \Delta V^{(0)} \end{bmatrix} = \begin{bmatrix} \Delta P^{(0)} \\ \Delta Q^{(0)} \end{bmatrix} \begin{bmatrix} \dfrac{\delta f_1}{\delta\delta} & \dfrac{\delta f_1}{\delta V} \\ \dfrac{\delta f_2}{\delta\delta} & \dfrac{\delta f_2}{\delta V} \end{bmatrix}^{-1} \tag{20.29}$$

Thus, $\Delta\delta^{(0)}$ and $\Delta V^{(0)}$ can be obtained from this equation.

The values determined for $\Delta\delta^{(0)}$ and $\Delta V^{(0)}$ are then added to the previous estimates of $V$ and $\delta$ to obtain new estimate of $V$ and $\delta$ to start next iteration. That is,

$$x^{(r+1)} = x^r - [j(x^r)]^{-1} f(x^r)$$

To apply the Newton–Raphson (NR) method in a power flow problem, let, in the $k$th bus,

$$V_k = |V_k| \angle \delta_k$$

$$V_n = |V_n| \angle \delta_n$$

$$Y_{kn} = |Y_{kn}| \angle \angle \phi_{kn}$$

The complex power expression is given by

$$P_k - jQ_k = \sum_{n=1}^{N} |Y_{kn}||V_n||V_k| \angle (\phi_{ik} + \delta_n - \delta_k)$$

where

$$P_k = \sum_{n=1}^{N} |Y_{kn}||V_n||V_k| \cos(\phi_{ik} + \delta_n - \delta_k)$$

and

$$Q_k = \sum_{n=1}^{N} |Y_{kn}||V_n||V_k| \sin(\phi_{ik} + \delta_n - \delta_k)$$

If $P$ and $Q$ are specified for every bus except the swing bus, it corresponds to specifying $C_1$ and $C_2$. Let us first estimate $V$ and $\delta$ for each bus except the slack or swing bus, for which they are known. Then we substitute these estimated values (which correspond to the estimated values for $x_1$ and $x_2$) in the power flow equation of $P_k$ and $Q_k$ above to calculate $Ps$ and $Qs$ that correspond $f_1\left(x_1^{(0)}, x_2^{(0)}\right)$ and $f_2\left(x_1^{(0)}, x_2^{(0)}\right)$.

In the next step, we compute,

$$\Delta P_k^{(0)} = P_{ks} - P_{kc}^{(0)}$$

$$\Delta Q_k^{(0)} = Q_{ks} - Q_{kc}^{(0)}$$

where the subscripts $s$ and $c$ are the specified and calculated values, respectively. These correspond to the values of $\Delta c_1^{(0)}$ and $\Delta c_2^{(0)}$ in Equation 20.25.

Thus, NR method in a three-bus power system model gives, similar to Equations 20.25 and 20.29.

$$
\begin{bmatrix} \Delta P_2^{(0)} \\ \Delta P_3^{(0)} \\ \Delta Q_2^{(0)} \\ \Delta Q_3^{(0)} \end{bmatrix} = 
\begin{bmatrix} 
\dfrac{\delta P_2}{\delta \delta_2} & \dfrac{\delta P_2}{\delta \delta_3} & \dfrac{\delta P_2}{\delta V_2} & \dfrac{\delta P_2}{\delta V_3} \\[2mm]
\dfrac{\delta P_3}{\delta \delta_2} & \dfrac{\delta P_3}{\delta \delta_3} & \dfrac{\delta P_3}{\delta V_2} & \dfrac{\delta P_3}{\delta V_3} \\[2mm]
\dfrac{\delta Q_2}{\delta \delta_2} & \dfrac{\delta Q_2}{\delta \delta_3} & \dfrac{\delta Q_2}{\delta V_2} & \dfrac{\delta Q_2}{\delta V_3} \\[2mm]
\dfrac{\delta Q_3}{\delta \delta_2} & \dfrac{\delta Q_3}{\delta \delta_3} & \dfrac{\delta Q_3}{\delta V_2} & \dfrac{\delta Q_3}{\delta V_3} 
\end{bmatrix}
\begin{bmatrix} \Delta \delta_2^{(0)} \\ \Delta \delta_3^{(0)} \\ \Delta V_2^{(0)} \\ \Delta V_3^{(0)} \end{bmatrix}
$$

By inverting the Jacobian, as done in Equation 20.25, the values of $\Delta \delta_k^{(0)}$ and $\Delta V_k^{(0)}$ can be calculated. The values can be utilized by adding $\Delta \delta_k$ and $\Delta V_k$ to the previous estimates of $V$ and $\delta$ to obtain new estimate in the iteration of next step. Iterations are continued till the values in either column matrix are as small as desired.

## 20.8 COMPARISON OF LOAD FLOW ANALYSIS METHODS

The choice of a particular method of load flow analysis depends upon the size of the system, rate of convergence, simplicity, computer memory, etc.

### 20.8.1 Advantages of GS Method

1. It can be easily programmed.

2. The solution technique is simple.

3. Computer memory requirements are smaller.

4. It takes less computational time per iteration.

### 20.8.2 Limitations of GS Method

1. The rate of convergence is slow and therefore, larger numbers of iterations are required. The GS method would take hundreds of iterations to converge, if a system with several hundred buses were to be analyzed.

2. The number of iterations increases directly with the number of buses in the system.

3. This method is sensitive to the choice of reference bus.

The GS method is used only for the system having small number of buses.

### 20.8.3 Advantages of NR Method

1. NR method possesses quadratic convergence characteristics. Therefore, the convergence is very fast.

2. The number of iterations are independent of the size of the system. Solution to a high accuracy is obtained nearly always in two to three iterations for both small and large systems.

3. The NR method convergence is not sensitive to the choice of the slack bus.

4. Overall there is a saving in computation time, since fewer numbers of iterations are required for convergence.

### 20.8.4 Limitations of NR Method

1. The solution technique is difficult.

2. It takes longer time as the elements of the Jacobian are to be computed for each iteration.

3. The computer memory requirement is large.

The NR method is more complicated than the GS method; however, it has advantages that far outweigh its shortcomings of complexity. It is the most reliable and powerful technique for solving load flow problems.

Although a large number of load flow methods are available in literature, it has been observed that only the NR and fast-decoupled load-flow methods are most popular. The fast decoupled load flow is definitely superior to the NR method from the point of view of speed and storage.

## WORKED EXAMPLES

### EXAMPLE 20.1

Network shown in Figure 20.2 has impedances in per unit (pu) as indicated. Determine the diagonal element $Y_{22}$ of the $Y_{bus}$ matrix of the network.

**Solution**

From the Figure 20.2, we get admittances connected to bus 2 is

$$Y_{22} = \frac{1}{j0.1} + \frac{1}{j0.1} + \frac{1}{-j20}$$
$$= -j10 - j10 + j0.05$$
$$= -j19.95_{pu}$$

### EXAMPLE 20.2

A three-bus network is shown in Figure 20.3, indicating pu impedances of each element. Find $Y_{bus}$ matrix of the network.

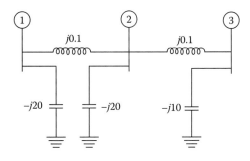

FIGURE 20.2   A three-bus power system network.

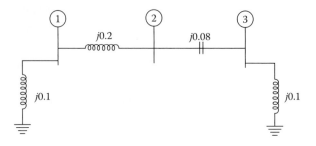

FIGURE 20.3   A three-bus network.

**Solution**

$$Y_{11} = \text{Admittance connected to the bus 1} = \frac{1}{j0.1} + \frac{1}{j0.2}$$

$$Y_{22} = \text{Admittance connected to the bus 2} = \frac{1}{j0.2} + \frac{1}{-j0.08}$$

$$Y_{33} = \text{Admittance connected to the bus 3} = \frac{1}{-j0.08} + \frac{1}{j0.1}$$

$Y_{12} = Y_{21}$ = Negative of the sum of all the admittances connected directly between bus 1 and bus 2 = $-Y_{12} = -(1/j0.2) = j5$.

$Y_{23} = Y_{32}$ = Negative of the sum of all the admittances connected directly between bus 2 and bus 3 = $-(1/-j0.08) = -j12.5$.

$Y_{13} = Y_{31}$ = Negative of the sum of all the admittances connected directly between bus 1 and bus 3 = 0.

$$Y_{bus} = \begin{bmatrix} \left(\frac{1}{j0.1} + \frac{1}{j0.2}\right) & j5 & 0 \\ j5 & \left(\frac{1}{j0.2} + \frac{1}{-j0.08}\right) & -j12.5 \\ 0 & -j12.5 & \left(\frac{1}{-j0.08} + \frac{1}{j0.1}\right) \end{bmatrix}$$

$$= \begin{bmatrix} -j15 & j5 & 0 \\ j5 & j7.5 & -j12.5 \\ 0 & -j12.5 & j2.5 \end{bmatrix}$$

## Example 20.3

Find the values of and for the following equations by NR method up to second iteration $x_1^2 - x_2 - 4 = 0$, $2x_1 - x_2 - 2 = 0$.

## Solution

Let the initial guess be such that

$$x_1^{(0)} = 1 \quad \text{and} \quad x_2^{(0)} = -1$$

Then

$$f_1(x_1^{(0)}, x_2^{(0)}) = 1 + 4 - 1 = 1$$
$$f_2(x_1^{(0)}, x_2^{(0)}) = 2 + 1 - 2 = 1$$

Also

$$\frac{\delta f_1}{\delta x_1} = 2x_1 = 2, \quad \frac{\delta f_2}{\delta x_1} = 2$$

$$\frac{\delta f_1}{\delta x_2} = -4, \quad \frac{\delta f_2}{\delta x_2} = -1$$

However,

$$f_1(x_1^{(0)}, x_2^{(0)}) + \Delta x_1^{(0)} \frac{\delta f_1}{\delta x_1} + \Delta x_2^{(0)} \frac{\delta f_1}{\delta x_2} = 0$$

$$f_2(x_1^{(0)}, x_2^{(0)}) + \Delta x_1^{(0)} \frac{\delta f_2}{\delta x_1} + \Delta x_2^{(0)} \frac{\delta f_2}{\delta x_2} = 0$$

Substitution yields,

$$1 + 2\Delta x_1 - 4\Delta x_2 = 0$$
$$1 + 2\Delta x_1 - \Delta x_2 = 0$$

Solution of these two equations yields

$$x_1^{(1)} = x_1^{(0)} + \Delta x_1 = 1 - 0.5 = 0.5$$
$$x_2^{(1)} = x_2^{(0)} + \Delta x_2 = -1 + 0 = -1.0$$

Proceeding in the same manner, the second iteration yields

$$x_1^{(2)} = 0.5357, \quad x_2^{(2)} = -0.9286$$

## EXERCISES

1. Develop the equations for real and reactive bus powers. Show that a diagonal element of a $Y_{bus}$ is equal to the sum of admittances directly connected to that bus and an off-diagonal element is equal to the negative of the sum of admittances directly connected between the buses.

2. What are the advantages of $Y_{bus}$ over $Z_{bus}$?

3. Compare the performance of Gauss–Siedel and Newton–Raphson methods for load flow solution. Explain the method of formation of $Y_{bus}$.

4. What is the significance of load flow analysis in a power system? Give the classification of various types of buses in a power system for load flow studies. Justify the classification.

# Bibliography

1. O.I. Elgerd, *Electric Energy System Theory—An Introduction*, McGraw-Hill, USA, 1971.
2. W.D. Stevenon Jr., *Elements of Power System Analysis*, McGraw-Hill, New York, 1962.
3. B.M. Weedy, *Electric Power Systems*, John Wiley & Sons, United Kingdom, 1974.
4. H. Cotton, *The Transmission and Distribution of Electrical Energy*, The English University Press Ltd., London, 1958.
5. H. Waddicor, *Principles of Electric Power Transmission*, Chapman and Hall, London, 5th edn., 1964.
6. J.D. Kraus and K.R. Carver, *Electromagnetics*, McGraw-Hill Inc., USA, 1991.
7. L.F. Woodruff, *Principles of Electric Power Transmission and Distribution*, John Wiley & Sons, New York, 1938.
8. A.T. Starr, *Generation, Transmission and Utilization of Electric Power*, Issac Pitman & Sons, London, 1957.
9. C.C. Barnes, *Electric Cables*, Pitman, London, 1964.
10. Westinghouse Electric Corporation, *Electric Transmission and Distribution Reference Book,* East Pittsburgh, Pennsylvania, 1964.
11. L. Emanueli, *High Voltage Cables*, Chapman and Hall, London, 1926.
12. D.J. Rhodes and A. Wright, Induced Voltages in the Sheaths of Cross Bonded a.c. Cables, *Proc. IEEE*, 113(1), January 1966, pp. 99–110.
13. H.A. Peterson, *Power System Transients*, Dover, New York, 1966.
14. L.W. Bewley, *Traveling Waves on Transmission Systems*, Dover, New York, 1961.
15. A. Greenwood, *Electrical Transients in Power System*, John Wiley & Sons, New York, 1971.
16. C.F. Wagner and R.D. Evans, *Symmetrical Components*, McGraw-Hill Inc., New York, 1933.
17. C.R. Mason, *The Arts and Science of Protective Relaying*, Wiley Easterns Ltd., New York, 1977.
18. A.R. Van and C. Warrington, *Protective Relays: Their Theory and Practice*, Vols. I and II, Chapman, New York, 1977.
19. The Electricity Council, *Power System Protection*, Vol. 1, 2, 3, MacDonald, London, 1969.

20. R.T. Lythall, *The J. & P. Switchgear Book*, Johnson & Phillips, London, 1969.
21. H. Trendam, *Circuit Breaking*, Butterworth, London, 1953.
22. E.W. Kimbark, *Power System Stability*, Vol. I, John Wiley & Sons, New York, 1974.
23. S.B. Crary, *Power System Stability*, Vol. I, John Wiley & Sons, New York, 1947.
24. I.J. Nagrath and D.P. Kothari, *Modern Power System Analysis*, Tata McGraw-Hill, New Delhi, 2008.
25. C.L. Wadhwa, *Electrical Power Systems*, New Age International Publishers, India, 2010.
26. A. Husain, *Electrical Power System*, CBS Publishers and Distributors Pvt. Ltd., India, 2012
27. B. Ram and D.N. Vishwakarma, *Power System Protection and Switchgear*, Tata McGraw-Hill, New Delhi, 2011.
28. A. Chakrabarti, M.L. Sony, P.V. Gupta, and U.S. Bhatnagar, *A Textbook on Power System Engineering*, Dhanpat Rai & Co., India, 2012.
29. J.B. Gupta, *A Course in Electrical Power*, S.K. Kataria & Sons, India, 2010.
30. D. Das, *Electrical Power System*, New Age International (P) Limited Publishers, India, 2006.
31. V.K. Mehta and R. Metha, *Principles of Power System*, S. Chand & Company Ltd., India, 2005.

# Index